In this ground-breaking book, John Garrick and his contributors make a most compelling argument for a rational approach to quantifying the risk associated with those low-probability but high-consequence occurrences known as catastrophes. Like the case studies in Quantifying and Controlling Catastrophic Risks, the publication of this book is itself one of those rare events that is of enormous consequence for science, technology, and society.

—Henry Petroski, PhD, Aleksandar S. Vesic
Professor of Civil Engineering, Duke University;
author of *To Engineer Is Human* and *Success through Failure.*

This is a foundational book. It brings the rigor and discipline of science to decision making – a sorely needed ability in a world faced with increasingly complex risks. John Garrick and his colleagues have armed us with a quantitative framework to address catastrophic events that are too often ignored or wishfully dismissed. I have little doubt that this book will become a standard resource for all who wrestle with complex, high-impact issues.

—Thomas O. Hunter, PhD,
President and Laboratories Director,
Sandia National Laboratories

This book is a must for practitioners and students alike in the field of quantitative risk assessment. John has elegantly laid out the methodology with a number of examples for evaluation of risks from complex systems and events.

—Vijay K. Dhir, PhD, Dean,
Henry Samueli School of Engineering & Applied Science,
University of California, Los Angeles.

Incredible! Arguably the "father" of quantitative risk assessment for nuclear power has now come forth with a book that extends this powerful technique to the inevitable global decisions related to potential catastrophes of our time. Since nuclear power remains the only carbon free technology that offers man any real hope of impacting climatic change, this is the perfect companion to his earlier work.

—Roger L. McCarthy, PhD, PE,
Chairman Emeritus, Exponent, Inc.

Quantifying and Controlling Catastrophic Risks

QUANTIFYING AND CONTROLLING CATASTROPHIC RISKS

B. John Garrick

Contributors
Robert F. Christie
George M. Hornberger
Max Kilger
John W. Stetkar

ELSEVIER

AMSTERDAM • BOSTON • HEIDELBERG • LONDON
NEW YORK • OXFORD • PARIS • SAN DIEGO
SAN FRANCISCO • SINGAPORE • SYDNEY • TOKYO
Academic Press is an imprint of Elsevier

Academic Press is an imprint of Elsevier
30 Corporate Drive, Suite 400, Burlington, MA 01803, USA
525 B Street, Suite 1900, San Diego, California 92101-4495, USA
84 Theobald's Road, London WC1X 8RR, UK

Library of Congress Cataloging-in-Publication Data
Application submitted

British Library Cataloguing-in-Publication Data
A catalogue record for this book is available from the British Library.

ISBN: 978-0-12-374601-6

For information on all Academic Press publications
visit our Web site at www.elsevierdirect.com

Printed and bound in the United Kingdom
Transferred to Digital Printing, 2010

Dedicated to Stan Kaplan for his many contributions to the risk sciences that made this book possible.

Contents

About the Author

B. John Garrick, PhD, PE, is a pioneer and prime mover in the development and application of the risk sciences and in particular quantitative risk assessment. He currently holds a White House appointment as Chairman of the US Nuclear Waste Technical Review Board and has testified before Congress on matters of nuclear safety, is an adjunct professor at Vanderbilt University, and is retired from PLG, Inc., an international applied science and engineering consulting firm, as President and CEO. Dr. Garrick is a Fellow of three professional societies and past president of the international society, the Society for Risk Analysis, receiving that society's most prestigious award, the Distinguished Achievement Award. He was elected to the National Academy of Engineering in 1993.

Acknowledgments

The motivation and encouragement for writing this book came from three principal sources, the late and distinguished Professor Norman C. Rasmussen of the Massachusetts Institute of Technology with whom I had three decades of professional association, my longtime friend and distinguished colleague Dr. Stan Kaplan and my former association with the consulting firm, PLG, Inc. Professor Rasmussen directed the seminal Reactor Safety Study for the U.S. Atomic Energy Commission (later the U.S. Nuclear Regulatory Commission) that will forever be the signature of the meaning of quantitative risk assessment (QRA) or as is better known in nuclear circles probabilistic risk assessment (PRA). My association with Dr. Kaplan enabled the development of the analytical algorithms that transitioned QRA and PRA into a form more applicable to any kind of natural or engineered system. At PLG the opportunity was provided primarily through private support for an energetic and highly motivated group of outstanding engineers and scientists that I was fortunate enough to lead, to advance the risk sciences while applying the results to very complex systems. The applications involved such systems as nuclear power plants, chemical and petroleum installations, the space shuttle, and transportation. The results were highly successful and have greatly influenced the path forward for quantitative risk assessment. These associations made this book possible.

Finally, I would like to acknowledge the very able assistance of my son Robert S. Garrick and my assistant Elizabeth M. Ward. Robert's advice and input on Chapter 7 were most helpful and Liz Ward's production skills greatly shortened the publishing time of the book.

Foreword

May 19, 2008

As this Foreword is being written, the front pages of the newspapers are filled with stories of the tragedies arising from the cyclone that struck Myanmar and the earthquakes that have ravaged China. Many tens of thousands of lives have been lost and the impacts on those who survived have been devastating. In light of the many destroyed homes, schools, factories and hospitals, the consequences will persist for years. These events raise the obvious question of whether we can do a better job in preparing for rare and catastrophic events. This book demonstrates that we can.

The author, B. John Garrick, is a leading practitioner of quantitative risk assessment (QRA). Dr. Garrick is one of the pathfinders in the use of probabilistic methods to understand complex systems. As a former Chairman of the Nuclear Regulatory Commission (NRC), I am very familiar with his work because the use of his techniques has greatly strengthened our understanding of nuclear reactors, with enormous benefit in enhancing nuclear safety. Indeed, the NRC is now embarked on the application of these techniques to refine the regulatory system in ways that enhance the focus on safety.

This book explores the application of QRA techniques more broadly. The key lesson, as demonstrated by the examples that are developed in the book, is that the systematic, integrated, and transparent examination of rare events can provide extraordinarily helpful insights that should inform policy. This is not to deny that there are inevitable limitations arising from uncertainty. But the book demonstrates that the systematic assembly of relevant information can help policy makers to make informed decisions about the allocation of resources and can help the general public to have realistic understanding of such events.

Perhaps the most striking example of this fact is the chapter concerning the risk that a major hurricane might strike New Orleans, overwhelming the dikes that protect the city. This chapter was written before the Katrina disaster, and it provides a remarkably accurate picture of the likelihood and consequences of the event that subsequently occurred. This chapter alone establishes that QRA can provide extraordinarily helpful advance information that can and should guide policy makers.

I anticipate that the need for such insights is growing. The climate change models indicate that extreme weather events may become much more frequent in the future. Moreover, the threat of terrorism presents the likelihood of human-induced events with extreme consequences. I served as Chairman of the NRC in the years before and after 9/11 and I know from first-hand experience that our decision-makers need tools that can provide a firmer foundation for societal preparation. This book opens the door to far more informed approaches than we have had in the past.

Richard A. Meserve, President
Carnegie Institution for Science
Washington, DC

Foreword

The author of this book dares to do something that very few of us in the risk assessment business would do. He proposes that major societal decisions involving catastrophic risks could, in fact, have a good dose of scientific analysis. I must say that, after I understood what the book is attempting to achieve, I felt apprehensive. On second thought, however, apprehension was replaced by admiration.

Whenever society faces threats with potentially catastrophic consequences, there are calls for prioritizing these threats so that resources to prevent or mitigate them would be allocated effectively and efficiently. A recent example is the threat of terrorism acts. Quantitative Risk Assessment (QRA) is the natural way to effect such prioritization. Yet, it is rarely utilized. The question is why?

The most commonly encountered arguments against the use of QRA are that it cannot be done and that the uncertainties are so large as to make the results useless. The message from this book is not that one can do a QRA for such major threats using 'cookbook' methods to produce precise numerical values for the risks. The main message is that there is an intellectual approach to these complex issues that has proven its worth in the assessment of risks from complex technological systems such as nuclear power plants and space systems. It is the scenario-based approach.

The systematic structuring of the set of scenarios (event sequences) that may start with a given threat and lead to undesirable consequences is an excellent first step in the understanding of that threat and the removal of some of the fear that may come with it. Completeness of the scenario list is always a problem, but this list provides a good starting point for the generation of additional scenarios.

The quantification of uncertainty is a second major theme in this book. The use of probability distributions to 'tell the truth' is what QRA is all about. The lack of (or complete absence of) strong statistical records should not be a deterrent. A probability distribution represents what we know now about an uncertain quantity (and what we know may be very little, in which case the distribution should reflect this fact). I know that this statement will sound like a truism to many readers, yet it is far from being common practice to employ probability curves in risk assessments rather than single ('point') values. For societal decisions, one would like to capture the knowledge of the relevant expert technical communities when these curves are derived. The structured elicitation of the judgments of experts is one way of evaluating what the scientific community knows about a particular subject.

I can imagine many readers shaking their heads and thinking that there are too many holes in the book's proposed approach. Completeness of the scenario list is a dream and the quantification of uncertainties is fraught with too many problems. My experience leads me to firmly believe that the proposed approach is intellectually sound. It provides the framework within which all the stakeholder concerns can be evaluated. If there are any disagreements, they can be placed in the right context and their significance can be evaluated.

Another point that is usually misunderstood about QRA is that its results and insights are intended to be inputs to the risk management process. Decisions are not to be QRA-based. The decision makers are expected to scrutinize the QRA results (especially their completeness) and to include other considerations, such as values, in their decisions. The alternative is to rely on ad hoc judgments and analyses without an overall intellectual framework for their evaluation.

Writing a book that promotes a way of thinking cannot be an easy task. The author has wisely chosen to limit the methodological sections to a minimum and to demonstrate the proposed approach using several case studies. I believe he has succeeded admirably.

George E. Apostolakis
Massachusetts Institute of Technology

Preface

This is a book about societal and environmental risks and how to quantify their likelihood of occurrence. It is about increasing our understanding of risks that could result in thousands or even millions of human fatalities and massive environmental damage. It is about obtaining the necessary knowledge to make good decisions on how to mitigate or reduce such risks.

In the past several decades there has been a growing awareness about the environment and the safety and security risks we face as a society. This awareness is especially relevant to that class of risks that rarely occur, but if they do, can have catastrophic consequences regionally and possibly even globally.

The excuse that is most often given for not quantifying the risk of rare and catastrophic events is that there is a lack of knowledge or too little data to do so. This book challenges the 'too little data' view on the basis of the progress that has been made in contemporary and 'evidence-based' probabilistic thinking. To be sure, there is seldom enough data about future events to be absolutely certain about when and where they will occur and what the consequences might be. But 'certainty' is seldom necessary to greatly improve the chances of making good decisions. Furthermore, while it may not be possible to predict *when* such location-specific events might actually occur, it is clearly possible to calculate, with uncertainty, their *frequency* of occurrence, thus providing important insights on the 'when' question. What is necessary is a process for organizing the information, however limited it may be, in such a way to maximize what can be inferred from it about specific catastrophic events at specific locations. The discipline of quantitative risk assessment (QRA), also known as probabilistic risk assessment, was developed to do just that.

The motivation for QRA was rooted in the inadequacies of past methods of risk analysis to provide answers on the frequency of rare events about which we know very little, but events of great concern to our welfare. For QRA to be successful, greater use had to be made of (1) analytical concepts capable of maximizing what can be inferred from very limited knowledge of specific events and (2) logic models of events about which we do have knowledge to the potentially catastrophic events of concern. For (1), this meant having a science of making 'uncertainty' an inherent part of the answer. For (2), it meant the development of logic models that could represent the course of scenarios involving events for which there is little to no direct experience. To illustrate these concepts and go beyond the theoretical elements of QRA, this book includes four case studies of potentially catastrophic events. Two of the case studies were carried to the point of the risk of catastrophic human fatalities and two others were limited to precursor events. The case studies were facilitated by contributions from distinguished experts in risk assessment, computer science, and earth science. The contributors to the case studies were Robert F. Christie, hurricanes and asteroids, John W. Stetkar and Max Kilger, terrorist attack of the national electrical grid, and George M. Hornberger, abrupt climate change.

The applications of the ideas constituting the new science of QRA have been enormously successful. Among the industries, services, and disciplines that have greatly benefited from this new way of thinking are nuclear, space, defense, chemical, health, and transportation. But the benefits could be much greater to society. It is the purpose of this book to present additional evidence of the need to provide our leaders with better information on the regional and global risks we face, and to provide assurance that methods of analysis exist to greatly enhance the likelihood of making the *right decisions* about possibly saving millions of lives and perhaps even the whole of society.

B. John Garrick, PhD, PE
2008

CHAPTER 1

Societal Risks in Need of Understanding and Action

1.1 The Target Risks

This book is about some of the health and safety risks that Hollywood loves to make movies about—global disasters brought about by events such as nuclear war, disease, climate changes, tsunamis, volcanic eruptions, asteroids, and terrorist attacks. Specifically, this book is about how to analyze such rare and catastrophic risks to better prepare for mitigating, controlling, or managing their consequences.

Why should we be interested in such risks? These are the risks that often don't get taken seriously and yet, they are the kind of risks that could greatly compromise or even terminate life. Many of these risks are beyond known human experience and generally tend not to raise much action from nations and their leaders. How can we get to the truth about risks that are rare and catastrophic and may be even irreversible unless anticipatory actions are taken? How can we possibly manage these risks if we don't really know what they are? Surely, there must be a better way than to rely on the fickle reports of the news media on the "threat of the day."

What is the basis for deciding what the priorities should be? Are our government leaders, who are primarily driven to deal with issues that coincide with election cycles, in a position to know how to lead the nation and the world on issues that have time constants of decades, centuries, and millennia and are dependent on scientific investigations and knowledge? Is the result of not understanding the risk of catastrophes a leadership that addresses symptoms, not root causes of problems? Is a news media that is primarily driven by the latest sensation a reliable source of information? Where can the public go to get the truth?

Certainly this book does not have all the answers to these questions, but it does suggest an important first step. That first step is to begin to identify, quantify, and prioritize the rare and catastrophic risks to society. The point of this book is that in order to make good decisions on the management and control of catastrophic risks, you first have to quantify those risks. But how can you quantify the risk of something so rare as a modern day global famine; an astronomical event leading to complete or partial extinction of life on Earth; a hundred or thousand year severe storm, earthquake, or volcanic eruption; a terrorist attack that can kill tens or hundreds of thousands of people; or a climate change that could lead to total extinction of life on Earth? The goal of this book is to provide a method with which to quantify such risks to support better decision-making to sustain human life on planet Earth.

The category of risks addressed in this book is a combination of "existential" risks and risks believed to be rare but of catastrophic consequences. An existential risk has been defined by Bostrom[1] as "one where an adverse outcome would either annihilate Earth-originating intelligent life or permanently and drastically curtail its potential." "Catastrophic risks" will be defined here as those risks that may threaten the existence of the human species globally, but also include risks that may have catastrophic consequences on just a local or regional basis. In this book, we assume an event is catastrophic if it results in 10,000 fatalities or greater. Examples of existential risks are global nuclear war, a runaway genetically engineered biological agent, a large asteroid, an irreversible atmospheric or climate change, or a runaway incurable disease. Catastrophic risks might involve a terrorist attack; an extremely severe storm; a super earthquake, tsunami, or volcanic eruption; a plague; or an industrial accident that could threaten the health and safety of thousands, but not necessarily result in extinction of the species. These are the types of risks that may occur only once in many lifetimes or even many millennia, but when they do occur the consequences can be cataclysmic and perhaps threatening to human existence on Earth. Risks having catastrophic consequences, either globally or regionally are further complicated by often having the property of irreversibility. Waiting for the occurrence of catastrophic consequences through direct experience may be too late to take risk-reducing actions.

Much has been written about risk and how to analyze it to support good decision-making and risk management. A plethora of textbooks, studies, and assessments of risk exists, several of which are referenced.[2–9] Beginning in the 1970s, the risk sciences started to come into their own—inspired by such issues as the safety of nuclear power plants and requirements for environmental impact statements.

We now have the capability to analyze risks better than at any other time in human history. A brief account of some of the roots and evolution of quantitative risk assessment (QRA) is provided in Appendix A.

One reason why we aren't doing much risk assessment on rare events with catastrophic consequences is the lack of quantitative risk information in the public domain. This sets the stage for another risk, namely complacency on the part of society about being able to control such risks. The public does not take rare but catastrophic risks seriously. Why is this? Besides the lack of good quantitative information on such risks, there is the perception that the risks that really threaten us are not the rare catastrophic events that may happen only once over many lifetimes, but risks such as disease, auto accidents, crime, airplane accidents, accidents around the home, and industrial accidents—risks that principally determine individual life expectancy, but not human life sustainability as a whole.

Most people are generally aware of their individual health risks. These health risks include being undernourished, unsafe sex, high blood pressure, tobacco use, alcohol consumption, high cholesterol, obesity, and the use of illicit drugs. Because of the experiential data, insurance companies for the most part already quantify these risks. The more subtle, but rare and catastrophic risks are not so much on people's minds because the present population has not experienced them. These risks have been talked about and in some cases even analyzed, but the results of the talks and analysis have not had meaningful impact. On the other hand, there is Hollywood's sensationalizing of such risks for the purpose of entertainment. The result is more of a "spoof" on such threats than any constructive dialog about their likelihood or consequence. The thesis of this book is that rare catastrophic events will most likely be the major threat to a life-sustaining planet for the centuries to follow and that proper analyses will result in better management of such events.

There is one sharp distinction between "individual risks" and "existential and catastrophic risks." That distinction is that in general we have excellent information on most personal risks and very poor information on the rare and catastrophic societal risks. We don't need to do more assessment on the risks we know. For example, in the United States of America (U.S.A.) we know[6] that up to about 40 years old, accidents (primarily auto accidents) are the major cause of death to humans. Somewhere between 40 and 50 years old, cancer and heart disease take over as the greatest threat; after 60 or 70 years old, the dominant individual risk is heart disease. These individual risks are generally not a threat to the society as a whole. Insurance companies best handle these types of risks. They mainly affect life expectancy, not the death or injury of very large numbers of people through a single event as might be the case for a terrorist attack, a super earthquake, a giant tsunami, or a planet-threatening asteroid.

The authors believe that current methods of analysis from the risk sciences exist with which to quantify rare and catastrophic risks and offer the possibility of either their mitigation or at least a reduction of the regional or global consequences. Had humans existed and the risk sciences been applied to rare events some 65 million years ago, we might have dinosaurs roaming our planet today. The same could probably be noted for many other species that do not exist in the 21st century. Thus, the real reason for focusing more on rare but catastrophic events is to be in a better position to mitigate them or at least to reduce their consequences so as to provide greater assurance of continued existence of the human species. Existential and catastrophic risks are where the highest payoff is for the application of contemporary quantitative risk assessment, and it is believed that such quantitative applications can result in extraordinary benefits to the society. "Quantification" in risk assessment is one of the central messages of this book.

Just what are the rare and catastrophic risks that we should be worrying about and why should we worry if they are, in fact, rare? Is it weapons of mass destruction? Is it overpopulation? Is it pollution? Is it lifestyle? Is it our inability to establish a world order for managing global risks? Just how much of a threat is environmental risk and how does it rank with other threats? How much of the risk is technological and how much is "other"? What can we do about religious radicalism and is that our greatest threat? What about emerging threats such as the risks associated with depleting resources, biological cloning, genetic alteration, atmospheric quality and opacity, oxygen depletion, and the buildup of carbon dioxide and other greenhouse gases?

What are the linkages between threats? For example, is energy production the most common link between such threats as global warming, alteration of the air we breathe, acid rain, resource depletion, etc., or is it technology in general and our inability to globally manage it? Or, are global risks really too deeply rooted in political and cultural factors to find workable and timely solutions? What are the risks we can do something about and what are the risks outside of our control? Presently, the most advertised rare and catastrophic risks are those associated with the environment such as global warming and the pollution or alteration of our air, water, and food chain. But as we shall see, environmental risk is just one of many risks that have the potential for possible catastrophic global consequences. Technological risk is perceived by many as a serious threat, but as pointed out by Lewis,[6] the benefits of technology have provided extensive compensation in terms of increased life expectancy.

1.2 The Quantitative Definition of Risk

What we mean by "risk" is what is described in the risk literature as the "set of triplets" definition of risk.[10] This triplet definition of risk and the theoretical basis of quantitative risk assessment are developed in detail in Chapter 2. The definition says that when we ask the question "What is the risk?" we are really asking three questions.

1. What can go wrong?
2. How likely is that to happen?
3. What are the consequences if it does happen?

We answer the first question in the form of a set of scenarios called the "risk scenarios." The "end states" of the scenarios are the consequences in terms of fatalities, injuries, etc., and the "likelihood" refers to the likelihood of the scenarios individually and collectively.

The triplet definition of risk is a more general definition than most found in the literature. For example, another definition often quoted is *probability times consequence.* The problem with this definition is that it loses the distinction between high likelihood-low consequence events and low likelihood-high consequence events. Also, this definition is often interpreted as an expected value for risk, which is far too restrictive in terms of the need to make uncertainty an inherent property of the calculated risk. The probability times consequence definition is vulnerable to the separation of the probability analysis from the consequence analysis and running the risk of losing the connection between the details of the consequence calculation and their impact on probability. That is, many of the assumptions in a consequence analysis, if not all, can be represented in probabilistic format to make the risk assessment more realistic and truly probabilistic. The triplet definition avoids these anomalies by characterizing the risk as the likelihood of scenarios and their attendant consequences. The issue of consequence assumptions and evidence become a basic part of the scenarios, thus guaranteeing that the boundary conditions for the end state of the scenario, that is the consequence, are an inherent property of the scenario likelihood calculation.

Some risk analysts choose to define risk to include value judgments or preferences in the form of a utility function. The position taken in this book is to make a distinction between "risk assessment" and "risk management." The idea is to focus the risk assessment on the "what can go wrong" scenarios, their consequences and likelihoods, and leave it to the decision analysts and decision-makers, often the citizenry, to apply value judgments and preferences to the results of the quantitative risk assessment.

1.3 The Process of Quantitative Risk Assessment

Although the scope, depth, and applications of quantitative risk assessments vary widely, they all follow the same basic steps:

Step 1. Define the system being analyzed in terms of what constitutes normal operation to serve as a baseline reference point.
Step 2. Identify and characterize the sources of danger, that is, the hazards (e.g., stored energy, toxic substances, hazardous materials, acts of nature, sabotage, terrorism, equipment failure, combinations of each, etc.).
Step 3. Develop "what can go wrong" scenarios to establish levels of damage and consequences while identifying points of vulnerability.
Step 4. Quantify the likelihoods of the different scenarios and their attendant levels of damage based on the totality of relevant evidence available.
Step 5. Assemble the scenarios according to damage levels, and cast the results into the appropriate risk curves and risk priorities.
Step 6. Interpret the results to guide the risk management process.

These steps provide the answers to the three fundamental questions of the triplet definition of risk.

Consider the simple example of wanting to know the risk of taking a hike in the well known primitive area of Idaho. Following the above six-step process we must first define what we are dealing with and what constitutes a normal or successful hike. We must understand what has to happen for the hike to be successful, which is Step 1. We call this the "success scenario" which now provides a reference for departures from success. Step 2 is to identify the sources of danger, that is, the hazards that might be encountered during the hike. Step 3 is to invoke the first question of the triplet, "what can go wrong." Asking this question conjures up all kinds of possibilities, such as being attacked by a wild animal, for example a bear, encountering an unexpected severe storm, being attacked by bandits, experiencing an earthquake or a forest fire, being crushed by a landslide, having a bad accident, a hiker going berserk, etc. The idea is to develop as complete a set as reasonable of the most important threat scenarios. The end point of the scenarios is the consequences we possibly face.

Step 4 is to quantify the levels of damage (consequence) resulting from each scenario. In the spirit of the risks targeted in this book, we are mainly interested in catastrophic events such as serious injuries, fatalities, or the loss of a hiker or of the entire hiking team. We evaluate the consequences of the scenarios and screen out the ones that do not meet our criteria of a catastrophe. During Step 4 we consider whatever evidence we can find bearing on the likelihood of each scenario. Evidence can be in the form of experience with similar hikes, the degree of difficulty of the hike, the experience and behavior of the members of the hiking team, and the susceptibility of the region to natural events, such as floods, storms, earthquakes, and fires. Based on the evidence, we assign a probability to each scenario. We now have a structured set of scenarios, their likelihoods, and consequences.

For our simple example, the structured set of scenarios may be small enough that we need not mathematically assemble the scenarios in order to make our decision and can skip Step 5 and proceed directly to Step 6. Of course, the decision does not have to be a "go" or "no go" decision. In fact, in many cases the decision is to get more information and reassess the risk. An example of that is given later. Now if the system is more complicated involving hundreds or thousands of scenarios as we shall see in the case studies to follow, it is necessary to implement Step 5 to facilitate the decision making process.

As we shall see in the case studies of Chapters 3 through 6, there are many intermediate steps in the six-step process when assessing catastrophic risks. But the principles are fundamental and very much rooted in the framework of the triplet. The table below identifies Steps 1 and 2 of the six-step process of quantitative risk assessment demonstrated in Chapters 3 and 4.

	Chapter 3	**Chapter 4**	
Step	Hurricanes	Asteroids	
1—System	City of New Orleans, LA	Contiguous 48 states of U.S.	City of New Orleans, LA
2—Hazard	Major hurricanes	Asteroids with high impact energy	Asteroids with high impact energy

The table below identifies the first two steps of the risk assessment scoping analyses for case studies 3 and 4 (Chapters 5 and 6).

	Chapter 5	**Chapter 6**
Step	Terrorism	Abrupt climate change
1—System	U.S. metropolitan region	North Atlantic states

2—Hazard	Terrorist attacks that cause loss of metropolitan region electrical grid for more than 24 h	Loss of the thermohaline circulation in the North Atlantic due to global warming

1.4 The Meaning of Quantification

If there were but one concept that we would like to communicate and clarify in this book, it is what we mean by being *quantitative* as in *quantitative risk assessment*. Webster's *New Collegiate Dictionary* offers one definition of quantitative as . . . *involving the measurement of quantity or amount*. The key word is *measurement*. Measurement is a good place to start as it is in tune with the basic notion of the scientific method—reducing observations to measurements. William Thomson, 1st Baron, usually known as Lord Kelvin, a 19th century British physicist, said it best.[11]

"I often say that when you can measure what you are speaking about, and express it in numbers, you know something about it; but when you cannot measure it, when you cannot express it in numbers, your knowledge is of a meager and unsatisfactory kind; it may be the beginning of knowledge, but you have scarcely, in your thoughts, advanced to the stage of science, whatever the matter may be."

Clearly, it is a scientific interpretation that the authors wish to give to the concept of quantification. But what we mean by quantification is still a bit of a mystery as it depends on what we mean by measurement. Generally, what is meant is a figure, or a number that indicates the amount of something. So, when we speak about the quantitative risk assessment of something, we are implying a process that results in a measure or an indication of the amount of risk associated with that something. This interpretation provides visions of real numbers that indicate levels of risk such as the likelihood of different consequences, that is, different outcomes of postulated risks. Postulated risks could be whatever we have fears of: flying in airplanes, automobile accidents, fires, earthquakes, radiation, climate changes, nuclear war, disease, severe storms, industrial accidents, terrorist attacks, Democrats, Republicans, or whatever.

One common perception of quantification is that you have to have *statistical quality* numbers about the frequency of risk scenarios in order to be quantitative. That is, you have to have a whole bunch of experience with the risk before you can quantify it. The authors of this book take exception to this common perception. We offer a different dimension to the meaning of quantification and this is another key message of this book. This different dimension is to embrace uncertainty in the risk measure. This frees one of the needs for so-called statistical quality data in order to make informed decisions. By doing this, the option exists to consider the evidence available as far as it can take you (but no farther); that is, tell the truth about the risk being considered based on the totality of available evidence. Of course, the more limited the data the greater the burden on the analyst to accurately model the risk. That modeling basically involves mapping from the events about which data does exist to the risk events of interest. So, *quantification* in this context is a measurement in a form that tells the truth about the accuracy of the measurement. *Quantification then, is our full state of knowledge about that measurement, including the uncertainty.*

Quantification represents the truth about a measurement based on the totality of *available* evidence. This view of quantification introduces different meanings to such descriptors as *precision*. The word precision conjures up such mind images as knowing a number out to so many decimal places. Precision in the context of quantification of a measurement involving uncertainty is not necessarily a precise number to so many decimal places. Precision in this case has to do with how well we have represented the truth about our knowledge of the measurement—how well we have quantified the number in terms of its uncertainty. Casting the number in a manner that expresses our state of knowledge about the number, that is the evidence supporting the number, enables us to make decisions based on the totality of available experience, not upon arbitrary and often opaque assumptions.

It should be noted that presenting risk measures in a form that communicates the uncertainty about them does not mean that when this is done it is because the risks being assessed are uncertain while other risks are not uncertain. The truth is that for the most part, we are always facing risks with uncertainty—it is just that we are generally not informed about the amount of uncertainty. We tend to be creatures of a binary world: yes, no; 1, 0; on, off; acceptable, not acceptable; go, no-go; safe, unsafe; etc., and do not respond well to the more frequent reality of "maybe" or "it's a possibility." That is, we live in a world of uncertainty; we just have a tendency to ignore it in many cases. Society is in trouble when the uncertainties are great and they are still ignored.

We are getting better at representing uncertainty. For example, we now require publication of the side effects of the medicines we use. In doing so, we are at least admitting uncertainty even though we have not advanced to the stage of quantifying it. Weather reports are now given on the basis of the "chance" of different types of weather. Clearly, these are steps towards embracing uncertainty, albeit small steps.

Adopting the definition of quantification as including uncertainty allows risks to be quantified where little or no direct experience exists, but there is evidence that such risks could come about. The key point is the quantification of what are called states of knowledge.

The state of knowledge about the risk of an event may be incomplete, but if properly presented, quantification of our state of knowledge can be extremely valuable in making decisions on how to mitigate the event. Just because we don't know the risk to several decimal places doesn't mean we are helpless, especially if we are able to put what we do know in its proper context. Quantification in this sense takes the form of measurements that clearly communicate their uncertainties—measurements that communicate what we *do* and *don't know*—measurements that tell the truth about what we know and how well we know it. So, when reference is made to quantification, what the authors mean is that we must present our numbers in a form that expresses uncertainty in the measurements and how the numbers and their uncertainties are connected to the supporting evidence.

As we build the case for what is meant by being quantitative, we have introduced another very important term, namely the term *evidence*. We choose the term evidence over such terms as "data" or "information" because of its broader implication. Too often terms like data are associated with statistics on failure rates and event frequencies—much too narrow of an interpretation for purposes of doing risk assessments. To be sure, such data is important and if it exists, it is very much a part of the supporting evidence for a risk assessment. The term information has a broader meaning, but doesn't quite have the punch of the term "evidence." Evidence carries with it a much broader meaning, including observations, intelligence, general experience, special investigations, and expert judgment. Of course, there is the need to invoke the *scientific method* for processing the evidence; that is, there must be a transparent process of reducing the evidence to measurements that can be used to make numerical calculations. The science of uncertainty allows this to be done.

Adopting the authors' meaning of quantification implies that we are no longer bound to knowing only those risks on which we have direct experience such as automobile accidents and flying in airplanes. We are now able to venture into the assessment of risks where we have either meager experience or no experience at all such as catastrophic terrorist attacks and global disease. We are able to infer from the evidence we already possess what can be concluded about the likelihood of events on which we may not have any direct experience.

The issue of quantification centers on calculating or presenting measurements that communicate what we know and what we don't know about the chosen risk measure. The triplet definition of risk says that what we measure are scenarios, likelihoods, and consequences. In particular, we calculate the *likelihood* of the different scenarios and their consequences. Likelihood becomes the parameter of the model. Following the thought process noted earlier, likelihood has to be defined in such a way that it accounts not only for what we do know about a particular scenario, but what we don't know as well. That is, when we are faced with uncertainty and we want to be quantitative, our measure of risk must reflect that uncertainty. Our risk measure, the likelihood function, must have a form that clearly communicates confidence and uncertainty in the results. Practitioners of quantitative risk assessment usually represent the likelihood of an event by one of three ways: (1) as a frequency, (2) as a probability, or (3) as a combination of the two, that is, as a *probability of frequency*. The meanings and distinctions of these interpretations of likelihood are covered in Section 2.3 of Chapter 2. The interpretation adopted here is the "probability of frequency" concept for capturing and quantifying the state of knowledge about a hypothesis. It is "evidence based" and embodies the first two interpretations as well. In particular, we use "frequency" to represent something that is measured and "probability" to communicate our uncertainty in that measurement.

If there could be but one word that best describes risk, it is "probability." What some call the "subjectivist" view of probability is best expressed by the physicist E. T. Jaynes:[12] "A probability assignment is 'subjective' in the sense that it describes a state of knowledge rather than any property of the 'real' world, but is 'objective' in the sense that it is independent of the personality of the user; two beings faced with the same background of knowledge must assign the same probabilities." The central idea of Jaynes is to bypass opinions and seek out the underlying evidence for the opinions. This distinction in evidence helps the analysis become more objective.

Jaynes' statement is correct, but we go a step further. We define "probability" as synonymous with "credibility." We can thus speak, and think, in terms of the "credibility" of a hypothesis based on all the evidence available. "Credibility" and thus our interpretation of probability is a positive number ranging from zero to one, and it obeys Bayes theorem (Chapter 2). As discussed in Chapter 2, Bayes theorem is the fundamental law governing what can be inferred from new information about the likelihood of a hypothesis or event. It tells us how the credibility of a hypothesis changes when new evidence occurs.

To illustrate Bayes theorem, suppose in our earlier hiking example the scenario of greatest concern was the risk of being attacked by a bear and that the decision to take the hike hinges on a better understanding of the risk of such an attack. In particular, the decision is to get more information on the risk of a bear attack before making the decision to go on the hike. The information desired is more specific data on humans encountering bears in the region of the planned hike. The prior analysis was based on forest ranger information indicating that in general there is a 50% probability of encountering a bear in the region of the desired hike. In Bayesian language, this is called the prior probability (prior to obtaining more information). The hikers know that the frequency of bear encounters depends on details that may not be evident at the level of the 50% information. Such details as the time of the hike, the route chosen, weather conditions, and the food supply all have an impact on (1) the likelihood of encountering a bear and (2) the temperament of the bear. The hiking planner goes to the forest ranger to obtain more detailed information on some of the above factors. Among the important pieces of new information is that the bears are still likely to be in hibernation at the time of the planned hike. The hikers update their estimate based on the new evidence. This updating is exactly what Bayes theorem does. The result of the new evidence about hibernation practices in the region of the hike as well as other factors considered is a reduction of a bear encounter by a factor of 10. Thus, updating our risk assessment indicates only a 5% chance of encountering a bear and the hikers decide to proceed with their plans.

1.5 Form of the Results of a Quantitative Risk Assessment

Risk can be thought of as a structured set of scenarios, their likelihoods and consequences. This structured set is generally depicted as an *event tree* or a number of event trees. Event trees as a risk assessment tool are discussed in Chapter 2 and applied in the case studies of Chapters 3–5. If the number of scenarios is small as in the case of the hiking example given earlier, then knowing their consequences and likelihoods may be all that is necessary to characterize the risk of the events being analyzed for decision-making. On the other hand, if the number of scenarios is large, it becomes clumsy to interpret the total risk at which point it is necessary to assemble the scenarios in an orderly fashion to convey the risk and the attendant uncertainty. How this is done analytically is covered in Chapter 2 and applied in the case studies.

Two types of presentation are popular for communicating risk—probability densities and cumulative probability distributions. The first type is usually used when wanting to present the risk of a specific consequence from a particular event, such as the risk of 10,000 fatalities from a hurricane at a particular location. That is, each consequence has its own "probability of frequency curve."

The second type of presentation treats consequence as a variable, thus presenting frequency as a function of consequence and probability of frequency as a family of curves, each with its own probability. These curves are known as "frequency of exceedance curves" or *complementary cumulative distribution functions.* Such curves can be developed for different types of consequence such as injuries, fatalities, and property damage. The technical basis of these types of presentations is discussed in Chapter 2.

As an illustration of an event tree and the two types of risk curves, we fast-forward to Chapters 3–5 and review selected results

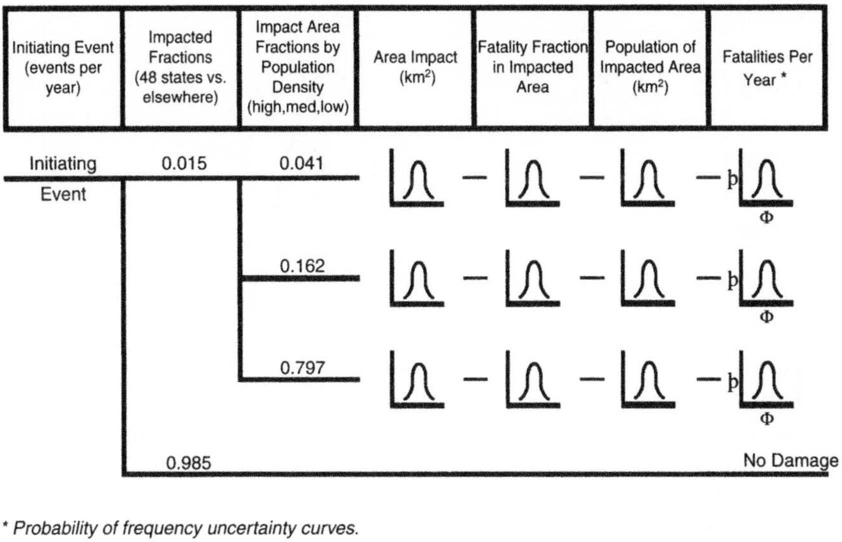

FIGURE 1.1. Asteroid scenarios for land impact.

that are developed and discussed on the risk of hurricanes, asteroids, and terrorism. We start with Figure 1.1 (Figure 4.2 from Chapter 4), which conceptualizes the logic in the model for analyzing the fatality risk of asteroids impacting the contiguous 48 states of the U.S.

In the case of a large asteroid impacting the land area of the 48 contiguous states of the U.S., we wish to know the frequency per year of impact of any asteroid hitting the 48 states, what type of population density exists at the impact site, what is the land area of the blast, the fatality fraction of the impacted population, and what are the resultant fatalities per year. Each fatality probability density function shown at the end of each sequence in Figure 1.1 may be the result of convoluting hundreds or thousands of individual probability density functions. The event tree depicts the logic of the asteroid risk assessment for land impact and allows for quantification of the fatalities per year, while accounting for the uncertainties involved. In Chapter 4, event trees also exist for asteroid impact in the Atlantic Ocean, the Gulf of Mexico, and the Pacific Ocean.

To illustrate how probability density functions are used to present risk information, we go to Figure 1.2 (Figure 5.11 from Chapter 5).

In Figure 1.2, probability density functions are used to display the risk of different damage states of the electrical grid system as a result of a terrorist attack. The damage states vary from local network damage to long-term damage to the regional grid.

Finally, to illustrate the complementary cumulative distribution function we present Figure 1.3 (Figure 3.2 from Chapter 3).

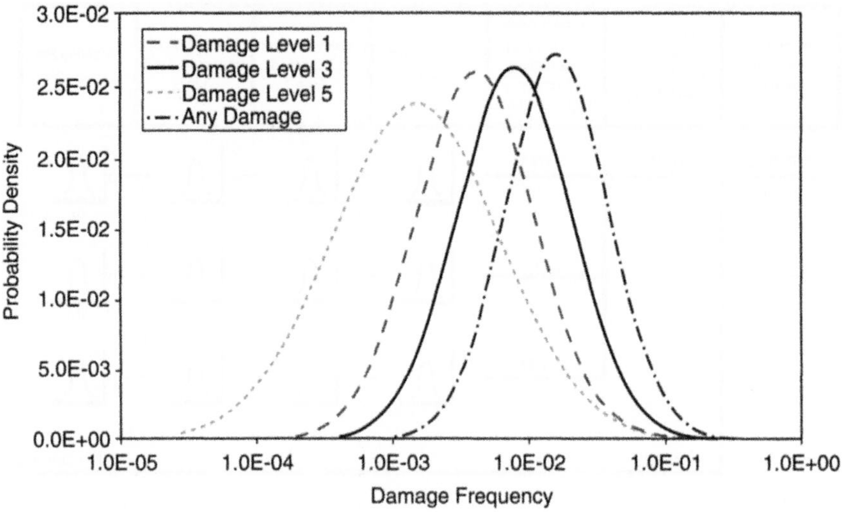

FIGURE 1.2. Results for top damage levels.

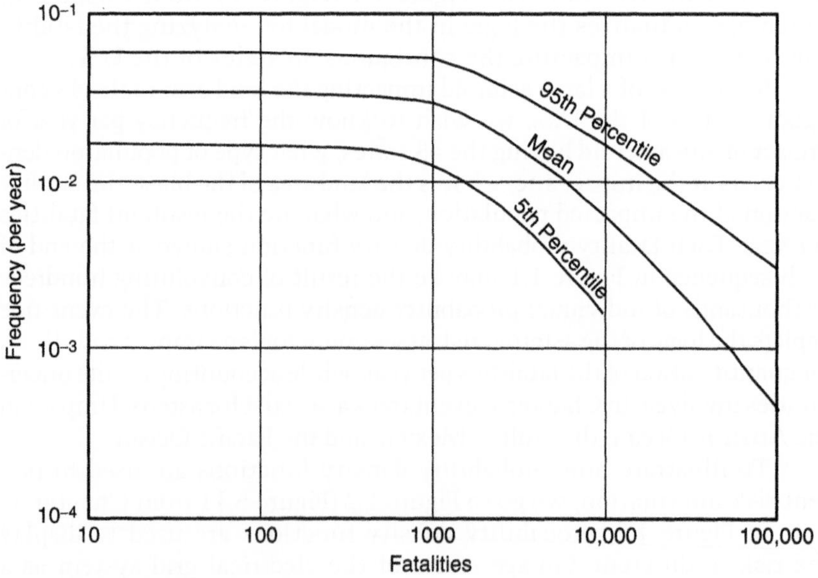

FIGURE 1.3. Hurricane fatality risk for New Orleans (based on U.S. hurricane data 1900–2004).

Figure 1.3 represents the risk of fatalities at New Orleans, LA, from hurricanes. These curves not only display the risk of different fatality consequences, but the uncertainty as a function of the number of fatalities. A quick inspection of the figure indicates that we can expect on average a 10,000-fatality hurricane event once every 130 years or so. But the curve tells us more than that about the 10,000-fatality event. It tells us how sure we are of our answer. In particular, we are 90% confident that the frequency of this event ranges from approximately one every 60 years to approximately one every 620 years. It is the mean value that is every 130 years.

In the material to follow, Chapters 3 (hurricanes) and 4 (asteroids) are case studies of two reasonably completed quantitative risk assessments involving events having the potential for catastrophic fatalities. Chapters 5 (terrorist attack) and 6 (abrupt climate change) are case studies of risk assessments that have not been developed to the point of quantifying the risk of catastrophic consequences. They have been developed to the point of quantifying the risk of events that may be "precursor events" to catastrophic consequences.

A number of factors were considered in choosing these four case studies. One factor was the state and accessibility of information on risks that are the target of this book. Another factor was the amount of risk assessment work that had been performed to accommodate the timeliness of getting quantitative risk assessment results. The authors were also interested in being sensitive to risks that might be of greatest interest to the public such as hurricanes as emphasized by Katrina, climate change because of global warming, and a terrorist attack because of 9/11/2001. The asteroid event was chosen for a variety of reasons including the fact that a considerable amount of risk-related work had already been performed, the fact that it is one of the few events that has the potential for the extinction of life, and the fact that asteroid risk is of growing concern to the public.

Chapter 7 presents numerous examples of risks targeted by this book as in need of greater understanding and action. Chapter 8 interprets and compares selected results and addresses the question of what it all means and how it fits in the grand scheme of the vision of the authors. Appendix A provides a historical perspective of quantitative risk assessment, including a comprehensive example and Appendices B and C provide the supporting evidence for the quantification steps in the case studies of Chapters 3 and 4.

The chapter immediately to follow, Chapter 2, presents the underlying methodology of quantitative risk assessment. For the reader who is not interested in the details of how to quantify risks, but only interested in the types of results that can be obtained by such methods, there is the choice of going directly to the case studies beginning with Chapter 3.

References

[1] Bostrom, N., "Existential Risks – Analyzing Human Extinction Scenarios and Related Hazards," *J. Evol. Technol.* 2002, *9*.

[2] Bernstein, P. L., *Against the Gods, The Remarkable Story of Risk*, Wiley, New York, 1996.

[3] Fullwood, R. F., *Probabilistic Safety Assessment in the Chemical and Nuclear Industries*, Butterworth-Heinemann, Oxford, 2000, 1968.

[4] Garrick, B. J., 1968. "Unified Systems Safety Analysis for Nuclear Power Plants," Ph.D. Thesis, University of California, Los Angeles.

[5] Haimes, Y. Y., *Risk Modeling, Assessment, and Management*, Wiley, New York, 1998.

[6] Lewis, H. W., *Technological Risk*, W.W. Norton & Company, New York, 1990.

[7] Molak, V. (Ed.), *Fundamentals of Risk Analysis and Risk Management*, CRC Press, Boca Raton, FL, 1997.

[8] NRC (National Research Council), *Improving Risk Communication*, National Academy Press, Washington, DC, 1989.

[9] U.S. Nuclear Regulatory Commission, 1975 "Reactor Safety Study: An Assessment of Accident Risks in U.S. Commercial Nuclear Power Plants," WASH-1400 (NUREG-75/014).

[10] Kaplan, S., Garrick, B. J., "On the Quantitative Definition of Risk," *Risk Anal.* 1981, *1*(1), 11–27.

[11] Thomson, W., *Lord Kelvin Quotations, Electrical Units of Measurement, PLA*, 1883, Vol. 1, 1883–05–03. http://zapatopi.net/kelvin/quotes/

[12] Jaynes, E. T., *Probability Theory: The Logic of Science*, Cambridge University Press, UK, 2003.

CHAPTER 2

Analytical Foundations of Quantitative Risk Assessment

One of the greatest challenges facing anyone attempting to discuss the analytical basis of quantitative risk assessment is using a language that is reasonably consistent within the risk science community and at the same time is understandable.[a] Consistency is the real problem. It is too easy to pick up two papers or two books on risk assessment and find the unsettling situation of both using the same words but with different meanings, or different words with the same meaning. This dilemma has been discussed elsewhere[1] and is only mentioned here as a reason for making a special effort to put forth clear definitions for such key terms as "risk," "scenario," "frequency," "likelihood," and "probability."

The elements of quantitative risk assessment as described in this Chapter are (1) a definition of risk that can serve as a general framework for what we mean by risk and can be applied to any type of risk, (2) a scenario approach that clearly links initial (*initiating events* (*IEs*) or *initial conditions* (*ICs*)) and final states (*consequences*) with well defined intervening events and processes, (3) the representation of uncertainty by a probability distribution (the *probability of frequency* concept), (4) a definition of probability that measures the credibility of a hypothesis based on the supporting evidence, and (5) the information processing according to Bayes theorem, the fundamental principle governing *inferential reasoning*.

[a] Appendix A to this book provides a brief account of the historical development of contemporary quantitative risk assessment.

2.1 Quantitative Definition of Risk

In order to properly support the process of decision-making in the face of large uncertainties, it is highly desirable, as well as highly illuminating, to quantify the risks associated with each of the decision options. For this purpose it is essential to define the concept of "risk" in such a way that it can be rigorously quantified. This purpose has led us to what is called the "set of triplets" definition of risk, which has its roots in the Refs. 2 and 3. That definition is,

$$R = \{\langle S_i, L_i, X_i \rangle\}_c,$$

where R denotes the risk attendant to the system or activity of interest. On the right, S_i denotes the ith risk scenario (a description of something that can go wrong), L_i denotes the likelihood of that scenario happening, and X_i denotes the consequences of that scenario if it does happen. The angle brackets $\langle\ \rangle$ enclose the risk triplets, the curly brackets $\{\ \}$ are MathSpeak for "the set of," and the subscript c denotes "complete," meaning that all, or at least all of the important scenarios, must be included in the set. The body of methods used to identify the scenarios (S_i) constitutes what we call the "Theory of Scenario Structuring."[4] Quantifying the L_i and the X_i is done, from the available evidence, using Bayes theorem, as we will illustrate later.

In accordance with this "set of triplets" definition of risk, the actual quantification of risk consists of answering the following three questions:

1. What can go wrong? (S_i)
2. How likely is that to happen? (L_i)
3. What are the consequences if it does happen? (X_i)

The first question is answered by describing a structured, organized, and complete set of possible risk scenarios. As above, we denote these scenarios by S_i. The second question requires us to calculate the "likelihoods," L_i, of each of the scenarios, S_i. Each such likelihood, L_i, is expressed either as a "frequency," a "probability," or a "probability of frequency" curve (more about this later).

The third question is answered by describing the "damage states" or "end states" (denoted X_i) resulting from these risk scenarios. These damage states are also, in general, uncertain. Therefore these uncertainties must also be quantified, as part of the QRA process. Indeed, it is part of the QRA philosophy to quantify all the uncertainties in all the parameters of the risk assessment.

Some authors have added other questions to the above definition such as "What are the uncertainties?" and "What corrective actions should be taken?" As already discussed and will be detailed later, the uncertainty question is embedded in the interpretation of "likelihood." The question about corrective actions is interpreted here as a matter of decision analysis and risk management, but not risk assessment per se. Therefore it is not considered a fundamental property of the definition of risk. Risk assessment does become involved to determine the impact of the corrective actions on the "new risk" of the affected systems.

2.2 The Scenario Approach to Quantitative Risk Assessment

We now describe the scenario approach to QRA. To do this we introduce the term "scenario structuring" to describe the process of identifying, categorizing, and portraying the risk scenarios S_i. Convenient categories, for example, are scenarios that originate as failures within a system such as equipment failures in an engineered facility, scenarios that originate as a result of natural threats external to a system such as earthquakes, and "terrorism scenarios" resulting from deliberate terrorist attacks or sabotage.

The first step in the process of structuring scenarios is to develop an event sequence diagram describing "the success" or "as planned," scenario (S_0). This scenario, after going through N events, leads to the successful, or "as planned" end-state, denoted ES_0. It describes the normal operating procedures for the system in the absence of any undesired initiating events (IEs). In other words, the success scenario describes the functioning of the system when it is working "as planned." It usually, but not necessarily, has a linear structure of N events as depicted in Figure 2.1.

Given the Figure 2.1 portrayal of the success scenario S_0, the possible failure scenarios can now be depicted as departures from S_0. For example, in Figure 2.2 a failure scenario is shown budding off the S_0 at node i into various end states (ES) or damage states.

A risk assessment asks, "What can go wrong?" with each part of the S_0 diagram, especially those parts considered most vulnerable to the risk being considered. The answers to this question are termed "the IEs" because they initiate the risk scenarios (S_i).

Given then that an IE has occurred at node i, as shown in Figure 2.2, a new event tree then emerges as shown. Each path through this tree represents a risk scenario and ends up at an end state (ES_i), as a result of the initiating event (IE_i).

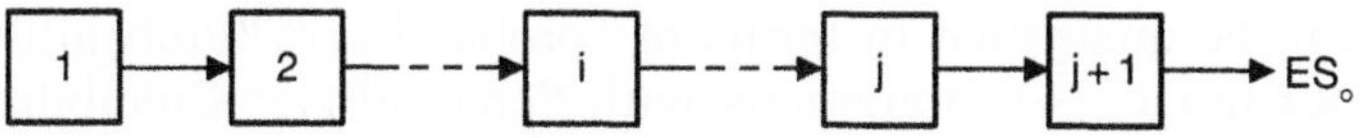

FIGURE 2.1. Diagram of a success scenario.

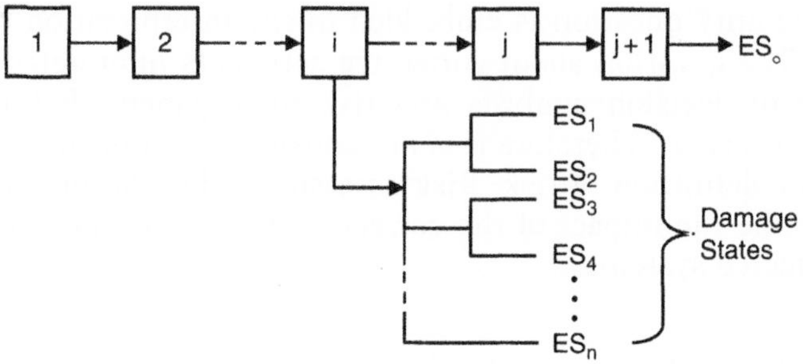

FIGURE 2.2. A "what can go wrong" event tree emerging from the success scenario.

The "what can go wrong" question has to be specialized to the nature of the IE. For example, for a sabotage or terrorism IE the question has to be given a different spin. Here the question becomes "How can the terrorist *make something go wrong* that will achieve his desired outcome?" The tasks of identifying such "IEs" and assessing their likelihoods are the major steps in applying QRA techniques to terrorism risk. The matter of using event trees and other logic diagrams to aid in the quantification of IEs and scenarios is covered in Section 2.4.3.

2.3 Interpretation of Probability and Likelihood

To quantify the likelihood of risk scenarios, it is first necessary to define the concept of likelihood in such a way that it can be quantified. So far, we have purposely used the term "likelihood" as a general, intuitive expression in the triplet definition of risk. Now we describe four explicit and quantitative interpretations of likelihood. These are "frequency," "probability," "credibility," and "probability of frequency." Two of the four, probability and credibility, in our usage are synonymous.

- *Frequency:* If the scenario is recurrent, that is, if it happens repeatedly, then the question "How frequently?" can be asked, and the answer can be expressed in occurrences per day, per year, per trial, per demand, etc.
- *Probability and credibility:* If the scenario is not recurrent (i.e., if it happens either once or not at all), then its likelihood can be quantified in terms of "probability." "Probability," in our usage, is synonymous with "credibility." Credibility is a scale that we humans have invented to quantitatively measure "degree of believability" of a hypothesis, in the same way that

we invented scales to measure distance, weight, temperature, etc. Thus, in our usage, *"probability" is the degree of credibility of the hypothesis in question, based on the totality of relevant evidence available.*

- *Probability of frequency:* If the scenario is recurrent (like a "hurricane" for example) and therefore has a frequency, but the numerical value of that frequency is not fully known, and if there is some evidence relevant to that numerical value, then Bayes theorem (as the fundamental principle governing the process of making inference from evidence) can be used to develop a probability curve over a frequency axis. This "probability of frequency" interpretation of "likelihood" is often the most informative, and thus is the preferred way of capturing/quantifying the state of knowledge about the likelihood of a specific scenario.

Having defined our meaning of "probability," it is of interest to note that it emerges also from what some call the "subjectivist" view of probability as best expressed by the physicist E.T. Jaynes:[5]

> *A probability assignment is 'subjective' in the sense that it describes a state of knowledge rather than any property of the 'real' world, but is 'objective' in the sense that it is independent of the personality of the user. Two rational beings faced with the same total background of knowledge must assign the same probabilities.*

The central idea of Jaynes is to bypass opinions and seek out the underlying evidence for the opinions, which thereby become more "objective" and less subjective.

We agree wholeheartedly with Jaynes' statement and, in our usage, go yet a step further. We define "probability" as synonymous with "credibility." We can thus speak, and think, in terms of the "credibility" of a hypothesis based on all the evidence available. This "credibility" interpretation of probability is thus a positive number ranging from zero to one, and it obeys Bayes theorem. Thus, if we write $p(H|E)$ to denote the credibility of hypothesis H, given evidence E, then

$$p(H|E) = p(H)\frac{p(E|H)}{p(E)} \tag{2.1}$$

which is Bayes theorem (derived in the next section), and which tells us how the credibility of hypothesis H changes when new evidence E, occurs. It does that without overt reference to a "user" or "sentient being"—it is completely objective as it is only evidence based, not opinion or personality based.

The debate between the so-called subjectivists, or Bayesians, and the frequentists, or classical statisticians, is legendary and has been going on for over 200 years. This debate has been the subject of textbooks and scientific articles on probability since the time of LaPlace and Bayes, a few of which are referenced.[6–8]

2.4 Quantification of the Scenarios

2.4.1 Bayes Theorem

As we have indicated, a central feature of "quantitative" risk assessment is making uncertainty an inherent part of the analysis. Uncertainty exists, to varying degrees, in all the parameters that are used to describe or measure risk. Of course there are other sources of uncertainty than parameter uncertainty such as the uncertainty that a particular phenomenon is being correctly modeled, that is, modeling uncertainty. A common approach to assessing modeling uncertainty is to apply different models to the same calculation in an attempt to expose modeling variability. Adjustments are made to the model to increase confidence in the results. The lack of confidence resulting from such an analysis can be a basis for assigning a modeling uncertainty component to parameter uncertainty for an improved characterization of the total uncertainty of the analysis.

In QRA, parameter uncertainties are quantified in the form of probability curves against the possible values of these parameters. This leads to the next question: How are these probability curves obtained? The answer is that they are inferred from all the available evidence, using the fundamental mathematical principle of logical inference, known as "Bayes theorem." This theorem has a long and bitterly controversial history, but in recent years has become widely understood and accepted, a view strongly supported by the authors of this book. For example, it has been characterized by Bernstein,[9] as:

> *"a striking advance in statistics by demonstrating how to make better informed decisions by mathematically blending new information into old information."*

Simply put, Bayes theorem is the fundamental logic principle governing the process of inferential reasoning. Specifically, the theorem answers the question: "How does the probability $p(H)$, of a given hypothesis H, change when we obtain a new piece of evidence E?"

The answer is very simply derived as follows: Let $p(H|E)$ denote the new probability of H, given that we now have the evidence E. Also, let $p(H \wedge E)$ denote the probability that both H and E are true. Then,

$$p(H \wedge E) = p(E)p(H|E) \tag{2.2}$$

This equation simply says that the probability of both H and E being true is equal to the probability that E is true times the probability that H is true, given the evidence E.

In the same way,

$$p(H \wedge E) = p(H)p(E|H) \tag{2.3}$$

Now note that the left hand sides of Equations (2.2) and (2.3) are the same. Therefore the two right hand sides must also be equal. Thus

$$p(E)p(H|E) = p(H)p(E|H) \tag{2.4}$$

and dividing Equation (2.4) by $p(E)$, we obtain

$$p(H|E) = p(H)\frac{p(E|H)}{p(E)} \tag{2.5}$$

This equation is Bayes theorem. It tells us how the probability (or, as we can also call it, the "credibility") of a hypothesis, H, changes when we learn a new piece of evidence, E.

An Example of Bayes Theorem in Action

As an example, suppose the hypothesis H stands for "terrorists preparing a bombing attack on the local high school." Suppose E stands for the evidence of *terrorists buying explosives.* Now, if the terrorists were preparing such an attack, H, then the likelihood that they would be acquiring explosives, E, goes up. Thus $p(E|H)$ is larger than $p(E)$ and from Equation (2.5), $p(H|E)$ would be larger than $p(H)$. Thus we see Equation (2.5) reflecting the common sense reasoning that if the terrorists are acquiring explosives, it probably means that they are planning to use them. (However, see footnotes.[b,c])

[b] Of course, any poker player would point out that it might be that the buying of explosives is a feint, a move to distract our attention while they really get ready to poison the water supply. Thus, we must act to protect ourselves against both the explosives and the poison. So goes the poker game of life.

[c] Because of the widespread confusion resulting from the many different definitions in use of the word "probability," we hereby, in this footnote, inform the reader that in the present document we use the word "probability" in the sense of "credibility," as in "the degree of credibility of a hypothesis based on all the relevant information available."

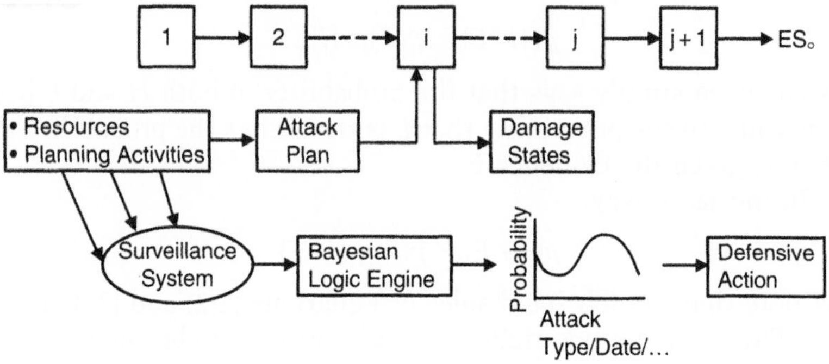

FIGURE 2.3. Surveillance of threats to systems.

The Use of Bayes Theorem for Surveillance

In this age of terrorism, Bayes theorem can be programmed into "surveillance computers" in such a way that when items of relevant evidence become noted, the computers will call attention to the fact that a terrorist action may be under preparation. The concept is illustrated in Figure 2.3. Given this forewarning, appropriate defensive action can be taken.

It should be noted in this connection that the "computers" mentioned above are not necessarily electronic machines. It could be a human being surveying the incoming evidence. Human brains also operate according to Bayes theorem, but they are not as fast or as reliable. It is best to combine the strengths of both. Thus, the computers, armed with Bayes theorem, help humans to "connect the dots" between different sources of information, and then to take corrective actions.

2.4.2 Initial Conditions and Initiating Events

Before the risk scenarios themselves can be quantified, the IEs or the initial conditions (ICs) of the risk scenarios must be identified and quantified.[d] The relationships between the initial states (IEs and ICs), the system being impacted, and the vulnerability of the system are illustrated in Figure 2.4.

That is, a QRA basically involves three analytical processes: (1) a system analysis that defines the system in terms of what constitutes success, (2) an IE and IC assessment that quantifies the threats to the system, and (3) a vulnerability assessment that quantifies the resulting risk scenarios and different damage states.

[d] Both IE and IC terminology are used, since for some systems such as the risk of a nuclear waste repository the issue is not so much an initiating event as it is a set of initial conditions such as annual rainfall.

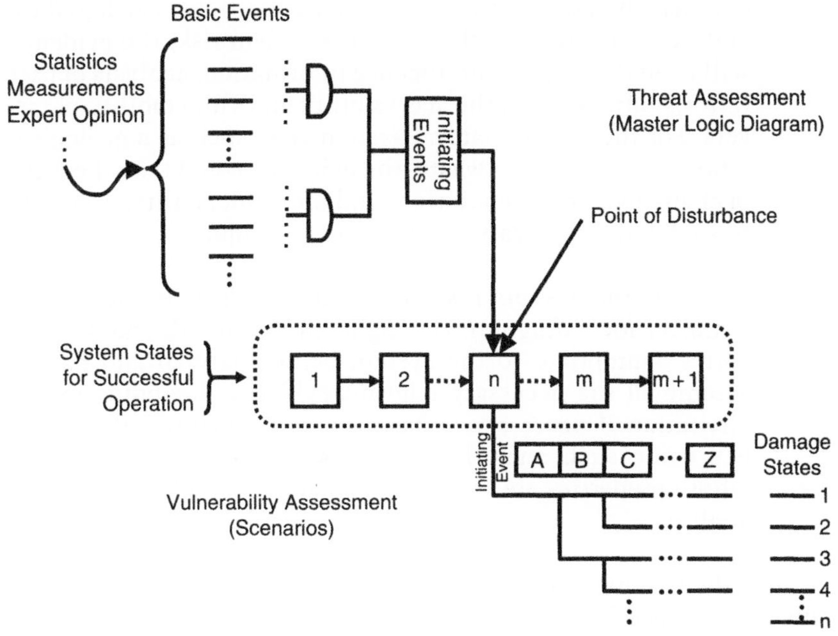

FIGURE 2.4. The concept of an integrated threat and vulnerability risk assessment.

A QRA-based IE or IC assessment involves the following:

1. *Identify IE or IC categories that could trigger catastrophic consequences.* This is a case of being able to identify the events that could disrupt an otherwise successfully operating system, be it a natural system, an engineered system, or a system that could be the target of such an event as a terrorist attack. The principal information needed for identifying such events is an understanding of how the system of interest works and its points of vulnerability to disruptions that could lead to catastrophic consequences. Criteria are necessary for screening out those ICs or IEs that would most likely not lead to catastrophic consequences.
2. *Develop the supporting evidence for the selected IEs and ICs.* For the case of technological system risk, the evidence takes the form of experience with similar systems in terms of system and equipment failure rates; construction, manufacturing, and maintenance practices; personnel training; success histories; and design basis. Location of the facility is also an important part of the evidence for determining IEs because of such

external threats as extreme weather conditions, earthquakes, and external fires. For the case of terrorism risk, the evidence will be in the form of intelligence information, analysis of past terrorist attacks, and the accessibility and vulnerability of targets. For the case of natural system risk, such as a geological repository for nuclear waste, the evidence could take the form of changing climate conditions and the vulnerability of the site to such natural threats as volcanoes, earthquakes, and human intrusion.

3. *Quantify the ICs and IEs using deductive logic in the form of a master logic diagram.* Having identified the IE and IC categories appropriate to the system being analyzed and having assembled the necessary relevant evidence, including expert elicitation if necessary, a basis exists to quantify the IEs or ICs using such analytical tools as a master logic diagram, whose properties are well established in the field of fault tree analysis.

A deductive logic model, that is, a fault tree or master logic diagram, is developed for each IE of the screened set. The structure of the logic model is to deduce from the "top events" that is, the selected set of hypothetical IEs or ICs the intervening events down to the point of "basic events." A "basic event" can be thought of as the initial input point for a deductive logic model of the failure paths of a system. For the case of accident risk, a basic event might be fundamental information on the behavior of structures, components, and equipment. For the case of a natural system such as a nuclear waste disposal site, a basic event could be a change in the ICs having to do with climate brought about by greenhouse gases. For the case of terrorism risk, the basic event relates to the *intentions* of the terrorist, that is, the *decision to launch* an attack. The intervening events of the master logic diagram for terrorism risk are representations of the planning, training, logistics, resources, activities, and capabilities of the terrorists. The intervening events of the master logic diagram for accident risk are the processes and activities that lead to the failure of structures, components, and equipment. The intervening events of the ICs for a nuclear waste disposal site could be factors that influence climate conditions.

2.4.3 The Event Tree Structure

The actual quantification of the risk scenarios is done with the aid of event trees like that in Figure 2.5. An event tree is a diagram that traces the response of a system to an IE, such as a terrorist attack, to different possible end points or outcomes (consequences). A single

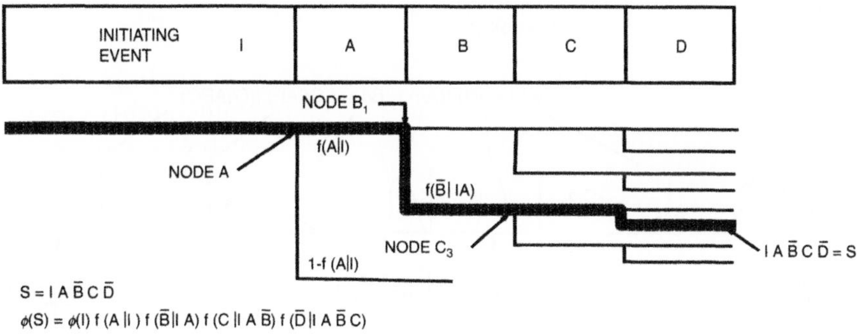

FIGURE 2.5. Quantification of a scenario using an event tree.

path through the event tree is a "scenario" or an "event sequence." The terms are sometimes used interchangeably. The event tree displays the systems, equipment, human actions, procedures, processes, etc., that can impact the consequences of an IE depending on the success or failure of intervening actions. In Figure 2.5, boxes with the letters A, B, C, and D represent these intervening actions. The general convention is that if the defensive action is successful, the scenario is mitigated. If the action is unsuccessful, then the effect of the IE continues as a downward line from the branch point as shown in Figure 2.5. For accident risk, an example of a mitigating system might be a source of emergency power. For terrorism risk, an action that could mitigate the hijacking of a commercial airliner to use it as a weapon to crash into a football stadium would be a remote takeover of the airplane by ground control. For a natural system, a mitigating feature might be an engineered barrier.

Each branch point in the event tree has a probability associated with it. It should be noted that the diagram shown in Figure 2.5 shows only two branches (e.g., success or failure). However, an event tree can have multiple branches to account for different degrees of degradation of a system. These branch points have associated "split fractions" that must be quantified based on the available evidence. The process involves writing an equation for each scenario (event sequence) of interest. For example, the path through the event tree that has been highlighted in Figure 2.5 could be a scenario that we wish to quantify. The first step is to write a Boolean equation, an algebraic expression, for the highlighted path. If we denote the scenario by the letter *S*, we have the following equation,

$$S = IA\bar{B}C\bar{D},$$

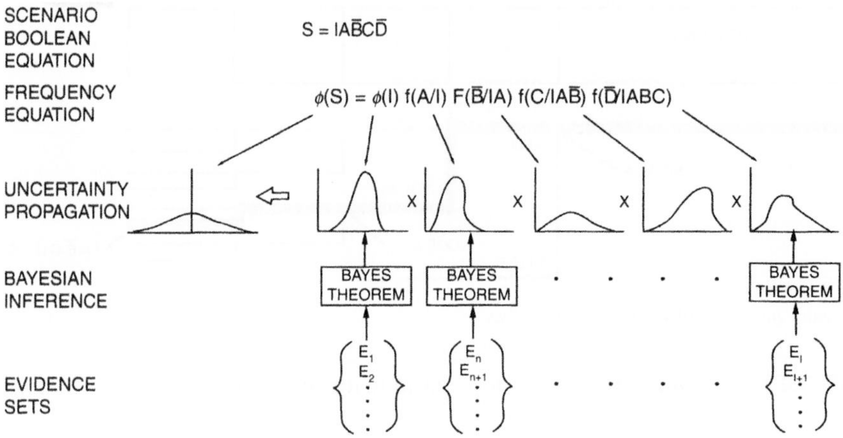

FIGURE 2.6. Bayes theorem used to process parameters.

where the bars over the letters indicate that the event in the box did not perform its intended function. The next step is to convert the Boolean equation into a numerical calculation of the frequency of the scenario. Letting φ stand for frequency and adopting the split fraction notation, $f(\ldots)$, of Figure 2.5, gives the following equation for calculating the frequency of the highlighted scenario,

$$\varphi(S) = \varphi(I)f(A|I)f(\bar{B}|IA)f(C|IA\bar{B})f(\bar{D}|IA\bar{B}C)$$

The remaining step is to communicate the uncertainties in the frequencies with the appropriate probability distributions. This is done using Bayes theorem to process the elemental parameters (Figure 2.6). The "probability of frequency" of the individual scenarios is obtained by convoluting the elemental parameters in accordance with the above equation.

2.5 Assembling the Results

Once the scenarios have been quantified, the results take the form of Figure 2.7. Each scenario has a probability-of-frequency curve in the form of a probability density function quantifying its likelihood of occurrence. The total area under the curve represents a probability of 1. The fractional area between two values of Φ represents the confidence, that is, the probability that Φ has the values over that interval (see below).

Figure 2.7 shows the curve for a single scenario or a set of scenarios leading to a single consequence. Showing different levels of damage, such as the risk of varying injuries or fatalities, requires a different type

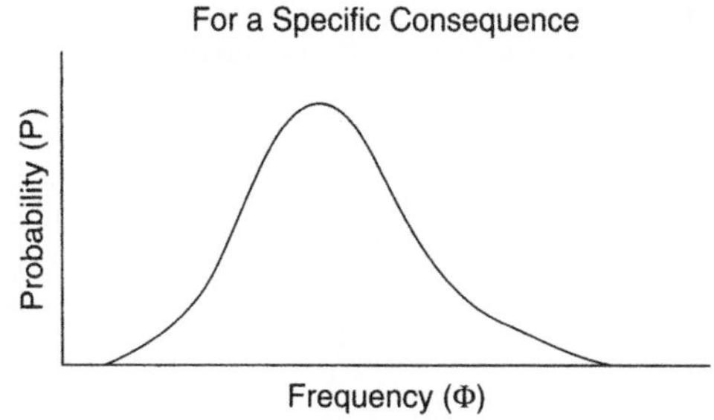

FIGURE 2.7. Probability of frequency curve.

of presentation. The most common form is the classical risk curve, also known as the frequency-of-exceedance curve, or the even more esoteric label, the complementary-cumulative-distribution-function. This curve is constructed by ordering the scenarios by increasing levels of damage and cumulating the probabilities from the bottom up in the ordered set against the different damage levels. Plotting the results on log–log paper generates curves as shown in Figure 2.8.

To illustrate how to read Figure 2.8, suppose P_3 has the value of 0.95, that is a probability of 0.95, and suppose we want to know the risk of an X_1 consequence at the 95% confidence level. According to the figure, we are 95% confident that the frequency of an X_1 consequence or greater is Φ_2. The family of curves (usually called percentiles) can include as many curves as necessary. The ones most often selected in practice are the 5th, 50th, and 95th percentiles. A popular fourth choice is the mean.

An often used method of communicating uncertainty in the risk of an event is to present the risk in terms of a confidence interval. To illustrate confidence intervals some notation is added to the above figures, which now become Figures 2.9 and 2.10. If the area between

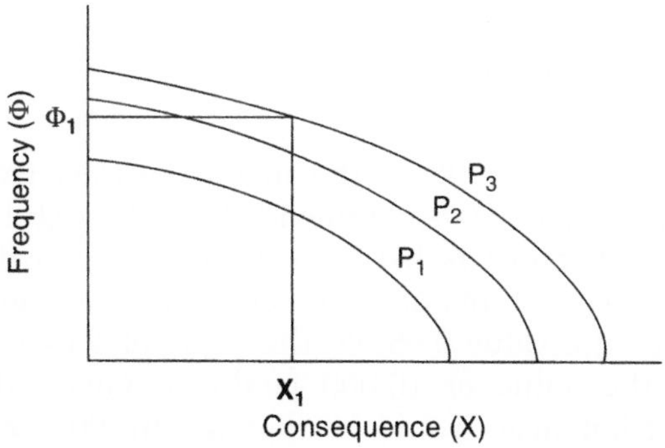

FIGURE 2.8. Risk curves for varying consequences.

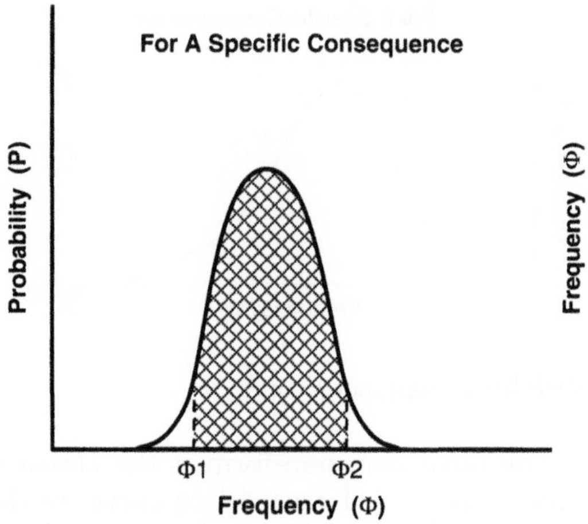

FIGURE 2.9. Probability density.

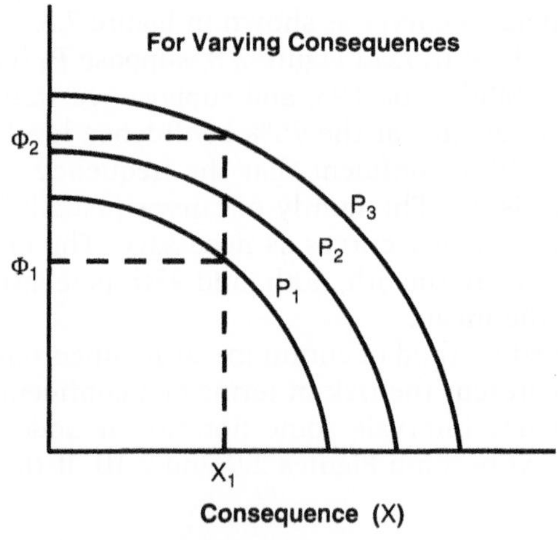

FIGURE 2.10. Cumulative probability.

Φ_1 and Φ_2 of Figure 2.9 is 90% of the area under the curve, the way to read this result is we are 90% confident (the 90% confidence interval) that the frequency range is between Φ_1 and Φ_2. To illustrate how to read Figure 2.10 in terms of a confidence interval, let P_1 have the value of 0.05, P_3 the value 0.95, Φ_1 the value of 1 in 10,000, Φ_2 1 in 1000, and X_1 the value of 10,000 fatalities. Given that $P_3 - P_1$ is 0.90 the proper language would be that we are 90% confident that the frequency of a 10,000-fatality consequence or greater varies from one every 10,000 years to as much as one every 1000 years.

Although risk measures such as those illustrated in Figures 2.9 and 2.10 answer the questions of what is the risk and how much confidence is there in the results, they are not necessarily the most important output of the risk assessment. Often the most important output is the exposure of the contributors to the risk—a critical result needed for effective risk management. The contributors are buried in the results assembled to generate the curves in Figures 2.9 and 2.10. Most risk assessment software packages contain algorithms for ranking the importance of contributors to the chosen risk measures.

Chapters 3 and 4 demonstrate explicitly the application of the methodology of this chapter to the risk of hurricanes in New Orleans, Louisiana, and the risk of high energy asteroids impacting the contiguous 48 states of the U.S. Chapters 5 and 6, while not as complete in scope and are referred to as scoping analyses, also demonstrate many of the above methods of risk assessment.

References

[1] Kaplan, S., "The Words of Risk Analysis," *Risk Analysis* 1997, *17*(4), 407–416.

[2] Garrick, B. J., *Unified Systems Safety Analysis for Nuclear Power Plants*, Ph.D. Thesis, University of California, Los Angeles, 1968.

[3] Kaplan, S., Garrick, B. J., "On the Quantitative Definition of Risk," *Risk Analysis* 1981, *1*(1), 11–27.

[4] Kaplan, S., Haimes, Y. Y., Garrick, B. J., "Fitting Hierarchical Holographic Modeling into the Theory of Scenario Structuring and a Resulting Refinement to the Quantitative Definition of Risk," *Risk Analysis* 2001, *21*(5), 807–819.

[5] Jaynes, E. T., *Probability Theory; The Logic of Science*, Cambridge University Press, 2003.

[6] Apostolakis, G., "The Concept of Probability in Safety Assessments of Technological Systems," *Science* 1990, *250*, 1359–1364.

[7] De Finetti, B., *Theory of Probability*, Vols 1 and 2, Wiley, New York, 1974.

[8] Lindley, D. V., *Making Decisions*, 2nd edition, Wiley, London, 1985.

[9] Bernstein, P. L., *Against the Gods: The Remarkable Story of Risk*, Wiley, New York, 1996.

CHAPTER 3

Case Study 1: Risk of a Catastrophic Hurricane in New Orleans, LA

The Katrina Hurricane

The input to the quantitative risk assessment in this chapter was completed in July 2005, prior to the landfall of hurricane Katrina in southeast Louisiana. In January 2006, the data available from the National Hurricane Center concerning hurricane Katrina was examined to consider revising our assessment. As a result of this examination, it was concluded that the Katrina hurricane would not significantly change the overall results of the quantitative risk assessment because of the strength of the evidence represented by the database that was used (1900–2004). Thus, the quantitative risk assessment was not changed except for minor edits. That is, the results of the risk assessment based on the history of hurricanes between 1900 and 2004 indicate that the occurrence of a hurricane of the severity of Katrina should not be a surprise. This in itself is an extremely important testament to the value of quantitative risk assessment. Certainly, any future risk assessment for New Orleans should include the Katrina event and data from all hurricanes after 2004. One important outcome of such an update might be to reduce the uncertainties in the results.

This chapter is a limited scope quantitative risk assessment of the fatalities from a major hurricane that directly impacts New Orleans, LA. To better calibrate how the hurricane risk changed over the 20th century, two risk assessments were actually performed. The first is based on hurricane data from 1900 to 2004, and the second is based on data from 1900 to only 1950. Prior to presenting the case study, it is appropriate to provide some background material and a summary of the study results.

Over the 20th century, the reduction in the number of fatalities following major hurricanes (defined as Category 3, 4, or 5) (see Major Hurricane section for a definition of hurricane categories) making landfall on the continental United States of America (U.S.A.) has been a success story as indicated in Table 3.1 and Figure 3.1.

In the last three decades of the 20th century there were approximately 600 human fatalities from hurricanes versus approximately 4000 in the first three decades.

The factors that impacted the reduction in actual fatalities include:

- Substantial improvements in the ability to forecast the track and intensity of hurricanes using improved equipment and information from both private and government organizations.
- The efforts of people in all levels of government and private organizations in the U.S. (national, state, and local) involved in mass evacuations before hurricane landfall.
- Substantial improvements in the technology of building structures to withstand hurricanes, including the work of professional engineering societies to define new building codes and construction practices.

Because of these factors, the actual number of fatalities from hurricanes in the U.S. has been reduced by almost a factor of 10 over the course of the 20th century. This has been accomplished in spite of a substantial increase in the population along the southeast Atlantic coast and the coast of the Gulf of Mexico.

TABLE 3.1
Atlantic hurricane fatalities in the U.S. during the 20th century

Year	*Past 10 Years*	*Past 20 Years*	*Past 30 Years*
1901	Start	Start	Start
1910	761	NA	NA
1920	1008	1769	NA
1930	2122	3130	3891
1940	1197	3319	4327
1950	216	1362	3486
1960	926	1110	2307
1970	531	1457	1641
1980	226	757	1683
1990	140	366	897
2000	242	382	608

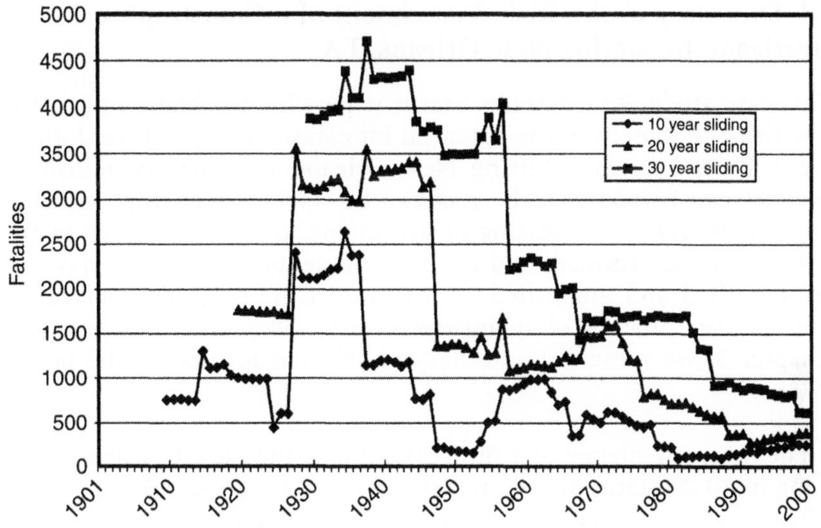

FIGURE 3.1. Fatalities—Atlantic hurricanes.

In the continental U.S., the last catastrophic hurricane (defined here as causing approximately 10,000 fatalities or greater) occurred in the year 1900 when a Category 4 hurricane made landfall at Galveston, Texas, and there were more than 8000 fatalities. In 1928, a Category 4 hurricane made landfall in southeast Florida killing 1836 people (the most deadly hurricane in the 20th century), but this event is not considered catastrophic since the number of fatalities did not approach our threshold for a catastrophe, namely 10,000 fatalities.

In spite of this reduction in the number of fatalities, there still exists the potential for a catastrophic hurricane in the U.S. given special circumstances. One such circumstance would occur if a major hurricane made landfall directly over a large metropolitan area combined with limited evacuation of the population.

It should be observed that the risk of fatalities is a much different issue than the risk of property damage. Unlike people, most personal property cannot be evacuated. For example, the impact of hurricanes on personal property increased substantially during the 20th century due to increased building and development along the southeast coast of the U.S. In the 21st century, it is expected that most major hurricanes making landfall in the continental U.S. will cause damage to personal property in the billions of dollars. For example, in September 2004, hurricane Ivan (Category 3 at landfall in Florida) resulted in only about 26 fatalities, but the property damage estimate was 10–15 billion dollars.

3.1 Summary of the Risk Assessment of a Catastrophic Hurricane Impacting New Orleans, LA

This case study demonstrates that it is possible to obtain a quantitative risk assessment of the potential for catastrophic fatalities following a major hurricane hitting New Orleans and that the following benefits can be realized: the risk can be tracked over time to determine if the risk is increasing or decreasing, the uncertainties in the risk can be determined, and the key elements that impact risk can be identified and quantified in importance. Prior to demonstrating the six-step process of quantitative risk assessment discussed in Chapter 1, we present a summary of selected results for the period 1900–2004.

- The mean frequency of a hurricane impacting New Orleans resulting in 10,000 fatalities or greater is 1 in 130 years (see Figure 3.2).
- The mean number of fatalities per year from major hurricanes impacting New Orleans is approximately 400 (see Table 3.2).
- The most likely hurricane to cause catastrophic fatalities is a Category 4 hurricane at landfall because of the combination of its frequency of occurrence and its consequence. The percentage of total

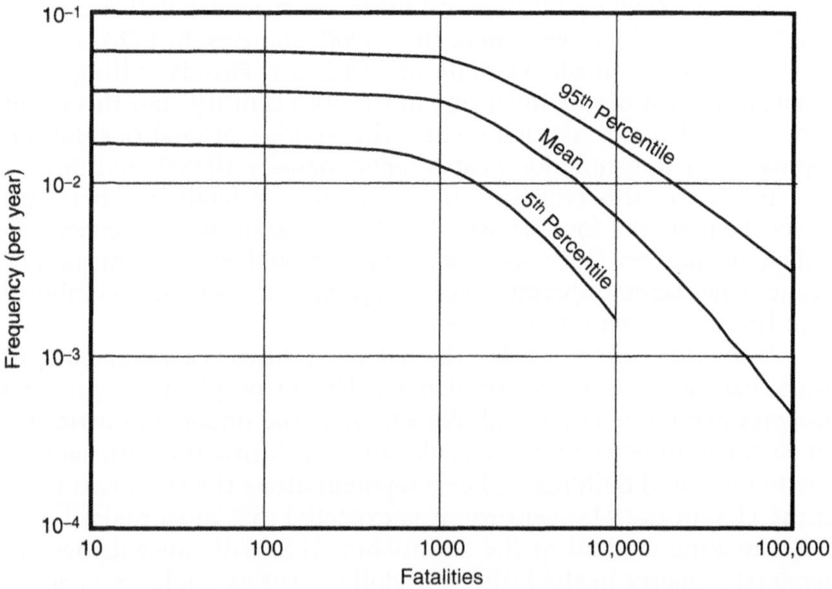

FIGURE 3.2. Hurricane fatality risk for New Orleans (based on U.S. hurricane data 1900–2004).

TABLE 3.2

Mean annual frequencies of U.S. hurricane risk scenarios and split fractions of intervening events[a] (based on U.S. hurricane data 1900–2004)

Risk Scenario									
Initiating Events (Frequency/ Year)	*Hurricane Category*[b]	*Evacuation Type*[c]	*New Orleans Impact Fraction*	*Category Fraction at Landfall*	*Evacuation Type Fraction*	*Damage State*[d]	*Risk Scenario (Frequency/ Year)*	*Fatalities of Risk Scenario*	*Annualized Fatalities*
Initiating Event 1									
0.176	5	Minimal	0.105	0.036	0.390	1	2.57E–04	104,000	27
0.176	5	Medium	0.105	0.036	0.590	2	3.89E–04	52,000	20
0.176	5	Full	0.105	0.036	0.000	4	0.00E+00	2600	0
0.176	4	Minimal	0.105	0.241	0.290	2	1.29E–03	52,000	67
0.176	4	Medium	0.105	0.241	0.690	3	3.07E–03	26,000	80
0.176	4	Full	0.105	0.241	0.000	5	0.00E+00	1300	0
0.176	3	Minimal	0.105	0.732	0.200	4	2.71E–03	2600	7
0.176	3	Medium	0.105	0.732	0.795	5	1.08E–02	1300	14
0.176	3	Full	0.105	0.732	0.000	6	0.00E+00	65	0
Initiating Event 1 total							1.85E–02		215
Initiating Event 2									
0.170	5	Minimal	0.085	0.046	0.290	1	1.94E–04	104,000	20
0.170	5	Medium	0.085	0.046	0.690	2	4.61E–04	52,000	24
0.170	5	Full	0.085	0.046	0.000	4	0.00E+00	2600	0
0.170	4	Minimal	0.085	0.251	0.200	2	7.23E–04	52,000	38
0.170	4	Medium	0.085	0.251	0.795	3	2.87E–03	26,000	75
0.170	4	Full	0.085	0.251	0.000	5	0.00E+00	1300	0

(Continued)

TABLE 3.2 (*Continued*)

Risk Scenario									
Initiating Events (Frequency/ Year)	*Hurricane Category*[b]	*Evacuation Type*[c]	*New Orleans Impact Fraction*	*Category Fraction at Landfall*	*Evacuation Type Fraction*	*Damage State*[d]	*Risk Scenario (Frequency/ Year)*	*Fatalities of Risk Scenario*	*Annualized Fatalities*
0.170	3	Minimal	0.085	0.682	0.110	4	1.08E–03	2600	3
0.170	3	Medium	0.085	0.682	0.890	5	8.74E–03	1300	11
0.170	3	Full	0.085	0.682	0.000	6	0.00E+00	65	0
Initiating Event 2 total							1.41E–02		171
Initiating Event 3									
0.054	5	Minimal	0.035	0.042	0.200	1	1.57E–05	104,000	1.63
0.054	5	Medium	0.035	0.042	0.690	2	5.41E–05	52,000	2.81
0.054	5	Full	0.035	0.042	0.110	4	8.63E–06	2600	0.02
0.054	4	Minimal	0.035	0.160	0.110	2	3.30E–05	52,000	1.71
0.054	4	Medium	0.035	0.160	0.690	3	2.07E–04	26,000	5.37
0.054	4	Full	0.035	0.160	0.200	5	5.99E–05	1300	0.08
0.054	3	Minimal	0.035	0.795	0.110	4	1.64E–04	2600	0.43
0.054	3	Medium	0.035	0.795	0.590	5	8.78E–04	1300	1.14
0.054	3	Full	0.035	0.795	0.290	6	4.32E–04	65	0.03
Initiating Event 3 total							1.85E–03		13
Total							**3.44E–02**		**399**

[a]See event tree Figures 3.3–3.5 for event sequence logic.
[b]Major hurricanes are Categories 3, 4, and 5 hurricanes as classified by the Saffir–Simpson hurricane scale.
[c]Three types of evacuation are defined: full, medium, and minimal corresponding to 1%, 20%, and 40% of the metropolitan population at risk, respectively. The not "at risk" population is assumed to be evacuated.
[d]Damage states are defined in terms of the number of fatalities incurred.

risk (fatalities) from each hurricane category was calculated from Table 3.2 and is shown below.

Hurricane Category at Landfall	*Approximate Percent of Total Risk (Fatalities)*
4	67
5	24
3	9

- The frequency of exceeding 10,000 fatalities from a major hurricane hitting New Orleans was essentially the same in 1950 as in 2004. This conclusion includes the fact that the population of the New Orleans metropolitan area has almost doubled from 1950 to 2004 (from 770,000 to 1.3 million). The number and intensity of hurricanes in the Gulf of Mexico along with the probability of impacting New Orleans for the first part of the 20th century was higher than for the second part of the 20th century. The smaller population in 1950 overcame the higher values for the number and intensity of hurricanes along with a higher probability of impacting New Orleans when the comparison was made for the two quantitative risk assessments (see Section 3.4).
- The frequency of exceeding 100,000 fatalities from a major hurricane impacting New Orleans was substantially smaller in 1950 than in 2004. The mean frequency of exceeding 100,000 fatalities in 1950 was approximately one event in 13,000 years. The mean frequency of exceeding 100,000 fatalities in 2004 was approximately 1 event in 2200 years. Again, the smaller population in 1950 was the dominant factor (see Section 3.4).

Figure 3.2 represents the bottom line risk assessment results for this period. For example, these results indicate that approximately every 130 years a major hurricane will hit New Orleans resulting in 10,000 fatalities or greater. The most likely hurricane to cause catastrophic fatalities is a Category 4 hurricane at landfall because of the combination of its frequency of occurrence and its consequence. In particular, 67% of the fatality risk derives from Category 4 hurricanes at landfall, 24% from Category 5, and 9% from Category 3. These results are based on the calculations shown in Table 3.2.

3.2 Risk Assessment (Based on Data from 1900 to 2004)

Chapter 1 (Section 1.3) identified a six-step process for performing a quantitative risk assessment using the framework of the triplet definition of risk. That process is followed in all the case studies.

The *step number* is identified in the parentheses following the appropriate section in the case study.

3.2.1 Definition of the System During Normal Conditions (Step 1)

For the case of the vulnerability of New Orleans to a catastrophic hurricane, the "system" is the city itself, its infrastructure, and its population. There are a number of unique features of New Orleans that make it a prime candidate for a catastrophic hurricane:

- A substantial portion of the city is below sea level.
- Major hurricanes striking southeast Louisiana or southwest Mississippi include a high portion of Category 4 or 5 hurricanes.
- New Orleans has a large metropolitan population (1.2–1.4 million people) with a substantial minority (100,000–250,000 people) of the population without personal transportation.
- Limited evacuation routes.
- A historical record of flawed evacuation during hurricanes.
- Barrier islands being eroded making the city more vulnerable.

New Orleans is bordered on the south and west by the Mississippi River (the largest river in the United States) and on the north by Lake Pontchartrain. Lake Pontchartrain is a large lake approximately 40 miles east to west and approximately 25 miles north to south. These major bodies of water around New Orleans make the effects of a hurricane more complicated compared to the effects of a hurricane for other large metropolitan areas on the coast of the Gulf of Mexico. In general, major hurricanes making landfall to the west of the center of New Orleans will drive water up the Mississippi River and then potentially into the city; major hurricanes making landfall to the east of the center of New Orleans will potentially drive water from Lake Pontchartrain into the city.

In past years, levees in New Orleans were constructed following certain destructive hurricanes. These levees exist today to protect against flooding from the Mississippi River and Lake Pontchartrain. These levees prevent all but major hurricanes from causing a catastrophic number of fatalities in New Orleans. It should be noted that the possibility of levee failure during a hurricane with a magnitude less than a major hurricane was not considered in the analysis.

If a major hurricane impacts New Orleans and water floods into the city from either the Mississippi River or Lake Pontchartrain due to either failure of the levee or flood surge over the levee, the water will be trapped in the city (bowl effect) until it is pumped out. Some of the New Orleans metropolitan area may remain submerged for many weeks (10 or more) since the large pumps that normally pump water out of the "bowl" during significant rainfall will not work with the entire "bowl" full of water. The flooding will impact many industrial facilities. Hazardous chemicals may be a significant factor in the recovery effort. The flooding will also impact the number of buildings available for use as long-term shelters.

The policy of downtown hotels in New Orleans regarding renting rooms to residents who desire to ride out the storm rather than evacuate is under review.[1] In a hurricane symposium for hotel operators at the Ritz Carlton New Orleans hotel on May 23, 2005, it was reported that public safety officials urged local hoteliers not to accept guests during hurricanes because they would not be able to accommodate them for some types of severe storms. Public safety officials said if a sufficiently severe hurricane does hit New Orleans, hotels might find themselves having to accommodate guests for a few months rather than a few days, which would require having appropriate amounts of food, water, and emergency electrical generators with large amounts of diesel fuel in reserve.

For New Orleans, evacuation is difficult because:

- Interstate 10, a principal evacuation route, is susceptible to flooding from Lake Pontchartrain.
- When wind exceeds 50 miles per hour (mph), the 24-mile Lake Pontchartrain Causeway is closed.
- Traffic gridlock on Interstate 10 is almost guaranteed for some portion of the evacuation period.
- It is estimated that approximately 20–25% of the citizens have no personal transportation. There are not enough buses to move all these people, and the city relies on the National Guard to help. Tourists are also a problem since many come in by airplane and the New Orleans airport is one of the first facilities closed on an approaching hurricane.

The evidence indicates that in the last two major New Orleans evacuations (in 1998 hurricane Georges made landfall near Biloxi, Mississippi, as a Category 2 hurricane and in 2004 hurricane Ivan made landfall just west of Gulf Shores, Alabama, as a Category 3 hurricane), traffic problems hindered evacuation. In the case of hurricane Georges in 1998, construction underway on Interstate 10 prior to the hurricane landfall impeded effective evacuation. In the case of hurricane Ivan in 2004, major portions of Interstate 10 became gridlocked for an extended time.

3.2.2 Identification and Characterization of System Hazards (Step 2)

The circumstances necessary for catastrophic fatalities from a hurricane at New Orleans are

- A major hurricane directly impacting New Orleans.
- Limited evacuation of a substantial portion of the population.

Major Hurricane

With respect to a major hurricane directly impacting New Orleans, the hurricanes considered in the analysis include hurricanes that make landfall as Category 5, Category 4, or selected Category 3 hurricanes on the Saffir–Simpson Hurricane Scale. These categories are defined as follows:

- *Category 5.* Winds greater than 155 mph [135 knots (kt) or 249 kilometers per hour (km/h)]. The storm surge is generally greater than 18 ft above normal. Low-lying escape routes are cut by rising water 3–5 h before arrival of the center of the hurricane. Major damage to lower floors of all structures located less than 15 ft above sea level and within 500 yards of shoreline. Massive evacuation of residential areas on low ground within 5–10 miles of the shoreline may be required.
- *Category 4.* Winds 131–155 mph (114–135 kt or 210–249 km/h). The storm surge is generally 13–18 ft above normal. Low-lying escape routes are cut by rising water 3–5 h before arrival of the center of the hurricane. Major damage to lower floors of all structures located near the shoreline. Terrain lower than 10 ft above sea level may be flooded, requiring massive evacuation of residential areas as far inland as 6 miles.
- *Category 3.* Winds 111–130 mph (96–113 kt or 178–209 km/h). The storm surge is generally 9–12 ft above normal. Low-lying escape routes are cut by rising water 3–5 h before arrival of the center of the hurricane. Terrain continuously lower than 5 ft above mean sea level may be flooded inland 8 miles or more. Evacuation of low-lying residences within several blocks of the shoreline may be required.

The criterion for including a Category 3 hurricane in the risk assessment was the listing of the hurricane in Table 4 of the report by Jarrell *et al.*[2]

Table 3.3 contains 41 major hurricanes included in the quantitative risk assessment ranked in order of minimum pressure at landfall. These 41 major hurricanes were derived from the list of 65 hurricanes in Table 4 of the report by Jarrell *et al.*[2] Thirty-nine of the 65 hurricanes were extracted from Table 4 of the Jarrell report as having made landfall in the Gulf of Mexico and having a chance of hitting New Orleans. Two hurricanes from 2004, hurricanes Charley and Ivan, were added to make the total 41 hurricanes listed in Table 3.3.

TABLE 3.3
Rank of most intense U.S. hurricanes (landfall) in Gulf of Mexico by minimum pressure (based on U.S. hurricane data 1900–2004)

Rank	*Hurricane*	*Year*	*Category at Landfall*	*Minimum Pressure (Millibars)*	*Fatalities*
1	Camille (Mississippi, Southeast Louisiana)	1969	5	909	256
2	Unnamed (Galveston, Texas)	1900	4	931	8000+
2	Unnamed (Grand Isle, Louisiana)	1909	4	931	350
2	Unnamed (New Orleans, Louisiana)	1915	4	931	275
2	Carla (North and Central Texas)	1961	4	931	46
6	Unnamed (Southeast Florida, Southeast Louisiana, Mississippi)	1947	4	940	51
7	Unnamed (North Texas)	1932	4	941	40
7	Charley (Southwest Florida)	2004	4	941	10
9	Opal (Northwest Florida, Alabama)	1995	3	942	59
10	Ivan (Alabama, Pensacola)	2004	3	943	26
11	Unnamed (Galveston, Texas)	1915	4	945	275
11	Audrey (Southwest Louisiana, North Texas)	1957	4	945	390
11	Celia (South Texas)	1970	3	945	15
11	Allen (South Texas)	1980	3	945	24
15	Frederic (Alabama, Mississippi)	1979	3	946	11
16	Unnamed (South Texas)	1916	3	948	
16	Unnamed (Mississippi, Alabama)	1916	3	948	
16	Betsy (Southeast Florida, Southeast Louisiana)	1965	3	948	75
19	Unnamed (South Texas)	1933	3	949	
20	Unnamed (Central Texas)	1942	3	950	
20	Hilda (Central Louisiana)	1964	3	950	38
20	Beulah (South Texas)	1967	3	950	10
23	Bret (South Texas)	1999	3	951	0

(Continued)

TABLE 3.3 (*Continued*)

Rank	*Hurricane*	*Year*	*Category at Landfall*	*Minimum Pressure (Millibars)*	*Fatalities*
24	Unnamed (Tampa Bay Florida)	1921	3	952	
24	Carmen (Central Louisiana)	1974	3	952	
26	Unnamed (Southwest Florida)	1910	3	955	30
26	Unnamed (Southwest Louisiana)	1918	3	955	34
26	Unnamed (Central Louisiana)	1926	3	955	
26	Eloise (Northwest Florida)	1975	3	955	
30	Unnamed (Mississippi, Alabama)	1906	3	958	134
30	Unnamed (North Texas)	1909	3	958	41
30	Unnamed (Northwest Florida)	1917	3	958	
30	Unnamed (North Texas)	1941	3	958	
30	Easy (Northwest Florida)	1950	3	958	
35	Elena (Mississippi, Alabama, Northwest Florida)	1985	3	959	4
36	Unnamed (Central Louisiana)	1934	3	962	
36	Unnamed (Southwest Florida, Northeast Florida)	1944	3	962	
36	Alicia (North Texas)	1983	3	962	
39	Unnamed (Northwest Florida)	1936	3	964	
40	Unnamed (Miami, Mississippi, Alabama, Pensacola)	1926	3	?	?
41	Andrew (Southeast Florida, Southeast Louisiana)	1992	3	?	?

Evacuation

The evacuation of people from New Orleans in advance of a hurricane has a significant impact on the number of fatalities that may occur. As indicated above, there are built-in problems for evacuation that are unique to New Orleans. Once the causeway over Lake Pontchartrain is closed because of high winds, there is no way to evacuate directly north from the city. Most people would have to go east using Interstate 10, U.S. 90, or west using Interstate 10 or U.S. 61 before they could turn north away from the Gulf. During actual hurricane evacuations in recent years, portions of these limited exit routes from New Orleans have become gridlocked for extended periods of time.

The Office of Homeland Security and Emergency Preparedness handles the overall coordination of evacuations in the state of Louisiana. A five-step process is used.

- Everyday activities (training) with no known threat of hurricane.
- Storm in the Gulf of Mexico.
- Storm might possibly hit Louisiana.
- Storm possible in 2–3 days to hit Louisiana. Declaration of emergency is issued by the governor.
- Recommendation to evacuate by local authorities.

If possible, for large hurricanes that are expected to hit New Orleans, evacuation is recommended at least 72 h before hurricane landfall in order to assure full evacuation. However, many hurricanes spend less than 72 h in the Gulf of Mexico.

Also, since knowledge about when the large hurricane is expected to hit landfall is imprecise, it is very difficult to assure 72 h of evacuation time. The decision to order an evacuation is left to local authorities. In the case of the city of New Orleans, the mayor makes the decision. For outlying regions, the head of the parish is generally the person to make the evacuation decision.

There are three levels of evacuation:

- Precautionary—decision to evacuate is left up to individual citizens.
- Recommended—decision to evacuate is left up to individual citizens.
- Mandatory—order is mandatory but difficult to enforce.

There are advantages and disadvantages to local officials if they order an evacuation. If the hurricane actually hits the local area and an evacuation was ordered prior to landfall, the local authorities receive credit for saving lives. If the hurricane does not hit the local area and an evacuation was ordered prior to landfall, there are a lot of costs with no benefits. Industry shuts down, traffic is a mess, extra police and emergency personnel are called into service, etc. Society is disrupted for at least one day to possibly a week at a cost of millions of dollars a day. Also, the possibility exists that the evacuation would move people into the path of the actual landfall and thus increases the risk of fatalities. If enough evacuations are ordered and no hurricane actually hits the areas evacuated, public confidence in the evacuation process is eroded and the probability of successful evacuation in the future is decreased.

At 2 h prior to expected hurricane landfall, an order to close all evacuation routes and evacuate traffic enforcement and news media to "last resort" refuges if necessary is made. Last resort refuges are set on all highways used for evacuation routes for people who might be on the highway as the hurricane approaches land. These "last resort" refuges are not guaranteed to survive a storm but are believed better than being in a car on the highway.

Information available to local officials as a hurricane approaches land consists of the strength, speed, projected path, and timing of landfall. This is provided by the National Hurricane Center approximately every 6 h. In addition, local authorities may have information from offshore facilities with respect to the amount of rainfall offshore and the offshore wind speed at sea level. Such information is vital to predicting local flooding when the hurricane hits land.

Counterflow operations (all lanes move out of the evacuated area) are possible and in place for certain Louisiana evacuation routes including New Orleans.

3.2.3 Structuring of the Risk Scenarios and Consequences (Step 3)

The key scenario parameters are

- The time the hurricane spends in the Gulf of Mexico before landfall.
- Whether the hurricane directly impacts New Orleans.
- The category of the hurricane at landfall.
- The type of evacuation prior to landfall.

The first parameter (the time the hurricane spends in the Gulf of Mexico before landfall) became the basis for the initiating events in the risk assessment. The other three parameters became top events (branch points) in the event trees used to define the logic of the sequence, that is, the scenarios leading to different damage states. Consideration of these four key parameters resulted in the definition of six damage states. The logic for calculating the likelihood of different scenarios and damage states is shown in three event trees: Figure 3.3 for Initiating Event 1, Figure 3.4 for Initiating Event 2, and Figure 3.5 for Initiating Event 3. The three event trees depict the sequences, that is, the scenarios that led to fatalities in New Orleans.

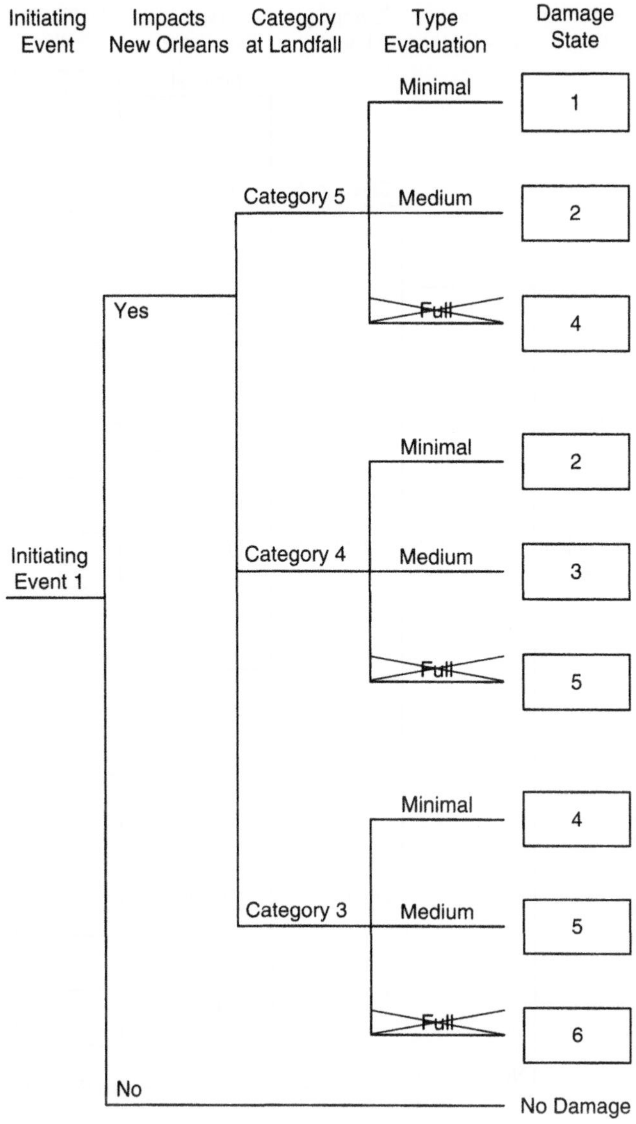

FIGURE 3.3. Hurricane scenarios for Initiating Event 1.

Initiating Events

For the risk assessment, it was decided to separate the 41 hurricanes in Table 3.3 into three categories of initiating events.

- *Initiating Event 1*—hurricanes residing 48 h or less in the Gulf of Mexico before landfall.

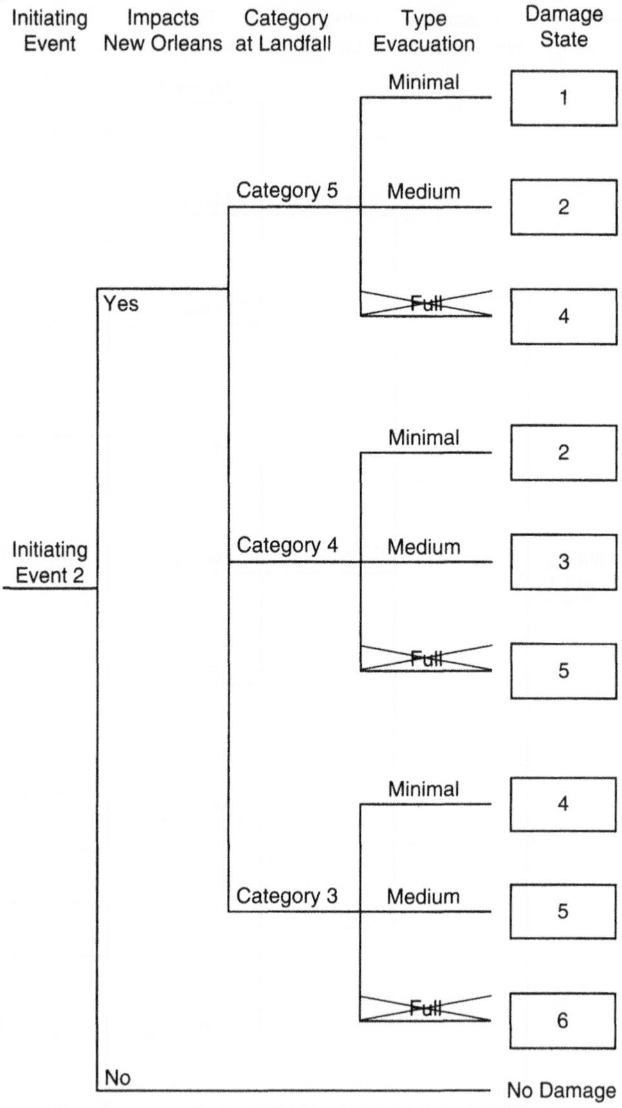

FIGURE 3.4. Hurricane scenarios for Initiating Event 2.

- *Initiating Event 2*—hurricanes residing between 48 and 72 h in the Gulf of Mexico before landfall.
- *Initiating Event 3*—hurricanes residing greater than 72 h in the Gulf of Mexico before landfall.

Knowing the residence times in the Gulf of Mexico facilitates the evaluation of evacuations prior to hurricane landfall. The procedures

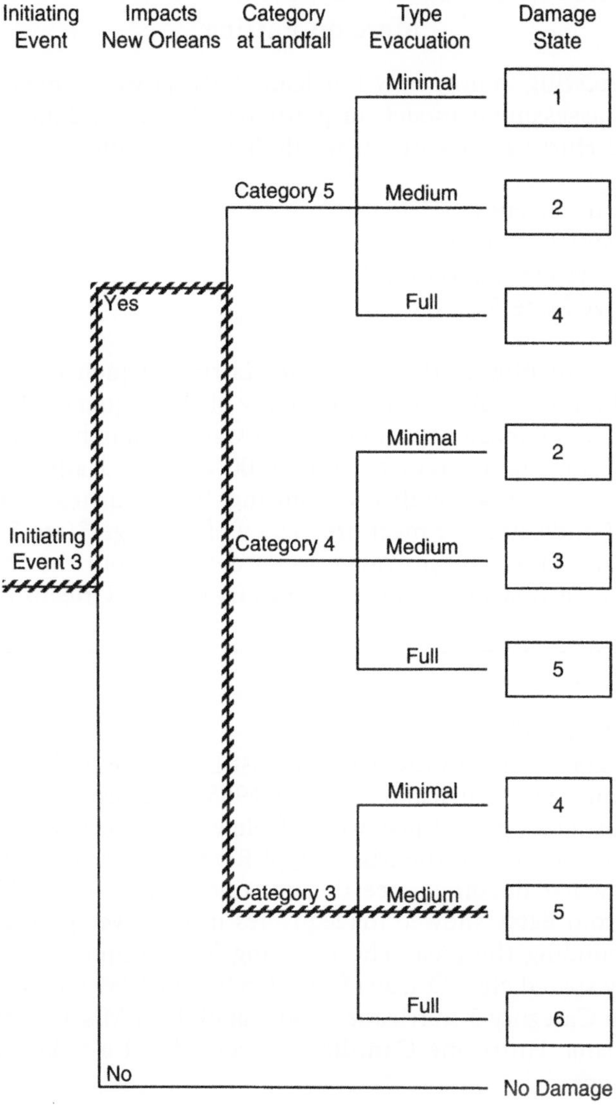

FIGURE 3.5. Hurricane scenarios for Initiating Event 3.

governing the evacuation of New Orleans recommend the commencement of evacuation at 72 h prior to hurricane landfall. This extended time allows the capability to evacuate the maximum number of people, including those people without private transportation. For times less than 72 h, there cannot be as complete an evacuation as the authorities would like.

More on Katrina

It is interesting to note how hurricane Katrina would have fit into our risk assessment model. In particular, based on data from the National Hurricane Center, it would have been input as follows:

1. Initiating Event 3.
2. Category 3 at landfall.
3. Moderate evacuation level.
4. Damage State 5.

Figure 3.5 highlights the hurricane Katrina sequence. Table 3.2 shows the mean value of the frequency of the sequence. For Damage State 5, the mean value of fatalities was taken to be 1300 deaths with the uncertainty ranging from 1100 to 1500 deaths, which is close to the actual fatalities resulting from hurricane Katrina. Table 3.6 indicates the mean frequency of Damage State 5 is once in 50 years. This is clear evidence that even without the data from hurricane Katrina such a threat to the city of New Orleans could be forecasted.

Impact on New Orleans

For the years covered in the data, the last hurricane to directly strike New Orleans was hurricane Betsy in 1965, a Category 3 hurricane. The eye of Betsy passed just west of New Orleans. During Betsy, a storm surge of 10 ft on the Mississippi River caused New Orleans to suffer its worst flooding since the unnamed hurricane of 1947. The flooding from Betsy showed inadequacies in the levee protection system surrounding the area. The resulting levee improvements made after Betsy spared New Orleans from flooding in 1969 when hurricane Camille, a Category 5 hurricane, made landfall in Mississippi, east of New Orleans. Hurricane Camille was considered a "close call" for New Orleans.

Category at Landfall

As described earlier, only three categories of hurricanes at landfall are considered: Category 5, Category 4, and Category 3.

Types of Evacuation

For the risk assessment, three types of evacuation were postulated: full, medium, and minimal.

Full evacuation was defined as moving 99% of the population to safety. Such an evacuation was assumed to require 72 h notification; good communication from the local authorities to the population; few if any complications in the roadways leading out of the city; and bus companies, national guard units, and other organizations able to handle the estimated 100,000–250,000 people without private transportation, etc.

Medium evacuation was defined as moving 80% of the population to safety. Some complications in the mass evacuation might arise but no major complications.

Minimal evacuation was defined as moving 60% of the population to safety.

It is expected that there will always be some mass evacuation from New Orleans before hurricane landfall. The people in New Orleans are well aware of the dangers of hurricanes, and it is expected most will evacuate given the possibility of a major hurricane. Therefore, the risk assessment does not include the possibility of a major hurricane making landfall at New Orleans with no evacuation prior to landfall.

For this case study, the total population was taken to be 1.3 million people in 2004.

Evacuation	*Percent at Risk*	*Population at Risk*
Full	1	13,000
Medium	20	260,000
Minimal	40	520,000

Damage States

In this case study it was estimated that a Category 5 hurricane at landfall would result in fatalities to 20% of the population at risk. A Category 5 hurricane that directly strikes the New Orleans area was assumed to cause the water in either the Mississippi River or Lake Pontchartrain to enter the city. The city is a bowl that will then fill up to an unknown height. A lot of people who were not evacuated will probably make it to safety in high-rise buildings, but it is not clear that this will be sufficient for long-term survival.

It was estimated in this case study that a Category 4 hurricane at landfall would result in fatalities to 10% of the population at risk. A Category 4 hurricane that directly strikes the New Orleans area was assumed to cause the water in either the Mississippi River or Lake Pontchartrain to enter the city. Essentially, the same conditions will occur as for a Category 5 hurricane.

It was estimated that a Category 3 hurricane at landfall would result in fatalities to 0.5% of the population at risk. It does not appear that a Category 3 hurricane that directly strikes the New Orleans area will cause the water in either the Mississippi River or Lake Pontchartrain to enter the city by overflowing the levees. Only certain Category 3 hurricanes are postulated to overflow the levees. However, this case study still considered fatalities to be possible from a Category 3 hurricane.

Hurricane Category at Landfall	*Evacuation*	*Fatalities (Mean Value)*	*Damage State*
5	Minimal	104,000	1
5	Medium	52,000	2
4	Minimal	52,000	2
4	Medium	26,000	3
5	Full	2600	4
3	Minimal	2600	4
4	Full	1300	5
3	Medium	1300	5
3	Full	65	6

3.2.4 Quantification of the Likelihood of the Scenarios (Step 4)

Initiating Events

As noted earlier, there are three initiating events used in the quantitative risk assessment. An evaluation of the 41 hurricanes resulted in the divisions shown in Table 3.4.

See Appendix B, Figures B.1–B.3, for details of the calculation of the frequency of initiating events and the accompanying uncertainty.

The mean frequency of the initiating events is shown in the table below.

Initiating Event	*Mean Frequency (Years)*
1	1 in 5.7
2	1 in 5.8
3	1 in 18

Impact on New Orleans

The impact on New Orleans is discussed in the "Impact on New Orleans" section under Section 3.2.3. The fraction of each initiating event impacting New Orleans is described in Appendix B. See Figures B.4–B.6 for the calculation of the fraction of initiating event hurricanes that directly impact New Orleans and the accompanying uncertainty.

TABLE 3.4
Most intense U.S. hurricanes (landfall) in Gulf of Mexico (based on U.S. hurricane data 1900–2004)

IE	*Hurricane*	*Year*	*Category at Landfall*	*Category Entering Gulf*	*Time in Gulf as Hurricane (h)*	*Category at 73 h from Landfall*	*Comments*
1	Unnamed (SE Florida, SE Louisiana, Mississippi)	1947	4	4	34	NA	Hit SE Florida as Cat 5; crossed Florida and entered Gulf as Cat 4.
1	Charley (SW Florida)	2004	4	3	14	NA	Entered Gulf from Cuba as Cat 3, dropped to Cat 2, landed as Cat 4.
1	Unnamed (New Orleans, Louisiana)	1915	4	2	42	NA	Entered Gulf from Yucatan channel as Cat 2.
1	Unnamed (North Texas)	1932	4	Storm	30	NA	Originated in Gulf. Went to Cat 4 and hit land in 30 hours.
1	Allen (South Texas)	1980	3	4	48	NA	Entered Gulf through Yucatan channel as Cat 4. Weakened from Cat 5 to Cat 3 before landfall.
1	Andrew (SE Florida, SE Louisiana)	1992	3	4	48	NA	Struck SE Florida as Cat 5. Went across Florida in 4 hours and exited as Cat 4. Weakened from Cat 4 to Cat 3 just before landfall.

(*Continued*)

TABLE 3.4 (*Continued*)

IE	*Hurricane*	*Year*	*Category at Landfall*	*Category Entering Gulf*	*Time in Gulf as Hurricane (h)*	*Category at 73 h from Landfall*	*Comments*
1	Unnamed (Miami, Mississippi, Alabama, Pensacola)	1926	3	3	36	NA	Hit Miami as Cat 4; crossed Florida and entered Gulf as Cat 3.
1	Unnamed (South Texas)	1916	3	3	36	NA	Entered Gulf from Yucatan channel as Cat 3.
1	Unnamed (Tampa Bay, Florida)	1921	3	3	36	NA	Entered Gulf from Yucatan channel as Cat 3.
1	Unnamed (SW Florida, NE Florida)	1944	3	2	24	NA	Entered Gulf from Cuba as Cat 2.
1	Betsy (SE Florida, SE Louisiana)	1965	3	2	35	NA	Entered Gulf from Florida Keys as Cat 2. Weakened from Cat 4 to Cat 3 just before landfall.
1	Unnamed (Central Texas)	1942	3	2	48	NA	Entered Gulf from Yucatan peninsula as Cat 2.
1	Unnamed (NW Florida)	1936	3	Storm	30	NA	Exits SW Florida as tropical storm.
1	Unnamed (SW Louisiana)	1918	3	Storm	30	NA	Passes over tip of Yucatan peninsula as tropical storm.
1	Eloise (NW Florida)	1975	3	Storm	36	NA	Entered Gulf from Yucatan peninsula as tropical storm.

1	Alicia (North Texas)	1983	3	Storm	36	NA	Originated in Gulf of Mexico.
1	Celia (South Texas)	1970	3	Storm	48	NA	Entered Gulf from Cuba as tropical storm.
1	Bret (South Texas)	1999	3	Storm	48	NA	Originated in Gulf of Mexico. Weakened from Cat 4 to Cat 3 just before landfall.
2	Camille (Mississippi, SE Louisiana)	1969	5	1	54	NA	Hit Cuba as Cat 3. Entered Gulf as Cat 1.
2	Unnamed (Galveston, Texas)	1915	4	3	60	NA	Entered Gulf from Cuba as Cat 3.
2	Unnamed (Grand Isle, Louisiana)	1909	4	1	72	NA	Entered Gulf from Cuba as Cat 1.
2	Audrey (SW Louisiana, North Texas)	1957	4	Storm	58	NA	Originated in Gulf of Mexico.
2	Unnamed (Galveston, Texas)	1900	4	Storm	72	NA	Entered Gulf from Cuba as tropical storm.
2	Ivan (Alabama, Pensacola)	2004	3	5	55	NA	Entered Gulf from Yucatan channel as Cat 5. Weakened from Cat 5 to Cat 3 before landfall.
2	Unnamed (NW Florida)	1917	3	3	72	NA	Hit Cuba as Cat 3; stayed as it entered the Gulf.
2	Unnamed (Mississippi, Alabama)	1916	3	2	54	NA	Entered Gulf from Yucatan peninsula as Cat 2.
2	Beulah (South Texas)	1967	3	2	66	NA	Entered Gulf from Yucatan peninsula as Cat 2.

(Continued)

TABLE 3.4 (*Continued*)

IE	*Hurricane*	*Year*	*Category at Landfall*	*Category Entering Gulf*	*Time in Gulf as Hurricane (h)*	*Category at 73 h from Landfall*	*Comments*
							Weakened from Cat 5 to Cat 3 just before landfall.
2	Unnamed (Mississippi, Alabama)	1906	3	1	60	NA	Entered Gulf from Yucatan channel as Cat 1.
2	Easy (NW Florida)	1950	3	1	54	NA	Entered Gulf from Cuba as Cat 1.
2	Unnamed (Central Louisiana)	1934	3	Storm	54	NA	Hit Yucatan peninsula as Cat 1; exit as tropical storm.
2	Opal (NW Florida, Alabama)	1995	3	Storm	58	NA	Entered Gulf from Yucatan peninsula as tropical storm. Weakened from Cat 4 to Cat 3 just before landfall.
2	Unnamed (North Texas)	1909	3	Storm	60	NA	Entered Gulf from Cuba as tropical storm. Went from Cat 1 to Cat 3 in last 12 hours.
2	Unnamed (Central Louisiana)	1926	3	Storm	60	NA	Entered Gulf from Yucatan channel as tropical storm.
2	Carmen (Central Louisiana)	1974	3	Storm	60	NA	Originated in Gulf. Weakened from Cat 4 to Cat 3 just before landfall.

2	Frederic (Alabama, Mississippi)	1979	3	Storm	64	NA	Entered Gulf from Cuba as tropical storm. Weakened from Cat 4 to Cat 3 just before landfall.
3	Carla (North and Central Texas)	1961	4	1	112	2	Entered Gulf from Yucatan channel as Cat 1. Weakened from Cat 5 to Cat 4 before landfall.
3	Unnamed (SW Florida)	1910	3	2	96	2	Entered Gulf from Cuba as Cat 2. Weakened from Cat 5 to Cat 4 before landfall.
3	Unnamed (South Texas)	1933	3	1	78	2	Entered Gulf from Florida-Cuba channel as Cat 1.
3	Hilda (Central Louisiana)	1964	3	Storm	90	2	Entered Gulf from Cuba as tropical storm. Weakened from Cat 4 to Cat 3 before landfall.
3	Elena (Mississippi, Alabama, NW Florida)	1985	3	Storm	96	2	Enters Gulf from Cuba as tropical storm; heads toward Florida then reverses direction.
3	Unnamed (North Texas)	1941	3	Storm	120	1	Crosses Florida as tropical storm. Spends approximately 3 days in Gulf as tropical storm.

IE, initiating event; Cat, category; SE, southeast; SW, southwest; NE, northeast; Gulf, Gulf of Mexico

The mean fraction of hurricanes impacting New Orleans is shown in the table below.

Initiating Event	*Fraction of Hurricanes Impacting New Orleans*
1	1 of 10
2	1 of 10
3	1 of 27

Category at Landfall

The Category at landfall is discussed in the "Category at landfall" section under Section 3.2.3. The uncertainties about the fraction of each initiating event hurricane making landfall by category is described in Appendix B, Figures B.7–B.15.

The mean value for the fraction of initiating event hurricanes making landfall by category is shown in the table below.

Initiating Event	*Category*	*Fraction of Hurricanes Making Landfall by Initiating Event and Category*
1	5	1 of 28
1	4	1 of 4
1	3	3 of 4
2	5	1 of 22
2	4	1 of 4
2	3	2 of 3
3	5	1 of 24
3	4	1 of 6
3	3	8 of 10

Types of Evacuation

The types of evacuation are discussed in the "Types of evacuation" section under Section 3.2.3. It is not likely that there will be a "full" evacuation for New Orleans prior to hurricane landfall. In order to get a "full" evacuation, the hurricane would have to enter the Gulf of Mexico from the Atlantic Ocean. If the hurricane originated in the Gulf of Mexico, it would generally strike land within 72 h precluding the time necessary for a "full" evacuation.

The hurricane would have to enter the Gulf of Mexico and then take more than 72 h to make landfall at New Orleans. In general, the stronger the hurricane, the faster it moves. If the hurricane were Category 1 or Category 2 when it entered the Gulf of Mexico and thus more likely to take more than 72 h to reach New Orleans, it is unlikely that people would be receptive to evacuating until it was very clear that New Orleans had a high probability of being directly impacted by the hurricane and that the hurricane was increasing in strength as it approached New Orleans.

Another obstacle to having a "full" evacuation is the tendency of some people in New Orleans to remain in their homes during hurricanes. As noted by Howell *et al.*,[3] only about a third of the residents of Jefferson and Orleans actually left the two-parish area during hurricane Georges. The survey also noted that a majority of those who evacuated waited until 24–30 h before the projected arrival of the storm.

It is not necessary for people to leave New Orleans for them to be safe from the impact of a major hurricane. This case study assumed that a "minimal" evacuation would be one similar to that during hurricane Georges in that perhaps 30% would leave the area and another 30% would move to places of safety in the New Orleans area. The "medium" evacuation was assumed to be between the "full" and the "minimal."

There can be no "full" evacuation for an Initiating Event 1 or 2 hurricane since these hurricanes reside 72 h or less in the Gulf of Mexico, which is less than the recommended 72 h.

For each initiating event, the uncertainty about the percentage of these hurricanes that will involve each particular evacuation is shown in Appendix B, Figures B.16–B.36.

The mean percentage of evacuation type for each initiating event is shown in the table below.

Category at Landfall	*Type Evacuation*	*Initiating Event 1 Percentage*	*Initiating Event 2 Percentage*	*Initiating Event 3 Percentage*
5	Full	Not applicable	Not applicable	11
5	Medium	59	69	69
5	Minimal	39	29	20
4	Full	Not applicable	Not applicable	20
4	Medium	69	79.5	69
4	Minimal	29	20	11
3	Full	Not applicable	Not applicable	29
3	Medium	80	89	59
3	Minimal	20	11	11

Damage States

Hurricane Betsy in 1965 was the last major hurricane that directly impacted New Orleans.[a] Hurricane Betsy entered the Gulf of Mexico through the Florida Keys as a Category 2 hurricane and made landfall approximately 35 h later as a Category 3 hurricane slightly to the west of New Orleans. The Mississippi River rose approximately 10 ft at New Orleans during the height of the storm. There were approximately 50 fatalities in Louisiana as a result of hurricane Betsy.

Specific data on fatality fractions as a function of hurricane category and typical New Orleans scenarios do not exist in the literature. Evidence from major hurricanes that impact areas along the coast in third world countries appears to indicate massive fatalities in the coastal populations due to ineffective evacuation caused by lack of knowledge of the incoming hurricane. In the case of New Orleans, there will be plenty of information concerning the hurricane prior to landfall, but even with such information the evacuation of New Orleans will not be complete.

The fatality fraction estimates used in this case study as a result of the hurricane directly impacting New Orleans are not based on any scientific data but are consistent with the estimates noted in the popular press by "experts." The estimated fatality percentages are 20% of the people at risk following a Category 5 hurricane, 10% following a Category 4 hurricane, and 0.5% following a Category 3 hurricane.

Damage states are discussed in the "Damage states" section under Section 3.2.3. The uncertainty about the number of fatalities for the damage states is described in Appendix B, Figures B.37–B.42.

Quantification Process

Propagating each initiating event through its respective event tree and compiling the results for all three initiating events quantified the risk model. For example, Initiating Event 1 was quantified through the event tree of Figure 3.3; Initiating Event 2 was quantified through the event tree of Figure 3.4; and Initiating Event 3 was quantified through the event tree of Figure 3.5. The top event (branch point) values in each event tree correspond to the fraction of each particular initiating event that impacts New Orleans, the hurricane category at landfall, and the type of evacuation that applies for each hurricane category.

An examination of Figures 3.3–3.5 shows that there are 10 possible sequences for each initiating event, including the "no damage" state. Table 3.2 summarizes the quantification results, without the three "no damage" event sequences. This spreadsheet indicates the

[a] As previously noted, this risk assessment was performed prior to hurricane Katrina.

quantitative values used for the initiating events, the branch points, and the damage states. The last column "Annualized Fatalities" is the mean number of fatalities per year for each sequence. It should be noted that the "Evacuation Type Fraction" is 0.0 for "full" evacuations during Initiating Events 1 and 2. Thus, those potential event sequences do not contribute numerically to the analysis results.

3.2.5 Assembly of the Scenarios into Measures of Risk (Step 5)

Appendix B summarizes the uncertainty distributions for all parameters in this risk assessment case study. The uncertainties in the initiating event frequencies, fractions of hurricanes that impact New Orleans, the hurricane category at landfall, and the type of evacuation for each hurricane were propagated through the event tree models to develop composite uncertainties for each damage state. The damage state uncertainty distributions were then combined with the corresponding fatality uncertainties and the results were assembled in the traditional complementary cumulative risk curve format shown in Figure 3.2.

The computer codes "Crystal Ball" and "EXCEL" were used for the calculations. These curves plot the cumulative frequency of exceeding a specific number of fatalities, including an explicit measure of the composite uncertainty. For example, Figure 3.2 shows that the mean frequency of exceeding 10,000 fatalities is approximately one in 130 years. This calculation is based on the historical hurricane evidence, the current metropolitan population, and the available evacuation estimates. Figure 3.2 also shows the uncertainty about these results. For example, our 90% confidence interval has a range of approximately 10, from about one in 60 years to about one in 620 years.

3.2.6 Interpretation of the Results (Step 6)

The information in Table 3.2 can be organized in many ways.

Table 3.5 rearranges Table 3.2 by the frequency of each individual risk scenario sequence in decreasing numerical order (mean values). The top two sequences are sequences where the hurricane makes landfall as a Category 3 hurricane with a medium evacuation prior to landfall. Thus, the most likely sequences to impact New Orleans (approximately one every 50 years) involve a Category 3 hurricane with medium evacuation and non-catastrophic fatalities.

Table 3.6 rearranges Table 3.2 by damage state.

The mean total damage state frequencies (in terms of a recurrence interval) and catastrophic fatalities are listed below.

TABLE 3.5
Mean annual frequencies of U.S. hurricane risk scenarios and split fractions of intervening events (based on U.S. hurricane data 1900–2004)

Risk Scenario									
Initiating Event No. (Frequency/ Year)	*Hurricane Category*	*Evacuation Type*	*New Orleans Impact Fraction*	*Category Fraction at Landfall*	*Evacuation Type Fraction*	*Damage State*	*Risk Scenario (Frequency/ Year)*	*Fatalities of Risk Scenario*	*Annualized Fatalities*
1 (0.176)	3	Medium	0.105	0.732	0.795	5	1.08E–02	1300	14
2 (0.170)	3	Medium	0.085	0.682	0.890	5	8.74E–03	1300	11
1 (0.176)	4	Medium	0.105	0.241	0.690	3	3.07E–03	26,000	80
2 (0.170)	4	Medium	0.085	0.251	0.795	3	2.87E–03	26,000	75
1 (0.176)	3	Minimal	0.105	0.732	0.200	4	2.71E–03	2600	7
1 (0.176)	4	Minimal	0.105	0.241	0.290	2	1.29E–03	52,000	67
2 (0.170)	3	Minimal	0.085	0.682	0.110	4	1.08E–03	2600	3
3 (0.054)	3	Medium	0.035	0.795	0.590	5	8.78E–04	1300	1.14
2 (0.170)	4	Minimal	0.085	0.251	0.200	2	7.23E–04	52,000	38
2 (0.170)	5	Medium	0.085	0.046	0.690	2	4.61E–04	52,000	24
3 (0.054)	3	Full	0.035	0.795	0.290	6	4.32E–04	65	0.03
1 (0.176)	5	Medium	0.105	0.036	0.590	2	3.89E–04	52,000	20
1 (0.176)	5	Minimal	0.105	0.036	0.390	1	2.57E–04	104,000	27
3 (0.054)	4	Medium	0.035	0.160	0.690	3	2.07E–04	26,000	5.37
2 (0.170)	5	Minimal	0.085	0.046	0.290	1	1.94E–04	104,000	20
3 (0.054)	3	Minimal	0.035	0.795	0.110	4	1.64E–04	2600	0.43
3 (0.054)	4	Full	0.035	0.160	0.200	5	5.99E–05	1300	0.08
3 (0.054)	5	Medium	0.035	0.042	0.690	2	5.41E–05	52,000	2.81
3 (0.054)	4	Minimal	0.024	0.160	0.110	2	3.30E–05	52,000	1.71
3 (0.054)	5	Minimal	0.034	0.042	0.200	1	1.57E–05	104,000	1.63
3 (0.054)	5	Full	0.035	0.042	0.110	4	8.63E–06	2600	0.02
1 (0.176)	5	Full	0.105	0.036	0.000	4	0.00E+00	2600	0
1 (0.176)	4	Full	0.105	0.241	0.000	5	0.00E+00	1300	0
1 (0.176)	3	Full	0.105	0.732	0.000	6	0.00E+00	65	0
2 (0.170)	5	Full	0.085	0.046	0.000	4	0.00E+00	2,600	0
2 (0.170)	4	Full	0.085	0.251	0.000	5	0.00E+00	1300	0
2 (0.170)	3	Full	0.085	0.682	0.000	6	0.00E+00	65	0
Total									**399**

TABLE 3.6

Mean annual frequencies of U.S. hurricane risk scenarios and split fractions of intervening events by damage state (based on U.S. hurricane data 1900–2004)

Risk Scenario									
Initiating Event No. (Frequency/Year)	*Hurricane Category*	*Evacuation Type*	*New Orleans Impact Fraction*	*Category Fraction at Landfall*	*Evacuation Type Fraction*	*Damage State*	*Risk Scenario (Frequency/Year)*	*Fatalities of Risk Scenario*	*Annualized Fatalities*
1 (0.176)	5	Minimal	0.105	0.036	0.390	1	2.57E–04	104,000	27
2 (0.170)	5	Minimal	0.085	0.046	0.290	1	1.94E–04	104,000	20
3 (0.054)	5	Minimal	0.035	0.042	0.200	1	1.57E–05	104,000	1.63
Total									**49**
1 (0.176)	4	Minimal	0.105	0.241	0.290	2	1.29E–03	52,000	67
2 (0.170)	4	Minimal	0.085	0.251	0.200	2	7.23E–04	52,000	38
2 (0.170)	5	Medium	0.085	0.046	0.690	2	4.61E–04	52,000	24
1 (0.176)	5	Medium	0.105	0.038	0.590	2	3.80E–04	52,000	20
3 (0.054)	5	Medium	0.035	0.042	0.690	2	5.41E–05	52,000	2.81
3 (0.054)	4	Minimal	0.035	0.160	0.110	2	3.30E–05	52,000	1.71
Total									**153**
1 (0.176)	4	Medium	0.105	0.241	0.690	3	3.07E–03	26,000	80
2 (0.170)	4	Medium	0.085	0.251	0.795	3	2.87E–03	26,000	75
3 (0.054)	4	Medium	0.035	0.160	0.690	3	2.07E–04	26,000	5.37
Total									**160**
1 (0.176)	3	Minimal	0.105	0.732	0.200	4	2.71E–03	2600	7
2 (0.170)	3	Minimal	0.085	0.682	0.110	4	1.08E–03	2600	3
3 (0.054)	3	Minimal	0.035	0.795	0.110	4	1.64E–04	2600	0.43
3 (0.054)	5	Full	0.035	0.042	0.110	4	8.63E–06	2600	0.02
Total									**10**
1 (0.176)	3	Medium	0.105	0.732	0.795	5	1.08E–02	1300	14
2 (0.170)	3	Medium	0.085	0.682	0.890	5	8.74E–03	1300	11
3 (0.054)	3	Medium	0.035	0.795	0.590	5	8.78E–04	1300	1.14
3 (0.054)	4	Full	0.035	0.160	0.200	5	5.99E–04	1300	0.08
Total									**27**
3 (0.054)	3	Full	0.035	0.795	0.290	6	4.32E–04	65	0.03

Damage State	*Recurrence Interval (Years)*	*Fatalities*
1	2200	104,000
2	340	52,000
3	160	26,000

As demonstrated with this case study, it is possible to perform a quantitative risk assessment of the risk of fatalities from major hurricanes impacting New Orleans using the techniques presented in Chapter 2. There are a number of analyses in this case study that would need more development prior to using it as a basis for decision-making on the risk of hurricanes in New Orleans. But the value of the process is clear. Examples of where more analysis would be necessary include more detailed modeling of evacuations and fatalities.

An important insight from the risk assessment was that Category 4 hurricanes at landfall dominate the risk of fatalities as noted in Table 3.2. While Category 5 hurricanes can cause more fatalities per event, the occurrence frequency of Category 5 is smaller than Category 4 and when one invokes the triplet definition of risk, Category 5 hurricanes at landfall have a smaller risk than Category 4. Category 3 hurricanes at landfall have minimal impact on the risk of fatalities.

The risk of fatalities in New Orleans from major hurricanes is not trivial. Spreading the consequences over time, the average number of fatalities turns out to be about 400 per year. On average, the risk of fatalities from hurricanes is greater than the vehicle accident risk for the 1.3 million people living in the New Orleans metropolitan area. However, the risk of fatality from major hurricanes does not come on a day-by-day basis as with vehicle accidents. The risk of fatalities from major hurricanes comes from rare events with catastrophic consequences.

3.3 Commentary on the New Orleans Hurricane Risk

These comments were written prior to the Katrina hurricane, but because of the overwhelming evidence that a Katrina hurricane is a credible event even before it actually happened, the comments are considered appropriate.

The risk of fatalities in New Orleans from major hurricanes is clearly recognized in a qualitative sense by government officials in Louisiana. The point of this case study and this book is to recognize the importance of saving lives by quantitatively assessing the risks to facilitate cost effective decision-making.

It is clear that the residents of New Orleans understand that a catastrophe can occur if a major hurricane hits their city. However, there almost seems to be a fatalistic attitude toward hurricanes among a large segment of the population. Government officials are focusing, rightfully so, on evacuation which is the key to reducing fatalities.

How can government officials make evacuations more successful? As noted earlier, full evacuation (99%) is not likely to occur in a population center like New Orleans. The quantitative risk assessment of this case study estimated that the minimal evacuation would be 60% (mean value) of the population and the medium evacuation would be 80% (mean value). It does not appear that these estimates can be significantly improved based on actual experience of hurricane evacuations. If these estimates are reasonable, then there will always be a considerable population at risk during a major hurricane in New Orleans.

It might be proper to take a look at what can be done for people who do not achieve safety before a Category 4 hurricane makes landfall. This is the most likely sequence for catastrophic fatalities. If people have not left low-elevation land or have not made it to a high-rise building, what can they do when the "bowl" is filling with water? Would a life raft be viable in a Category 4 hurricane? Would it be feasible to have large boats stored at public places capable of riding out the flooding similar to lifeboats on large ships? It might be argued that such measures would make people less likely to evacuate, but what is the alternative? Should one just accept the fact that not everyone can move to safety and accept the resultant fatalities as being acts of nature?

It also might be proper to take a look at the recovery efforts if the "bowl" fills. What actions are planned to get life rafts or boats into the bowl after the bowl fills? The Mississippi River and Lake Pontchartrain will recede to normal levels after the hurricane moves inland. How does one get equipment and boats into the bowl after the flooding? What actions are planned to use portable pumps to pump water out of the bowl?

3.4 Risk Assessment (Based on Data from 1900 to 1950)

The data presented in Figure 3.1 and Table 3.1 clearly indicates that Atlantic hurricanes for the first half of the 20th century were more damaging in terms of human fatalities than for the latter half of the century. The data further indicates higher frequencies of hurricanes impacting the New Orleans area during that period. In order to determine if the fatality risk was rising or falling over the course of the entire 20th century, it was decided to extend the case study to quantify the risk of major hurricanes during the first half of the century. In particular, the risk assessment was conducted as if the analyst had hurricane data only through December 31, 1950.

3.4.1 Overview

The assessment again followed the six-step process of quantitative risk assessment and is detailed in Appendix B, Section B.2. Basically, there was no change from the above assessment in Steps 1, 2, and 3. The same prior distributions were used in the Bayesian updates for the frequency and intensity of hurricanes as were used for the years 1900–2004. The evacuation model remained the same, as did the event tree logic for structuring the scenarios. The quantification was performed exactly the same as for the data from 1900 to 2004.

The changes that were made for the 1900–1950 case included a change in the population (770,000 as opposed to 1.3 million), the values for the initiating events, the frequency and category of hurricanes impacting New Orleans, the fatality distributions due to the change in population, the number of major hurricanes, and the damage state distributions to reflect the smaller population.

3.4.2 Summary

The results of this case study are summarized below.

- The mean frequency of a major hurricane hitting New Orleans resulting in 10,000 fatalities or greater is one in 130 years. The full range of frequencies and consequences is presented in Figure 3.6. The 90% confidence interval varies from approximately one in 50 years to about one in 2700 years.
- The differences between the mean values for the period 1900–1950 and the period 1900–2004 are shown in Figure 3.7.
- The mean frequency of exceeding 10,000 fatalities from a major hurricane hitting New Orleans was essentially the same in 1950 as in 2004. The mean frequency of exceeding 100,000 fatalities from a major hurricane hitting New Orleans was substantially smaller in 1950 than in 2004.
- The mean value of fatalities per year in 1950 from major hurricanes hitting New Orleans was approximately 295 per year. See Table 3.7.

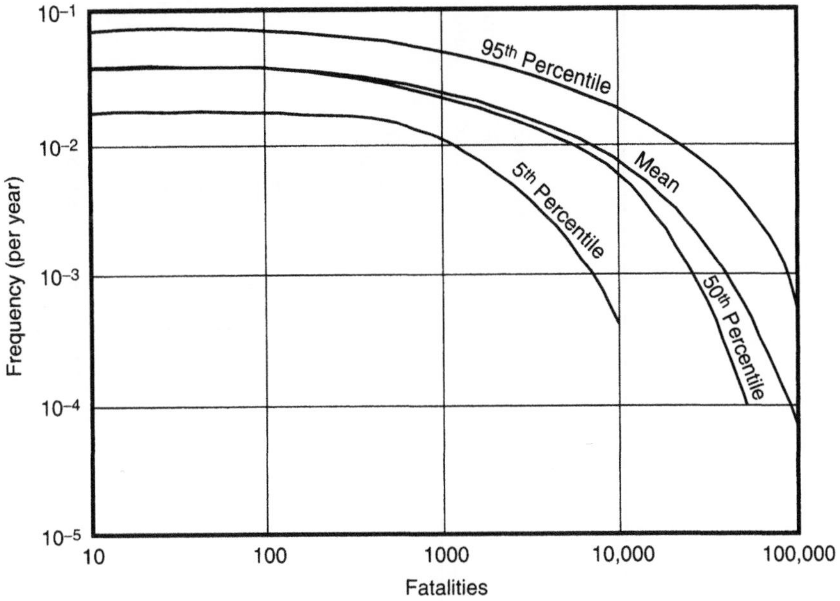

FIGURE 3.6. Hurricane fatality risk for New Orleans (based on U.S. hurricane data 1900–1950).

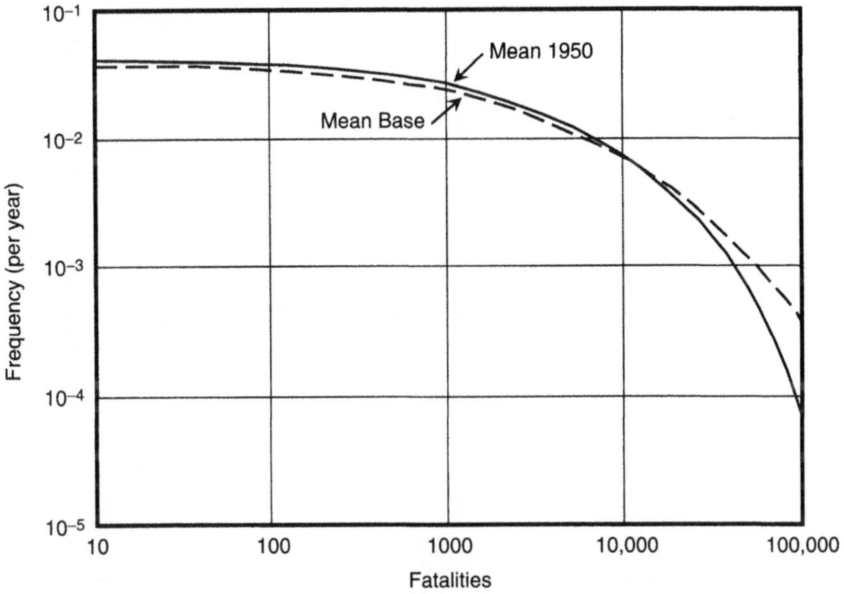

FIGURE 3.7. Hurricane fatality risk for New Orleans: Comparison of results based on different data sets (1900–1950 vs 1900–2004).

TABLE 3.7
Mean annual frequencies of U.S. hurricane risk scenarios and split fractions of intervening events[a] (based on U.S. hurricane data 1900–1950)

Risk Scenario									
Initiating Events (Frequency/ Year)	*Hurricane Category*[b]	*Evacuation Type*[c]	*New Orleans Impact Fraction*	*Category Fraction at Landfall*	*Evacuation Type Fraction*	*Damage State*[d]	*Risk Scenario (Frequency/ Year)*	*Fatalities of Risk Scenario*	*Annualized Fatalities*
Initiating Event 1									
0.184	5	Minimal	0.100	0.039	0.390	1	2.81E–04	61,600	17
0.184	5	Medium	0.100	0.039	0.590	2	4.26E–04	30,800	13
0.184	5	Full	0.100	0.039	0.000	4	0.00E+00	1540	0
0.184	4	Minimal	0.100	0.296	0.290	2	1.58E–03	30,800	49
0.184	4	Medium	0.100	0.296	0.690	3	3.76E–03	15,400	58
0.184	4	Full	0.100	0.296	0.000	5	0.00E+00	770	0
0.184	3	Minimal	0.100	0.674	0.200	4	2.48E–03	1540	4
0.184	3	Medium	0.100	0.674	0.795	5	9.86E–03	770	8
0.184	3	Full	0.100	0.674	0.000	6	0.00E+00	38	0
Initiating Event 1 total							1.84E–02		148
Initiating Event 2									
0.184	5	Minimal	0.100	0.039	0.290	1	2.09E–04	61,600	13
0.184	5	Medium	0.100	0.039	0.690	2	4.98E–04	30,800	15
0.184	5	Full	0.100	0.039	0.000	4	0.00E+00	1540	0

0.184	4	Minimal	0.100	0.296	0.200	2	1.09E–03	30,800	34
0.184	4	Medium	0.100	0.296	0.795	3	4.33E–03	15,400	67
0.184	4	Full	0.100	0.296	0.000	5	0.00E+00	770	0
0.184	3	Minimal	0.100	0.674	0.110	4	1.36E–03	1540	2
0.184	3	Medium	0.100	0.674	0.890	5	1.10E–02	770	8
0.184	3	Full	0.100	0.674	0.000	6	0.00E+00	38	0
Initiating Event 2 total							1.85E–02		139
Initiating Event 3									
0.054	5	Minimal	0.038	0.045	0.200	1	1.81E–05	61,600	1
0.054	5	Medium	0.038	0.045	0.690	2	6.26E–05	30,800	2
0.054	5	Full	0.038	0.045	0.110	4	9.98E–06	1540	0
0.054	4	Minimal	0.038	0.113	0.110	2	2.53E–05	30,800	1
0.054	4	Medium	0.038	0.113	0.690	3	1.59E–04	15,400	2
0.054	4	Full	0.038	0.113	0.200	5	4.60E–05	770	0
0.054	3	Minimal	0.038	0.866	0.110	4	1.94E–04	1540	0
0.054	3	Medium	0.038	0.866	0.590	5	1.04E–03	770	1
0.054	3	Full	0.038	0.866	0.290	6	5.11E–04	38	0
Initiating Event 3 total							2.06E–03		7
Total							**3.90E–02**		**295**

[a]See event tree Figures 3.3–3.5 for event sequence logic.
[b]Major hurricanes are Categories 3, 4, and 5 hurricanes as classified by the Saffir–Simpson hurricane scale.
[c]Three types of evacuation are defined: full, medium, and minimal corresponding to 1%, 20%, and 40% of the metropolitan population at risk, respectively. The not "at risk" population is assumed to be evacuated.
[d]Damage states are defined in terms of the number of fatalities incurred.

- The most likely hurricane to cause catastrophic fatalities is a Category 4 hurricane at landfall. The percentage of total risk (fatalities) from each hurricane category was calculated from Table 3.7 and is shown below.

Hurricane Category at Landfall	*Approximate Percent of Total Fatality Risk*
4	71
5	21
3	8

Using the data in Table 3.3, Table 3.8 describes the 23 major hurricanes making landfall in the Gulf of Mexico in the years 1900–1950. Table 3.8 also shows the separation of the hurricanes into the three initiating events similar to Table 3.4.

The information in Table 3.7 can be organized in many ways.

Table 3.9 rearranges Table 3.7 by the frequency of each individual sequence in decreasing numerical order (mean values). The top two sequences are sequences where the hurricane makes landfall as a Category 3 hurricane with a medium evacuation prior to landfall. Thus, the most likely sequences to impact New Orleans (approximately once every 50 years) involve a Category 3 hurricane with medium evacuation and non-catastrophic fatalities.

Table 3.10 rearranges Table 3.7 by the damage states.

The total damage state mean frequencies and fatalities are listed in the table below.

Damage State	*Frequency*	*Fatalities*
1	1 in 2000 years	61,600
2	1 in 270 years	30,800
3	1 in 120 years	15,400

TABLE 3.8
Most intense U.S. hurricanes (landfall) in Gulf of Mexico (based on U.S. hurricane data 1900–1950)

IE	*Hurricane*	*Year*	*Category at Landfall*	*Category Entering Gulf*	*Time in Gulf as Hurricane (Hours)*	*Category at 73 h from Landfall*	*Comments*
1	Unnamed (North Texas)	1932	4	Storm	30	NA	Originated in Gulf of Mexico. Went to Cat 4 and hit land in 30 hours.
1	Unnamed (SE Florida, SE Louisiana, Mississippi)	1947	4	4	34	NA	Hit SE Florida as Cat 5; crossed Florida and entered Gulf as Cat 4.
1	Unnamed (New Orleans, Louisiana)	1915	4	2	42	NA	Entered Gulf from Yucatan channel as Cat 2.
1	Unnamed (SW Florida, NE Florida)	1944	3	2	24	NA	Entered Gulf from Cuba as Cat 2.
1	Unnamed (SW Louisiana)	1918	3	Storm	30	NA	Passes over tip of Yucatan peninsula as tropical storm.
1	Unnamed (NW Florida)	1936	3	Storm	30	NA	Exits SW Florida as tropical storm.
1	Unnamed (South Texas)	1916	3	3	36	NA	Entered Gulf from Yucatan channel as Cat 3.
1	Unnamed (Tampa Bay, Florida)	1921	3	3	36	NA	Entered Gulf from Yucatan channel as Cat 3.
1	Unnamed (Miami, Mississippi, Alabama, Pensacola)	1926	3	3	36	NA	Hit Miami as Cat 4; crossed Florida and entered Gulf as Cat 3.
1	Unnamed (Central Texas)	1942	3	2	48	NA	Entered Gulf from Yucatan peninsula as Cat 2.
2	Unnamed (Galveston, Texas)	1915	4	3	60	NA	Entered Gulf from Cuba as Cat 3.

(*Continued*)

TABLE 3.8 (*Continued*)

IE	*Hurricane*	*Year*	*Category at Landfall*	*Category Entering Gulf*	*Time in Gulf as Hurricane (Hours)*	*Category at 73 h from Landfall*	*Comments*
2	Unnamed (Galveston, Texas)	1900	4	Storm	72	NA	Entered Gulf from Cuba as tropical storm.
2	Unnamed (Grand Isle, Louisiana)	1909	4	1	72	NA	Entered Gulf from Cuba as Cat 1.
2	Unnamed (Mississippi, Alabama)	1916	3	Storm	54	NA	Entered Gulf from Yucatan peninsula as Cat 2.
2	Unnamed (Central Louisiana)	1934	3	Storm	54	NA	Hit Yucatan peninsula as Cat 1; exit as tropical storm.
2	Easy (NW Florida)	1950	3	1	54	NA	Entered Gulf from Cuba as Cat 1.
2	Unnamed (Mississippi, Alabama)	1906	3	1	60	NA	Entered Gulf from Yucatan channel as Cat 1.
2	Unnamed (North Texas)	1909	3	Storm	60	NA	Entered Gulf from Cuba as tropical storm. Went from Cat 1 to Cat 3 in last 12 hours.
2	Unnamed (Central Louisiana)	1926	3	Storm	60	NA	Entered Gulf from Yucatan channel as tropical storm.
2	Unnamed (NW Florida)	1917	3	3	72	NA	Hit Cuba as Cat 3; stayed 3 as it entered the Gulf.
3	Unnamed (South Texas)	1933	3	1	78	2	Entered Gulf from Florida-Cuba channel as Cat 1.
3	Unnamed (SW Florida)	1910	3	2	96	2	Entered Gulf from Cuba as Cat 2. Stalled for 2 days just north of Cuba. Weakened from Cat 4 to Cat 3 just before landfall.
3	Unnamed (North Texas)	1941	3	Storm	120	1	Crosses Florida as tropical storm. Spends approximately 3 days in Gulf as tropical storm.

IE, initiating event; Cat, category; SE, southeast; SW, southwest; NE, northeast; Gulf, Gulf of Mexico

TABLE 3.9
Mean annual frequencies of U.S. hurricane risk scenarios and split fractions of intervening events by decreasing frequency of the risk scenario (based on U.S. hurricane data 1900–1950)

Risk Scenario									
Initiating Event No. (Frequency/ Year)	*Hurricane Category*	*Evacuation Type*	*New Orleans Impact Fraction*	*Category Fraction at Landfall*	*Evacuation Type Fraction*	*Damage State*	*Risk Scenario (Frequency/ Year)*	*Fatalities of Risk Scenario*	*Annualized Fatalities*
2 (0.184)	3	Medium	0.100	0.674	0.890	5	1.10E–02	770	8.5
1 (0.184)	3	Medium	0.100	0.674	0.795	5	9.86E–03	770	7.6
2 (0.184)	4	Medium	0.100	0.296	0.795	3	4.33E–03	15,400	66.7
1 (0.184)	4	Medium	0.100	0.296	0.690	3	3.76E–03	15,400	57.9
1 (0.184)	3	Minimal	0.100	0.674	0.200	4	2.48E–03	1540	3.8
1 (0.184)	4	Minimal	0.100	0.296	0.290	2	1.58E–03	30,800	48.6
2 (0.184)	3	Medium	0.100	0.674	0.110	4	1.36E–03	1540	2.1
2 (0.184)	4	Minimal	0.100	0.296	0.200	2	1.09E–03	30,800	33.5
3 (0.054)	3	Medium	0.038	0.866	0.590	5	1.04E–03	770	0.8
3 (0.054)	3	Full	0.038	0.866	0.290	6	5.11E–04	38	0.0
2 (0.184)	5	Medium	0.100	0.039	0.690	2	4.98E–04	30,800	15.3
1 (0.184)	5	Medium	0.100	0.039	0.590	2	4.26E–04	30,800	13.1
1 (0.184)	5	Minimal	0.100	0.039	0.390	1	2.81E–04	61,600	17.3
2 (0.184)	5	Minimal	0.100	0.039	0.290	1	2.09E–04	61,600	12.9
3 (0.054)	3	Minimal	0.038	0.866	0.110	4	1.94E–04	1540	0.3
3 (0.054)	4	Medium	0.038	0.113	0.690	3	1.59E–04	15,400	2.4
3 (0.054)	5	Medium	0.038	0.045	0.690	2	6.26E–05	30,800	1.9
3 (0.054)	4	Full	0.038	0.113	0.200	5	4.60E–05	770	0.0
3 (0.054)	4	Minimal	0.038	0.113	0.110	2	2.53E–05	30,800	0.8
3 (0.054)	5	Minimal	0.038	0.045	0.200	1	1.81E–05	61,600	1.1
3 (0.054)	5	Full	0.038	0.045	0.110	4	9.98E–06	1540	0.0
1 (0.184)	5	Full	0.100	0.039	0.000	4	0.00E+00	1540	0.0
1 (0.184)	4	Full	0.100	0.296	0.000	5	0.00E+00	770	0.0
1 (0.184)	3	Full	0.100	0.674	0.000	6	0.00E+00	38	0.0
2 (0.184)	5	Full	0.100	0.039	0.000	4	0.00E+00	1540	0.0
2 (0.184)	4	Full	0.100	0.296	0.000	5	0.00E+00	770	0.0
2 (0.184)	3	Full	0.100	0.674	0.000	6	0.00E+00	38	0.0
Total									**295**

TABLE 3.10
Mean annual frequencies of U.S. hurricane risk scenarios and split fractions of intervening events by damage state (based on U.S. hurricane data 1900–1950)

Risk Scenario									
Initiating Event No. (Frequency/ Year)	*Hurricane Category*	*Evacuation Type*	*New Orleans Impact Fraction*	*Category Fraction at Landfall*	*Evacuation Type Fraction*	*Damage State*	*Risk Scenario (Frequency/ Year)*	*Fatalities of Risk Scenario*	*Annualized Fatalities*
1 (0.184)	5	Minimal	0.100	0.039	0.390	1	2.81E–04	61,600	17.3
2 (0.184)	5	Minimal	0.100	0.039	0.290	1	2.09E–04	61,600	12.9
3 (0.054)	5	Minimal	0.038	0.045	0.200	1	1.81E–05	61,600	1.1
Total									**31**
1 (0.184)	4	Minimal	0.100	0.296	0.290	2	1.58E–03	30,800	48.6
2 (0.184)	4	Minimal	0.100	0.296	0.200	2	1.09E–03	30,800	33.5
2 (0.184)	5	Medium	0.100	0.039	0.690	2	4.98E–04	30,800	15.3
1 (0.184)	5	Medium	0.100	0.039	0.590	2	4.26E–04	30,800	13.1
3 (0.054)	5	Medium	0.038	0.045	0.690	2	6.26E–05	30,800	1.9
3 (0.054)	4	Minimal	0.038	0.113	0.110	2	2.53E–05	30,800	0.8
Total									**113**
2 (0.184)	4	Medium	0.100	0.296	0.795	3	4.33E–03	15,400	66.7
1 (0.184)	4	Medium	0.100	0.296	0.690	3	3.76E–03	15,400	57.9
3 (0.054)	4	Medium	0.038	0.113	0.690	3	1.59E–04	15,400	2.4
Total									**127**
1 (0.184)	3	Minimal	0.100	0.674	0.200	4	2.48E–03	1540	3.8
2 (0.184)	3	Minimal	0.100	0.674	0.110	4	1.36E–03	1540	2.1
3 (0.054)	3	Minimal	0.038	0.866	0.110	4	1.94E–04	1540	0.3
3 (0.054)	5	Full	0.038	0.045	0.110	4	9.98E–06	1540	0.02
Total									**6**
2 (0.184)	3	Medium	0.100	0.674	0.890	5	1.10E–02	770	8.5
1 (0.184)	3	Medium	0.100	0.674	0.795	5	9.86E–03	770	7.6
3 (0.054)	3	Medium	0.038	0.866	0.590	5	1.04E–03	770	0.8
3 (0.054)	4	Full	0.038	0.113	0.200	5	4.60E–05	770	0.04
Total									**17**
3 (0.054)	3	Full	0.038	0.866	0.290	6	5.11E–04	38	0.02

References

[1] Mowbray, R., Hotels to evacuate for hurricanes, Business writer, Times—Picayune Updates, NOLA.com, 2006.

[2] Jarrell, J. D., Mayfield, M., Rappaport, E. N., Landsea, C. W., Deadliest, costliest, and most intense United States hurricanes from 1900 to 2000 (and other frequently requested hurricane facts), NOAA Technical Memorandum NWS TPC-1. NOAA/NWS/Tropical Prediction Center, Miami, Florida, NOAA/AOML/Hurricane Research Division, Miami, Florida, 2001.

[3] Howell, S. E., McLean, W., Haysley, V., Evacuation Behavior in Orleans and Jefferson Parishes—Hurricane Georges, University of New Orleans Survey Research Center, November, 1998.

CHAPTER 4

Case Study 2: Risk of Asteroids Impacting the Earth

This chapter is a limited scope quantitative risk assessment of the fatalities from high energy asteroids impacting[a] the 48 contiguous states of the United States (48 States).

Two risk assessments were performed. The first is based on high energy asteroids impacting the 48 states and is considered the case study. The second is based on high energy asteroids impacting New Orleans, LA, and was included to demonstrate the value of being able to compare location-specific risks.

In the Earth's solar system, there are thousands of objects called asteroids (both stone and metal), planetoids (very small particles), and comets (rock, dust, and ice) revolving around the sun. These objects vary in composition with almost all of them being very small in size. In this chapter, all of these objects in the Earth's solar system will be referred to collectively as asteroids. Some of these asteroids impact the Earth's atmosphere on a daily basis. In all but very rare cases, the asteroid burns up in the Earth's atmosphere with no impact on the Earth or its inhabitants. The Earth's atmosphere acts as a protective shield to prevent damage. In order for the blast effects to reach the ground, the energy released must be equivalent to a few megatons of TNT (~50–60 m in diameter[b]

[a] Impact energies are represented in terms of TNT (trinitrotoluene) equivalences (TNTE), similar to nuclear explosives. A 10-megaton energy impact implies that the impact energy is equivalent to 10 million tons of TNT.

[b] Since there is not a direct relationship between the diameter and energy impact of an asteroid because of differences in their densities, the authors have chosen impact-energy as the primary descriptor of an asteroid.

depending on the composition of the asteroid). The 1908 Tunguska airburst of an asteroid impacting in Siberia is an example of such a threshold event.[1] It is estimated that the Tunguska asteroid impact energy was about 15 megatons of TNTE and was approximately 60 m in diameter. The remoteness of the impact area prevented any catastrophic consequences.

Geological observations on Earth indicate that there have been asteroids impacting the Earth similar to the Tunguska event capable of causing fatalities had people been in the impact area. Based on the geological record of the Earth and its increasing human population, there presently exists the potential for a catastrophic event (defined here as an event involving 10,000 human fatalities or more) when a high energy asteroid impacts the Earth. In fact, it is estimated that there are thousands of asteroids in space large enough to cause fatalities should they impact the Earth.[2,3] Even with this evidence it has been observed by one expert[4] that, "in the modern world, no one has been killed by an impact explosion. Indeed, there are no reliable historical examples of mass mortality due to impacts."

So, what's the problem? There are two issues. The first issue is that modern society has become very conscious of the fragility of the planet and is much more on the alert to manmade and natural events that may alter the course of the habitability of our environment. That higher level of consciousness has resulted in much more serious and scientifically based assessments of threats to the health and safety of the planet's inhabitants. The result is we are now assessing more diligently our risks over long time periods, while searching for benchmarks to put the different risks in proper context for taking protective actions, if possible. The second issue is the short time of reliable human historical records in comparison to the desire and need to assure human and species survival for many future millennia. The risk of catastrophic consequences from asteroids colliding with Earth clearly exists, and it is incumbent on the Earth's inhabitants to manage this risk to a level commensurate with the resources being allocated to address other similar risks.

What sets the asteroid risk apart from many other risks is the potential for global impacts. As argued by Chapman[5,6] asteroid "impacts must be exceptionally more lethal globally than any other proposed terrestrial causes for mass extinctions because of two unique features: (1) their environmental effects happen essentially instantaneously (on timescales of hours to months, during which species have little time to evolve or migrate to protective locations) and (2) there are compound environmental consequences (e.g., broiler-like skies as ejecta re-enter the atmosphere, global firestorm, ozone layer destroyed, earthquakes and tsunamis, months of ensuing 'impact winter,' centuries of global warming, poisoning of the oceans)." As will be discussed later, global impacts are not necessary for there to be regional and local catastrophes as previously defined—a target of this book as well.

It is interesting to observe that only relatively recently has the asteroid risk been taken seriously. The principal reason is that there have been no catastrophic events involving asteroids *that have been clinically recorded*. Many planetary scientists believe that the turning point in the interest in the asteroid risk was the pioneering work of Alvarez and his colleagues.[7] Alvarez showed that the extinction of numerous species at the Cretaceous—Tertiary (K–T) geologic boundary was almost certainly caused by the impact of a massive asteroid. The site of the impact was later identified with the Chiexulub crater in the Yucatan peninsula and is the event most often associated with the demise of the dinosaurs some 65 million years ago. Other work and events that greatly stimulated interest and awareness of the asteroid risk was that of Shoemaker[8] and a NASA sponsored Snowmass, Colorado, Workshop on "Collision of Asteroids and Comets with the Earth: Physical and Human Consequences," chaired by Shoemaker in the early 1980s.

A further distinction of the asteroid risk than just its possible existential consequences is that unlike many other natural hazards such as earthquakes, hurricanes, and volcanoes, an asteroid collision with Earth may be something that can be prevented.[9] In particular, timely detection of the course of a threatening asteroid may allow for actions to prevent it from colliding with Earth. For example, if it is possible to have several decades' advance notice of an impact of a high energy asteroid (greater than 1 km in diameter) at a specific location, the opportunity may exist for a variety of mitigating actions to deflect the asteroid and completely prevent any collision with Earth. In this regard, at the request of the U.S. Congress,[10] the National Aeronautics and Space Administration (NASA) has an asteroid survey program known as Spaceguard for the purpose of identifying asteroids larger than 1 km in diameter (equivalent). In particular, the Spaceguard goal is to identify 90% of all near Earth orbit (NEO) asteroids by 2008 with absolute magnitude $H < 18$ (approximate diameters greater than 1 km). It is estimated that there are some 1100 asteroids in this size range and that as of August 2006 over 800 of them or approximately 80% had already been identified.

NASA's Spaceguard program is a major step forward in assessing the threat of a high energy asteroid impacting Earth. In the spirit of this book, what about the less than global consequences that are still in the catastrophic category (10,000 or more fatalities)? That is, what about the risk of all of the asteroids between the limit of atmospheric penetration (a few megatons and in the 50–60 m diameter range) and the so-called K–T (mass extinction) scale? How do we cast the asteroid risk in a form that allows comparison with other risks such that a rational basis can exist for the fair allocation of public funds and other resources for the risk management of all natural events? Clearly, it is not necessary for an asteroid to have a diameter greater than 1 km for there to be a catastrophic impact.

For sub-km impacts planetary scientists suggest that the most property and life threatening phenomena are fires, tsunamis, and airbursts over land. Numerous studies have been made on the merits of a Spaceguard-like program for asteroids in the sub-km range.[11–16] The debate continues on the merits of a sub-km search program for asteroids. Clearly, a survey program of asteroids in the sub-km range could greatly reduce the risk of a surprise impact that could result in catastrophic consequences. There is a size below which the cost of surveying greatly exceeds the benefit. That size is most likely in the range of the threshold for atmospheric penetration. For example, Lewis[3] points out that the cost of surveying small (10 m) near Earth asteroids "probably surpasses the expected benefits of finding them by roughly a factor of 10,000."

There have been many important scientific developments in recent decades contributing to an increased understanding of the threat of asteroids, especially asteroids whose impact could lead to global consequences. The emphasis has been on large diameter and very high energy asteroids (1 km and greater). To set the stage for our case study on high energy asteroids, we have attempted in Table 4.1 to summarize qualitatively the current state of knowledge of the asteroid threat. In particular, Table 4.1 highlights various thresholds for events such as atmospheric penetration and events that could have regional and global consequences. Two known asteroid impacts are included to provide benchmarks for two of the event categories. Ranges of numbers are given in Table 4.1 to partly account for the uncertainties involved, including the dependencies on asteroid composition and the uncertainties in the inventory of asteroids in space that could collide with Earth.

Of course, as will be noted later, the information of Table 4.1 is insufficient as a basis for decision-making on the allocation of resources for the purpose of effective risk management of multiple threats to society. Location-specific quantitative risk assessments are required to have a scientific basis for decision-making on how to manage the asteroid risk in context with other risks facing society. Such an assessment provides a scientific basis for allocating risk management resources. Lewis[3] is among those who tackled the problem of providing some basis for decision-making on actions to manage the risk of asteroids.

There is evidence, as will be shown later, that the case is stronger for managing the risk of sub-km size asteroids than is reflected

TABLE 4.1
Summary of the asteroid threat

Event	*Energy Release (Megatons TNTE)*	*Diameter (Meters)*	*Recurrence Interval (Years)*
Atmospheric Penetration Threshold	Several	50–100	Hundreds to Thousands
Tunguska	~15	~60	Similar Range
Regional Catastrophe	100–1000	100–400	Thousand to Several Thousand
Global Catastrophe	Thousand to Millions	Hundreds to Several Thousand	Several Thousand to Millions
Species Extinction	Ten to Fifty Million	Four to Six Thousand	Greater than Ten Million
K-T Mass Extinction	~100 Million	Ten to Twenty Thousand	Fifty to One Hundred Million

in most current programs because of their increased frequency of occurrence. In recent years more attention is being given to sub-km asteroids. For example, Stokes[13] has studied sub-km asteroids and observes that from about 50 to 150 m diameter, the impacts are primarily airbursts and the larger asteroids "reach the ground and produce craters." Stokes and Morrison indicate that the most life-threatening hazard from sub-km impacts is associated with airbursts over land. Stokes observes that at about 300 m diameter (~1000 to a few thousand megaton airburst), the damage area may be as "large as a U.S. state or small European country." According to Stokes, the greatest hazard for sub-km land impacts is from asteroids 100–200 m in diameter. The greatest hazard for tsunamis comes from impacts in the ocean of asteroids having diameters in the 200–500 m range. The smaller diameter asteroids generally result in airbursts that do not generate tsunamis. Asteroids greater than 500 m tend to bottom out even in deep oceans, and no longer make increasingly larger waves. Chesley and Ward[17] in their studies of asteroid-induced tsunamis estimate that one near Earth asteroid will impact somewhere into the Earth's oceans every 5880 years. While tsunamis are a serious collateral effect of an ocean impact, in many cases there will be adequate warning time for evacuation of threatened coastlines.

Much has been learned about the asteroid collision threat to planet Earth during the past three decades. The Spaceguard program has contributed to the knowledge base of large asteroids whose impact could result in global consequences. Other research projects and studies have provided insights on the general threat of asteroids. But there is much more risk information needed for nations to be in a position to make the right decisions on the management of the asteroid risk. Among the questions whose answers are necessary for such decision-making are (1) What is the risk over a wide range of consequences for specific regions? (2) What is a form of the asteroid risk that allows direct comparison with other risks facing society? and (3) What is the range of the asteroid risk uncertainty as a function of consequences? It is the purpose of this case study to demonstrate how to answer these questions. Of course, location is not necessarily an issue for impacts that have global consequences, but regional catastrophic consequences occur more frequently and need quantification to support a rational societal risk management program. For example, as is shown later in this case study, over 70% of the asteroid fatality risk to people living in the 48 states comes from impacts involving energy releases below the global threshold.

This case study demonstrates that it is possible to obtain a quantitative risk assessment of the potential for catastrophic fatalities following a high energy impact of an asteroid on the 48 states. The benefits of such analyses are many, including a quantification of (1) the magnitude of the risk of catastrophic fatalities, (2) the uncertainties in the risk, and (3) the key risk contributors.

The approach is to follow the six-step process outlined in Chapter 1, as it was for the hurricane case study. Also, to support the point of the importance of location-specific risk assessments, this case study selects New Orleans for demonstrating an effective form for comparing different risks to a specific region.

Prior to presenting the details of the case study, we will fast-forward to a summary of selected results.

4.1 Summary of the Fatality Risk of an Asteroid Impacting the 48 States

Figure 4.1 is a graphical display of the risk of asteroid events causing various levels of fatalities in the 48 states.

- Figure 4.1 shows that the mean frequency of an asteroid impacting the 48 states that could result in 10,000 fatalities or more is one in 6800 years. There is considerable uncertainty in this result. In particular, the 90% confidence interval for 10,000 fatalities or more has a frequency range from one in 1900 years to one in 520,000 years. The population at risk is approximately 280 million people

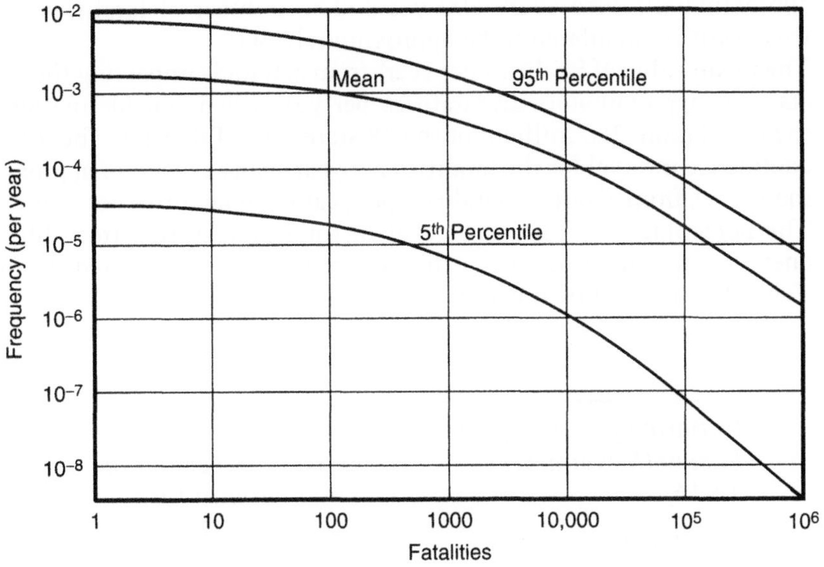

FIGURE 4.1. Asteroid fatality risk for the 48 states.

and the target area includes land impact and offshore impacts that could generate tsunamis.

Other key results from the case study follow.

- The mean frequency of a very large asteroid with a global impact is one in 50 million years. Again, there is large uncertainty in this value. The 90% confidence interval ranges from a frequency of one in 20 million years to one in a billion years.
- The most likely asteroid to cause catastrophic fatalities in the 48 states is an asteroid with an impact energy between 0.1 and 10 megatons TNTE that strikes land over a high population density area with a mean frequency of approximately once every 12,000 years. The next most likely asteroid to cause catastrophic fatalities in the 48 states is an asteroid with impact energy between 10 and 10,000 megatons TNTE that strikes the Atlantic Ocean between 500 km and 2500 km from the Atlantic shoreline. The mean frequency of this event is approximately one every 16,000 years.
- Asteroids causing fewer fatalities than 10,000 (the chosen threshold for a catastrophic event) of course occur more frequently. The most likely asteroid to impact the 48 states is an asteroid with an impact energy between 0.1 and 10 megatons TNTE that strikes land over a low population density area with a mean frequency of

approximately once every 600 years. The number of fatalities for this event is calculated to be approximately 500.

- The mean value of fatalities per year[c] from asteroids impacting the 48 states is approximately 22 fatalities per year when considering just the population (280 million) of the 48 states. See Table 4.2. The 90% confidence interval for the fatalities per year spans a factor of approximately 50, from about 1.5 fatalities per year to 86 fatalities per year.
- The percentage by initiating event of total fatality risk from high energy asteroids impacting the 48 states was calculated from Table 4.2 and is shown below.

Initiating Event Impact Energy (Megatons of TNTE)	*Approximate Percent of Total Fatality Risk*
10–10,000	54
Global	25
0.1–10	19

- The percentage by phenomenon that causes fatalities from high energy asteroids impacting the 48 states was calculated from Table 4.2 and is shown below.

Phenomenon	*Approximate Percent of Total Fatality Risk*
Tsunami	47
Land Blast	28
Global	25

4.2 Risk Assessment

The methods of risk assessment employed are those presented in Chapters 1 and 2. The methods can be applied to any location and can address different measures of risk if desired including

[c] This is not to suggest that we expect this many fatalities each year from asteroid impacts. Large asteroid impacts are rare events with no fatalities between events. This is simply a matter of putting the fatalities from a high energy asteroid impact on a per year basis. This type of metric is often used for making risk comparisons.

TABLE 4.2
Mean annual frequencies of asteroid risk scenarios by initiating event for the 48 states

Risk Scenarios													
Initiating Event No./Frequency/Year	Type of Impact (Land-Population Density) (Water-depth/Region)	Fraction of Earth Area Impacted	Fraction of Land by Population Density (High, Medium, Low)	Impact Fraction by Water Depth/Region	Risk Scenario Frequency/Year	Coastline Impacted (km)	Amplitude of Wave at Landfall (Meters)	Distance Water Travels Inland (km)	Blast Damage Area (km^2)	Population Density (per km^2)	Fatality Fraction	Population at Risk	Fatalities/Year
1/1.34E−01	Land–High	0.0150	0.0410		8.24E−05				176	374	0.332	21,854	1.80E+00
1/1/34E−01	Land-Medium	0.0150	0.1620		3.26E−04				176	87	0.332	5095	1.66E+00
1/1.34E−01	Land-Low	0.0150	0.7970		1.60E−03				176	9	0.332	526	8.42E−01
2/3.82E−03	Land-High	0.0150	0.0410		2.35E−06				1420	374	0.570	302,716	7.11E−01
2/3.82E−03	Land-Medium	0.0150	0.1620		9.28E−06				1420	87	0.570	70,580	6.55E−01
2/3.82E−03	Land-Low	0.0150	0.7970		4.57E−05				1420	9	0.570	7285	3.33E−01
2/3.82E−03	Gulf-Shallow	0.0024		0.5260	4.82E−06	2610	0.210	0.067		71	0.049	610	2.94E−03
2/3.82E−03	Gulf-Medium	0.0024		0.2100	1.93E−06	2610	1.500	1.017		71	0.150	28,428	5.47E−02
2/3.82E−03	Gulf-Deep	0.0024		0.2630	2.41E−06	2610	2.275	1.592		71	0.280	83,069	2.00E−01
2/3.82E−03	Pacific-Reg I	0.0147		0.0289	1.62E−06	200	12.085	14.800		152	1.000	449,920	7.30E−01
2/3.82E−03	Pacific-Reg II	0.0147		0.1440	8.09E−06	1200	2.048	1.398		152	1.000	254,995	2.06E+00
2/3.82E−03	Pacific–Reg III	0.0147		0.8270	4.64E−05	2070	0.520	0.231		152	0.151	10,975	5.10E−01
2/3.82E−03	Atlantic–Reg I	0.0196		0.0337	2.52E−06	200	12.085	14.800		203	1.000	600,880	1.52E+00
2/3.82E−03	Atlantic–Reg II	0.0196		0.1540	1.15E−05	1200	2.048	1.398		203	1.000	340,553	3.93E+00
2/3.82E−03	Atlantic-Reg III	0.0196		0.8120	6.08E−05	3310	0.520	0.231		203	0.151	23,438	1.42E+00
3/1.24E−05	Land-High	0.0150	0.0410		7.63E−09				26,000	208	1.000	5,408,000	4.12E−02
3/1.24E−05	Land-Medium	0.0150	0.1620		3.01E−08				26,000	43	1.000	1,112,800	3.35E−02
3/1.24E−05	Land-Low	0.0150	0.7970		1.48E−07				26,000	9	1.000	234,000	3.47E−02
3/1.24E−05	Gulf-Shallow	0.0024		0.5260	1.57E−08	2610	0.210	0.067		71	0.049	610	9.56E−06
3/1.24E−05	Gulf-Medium	0.0024		0.2100	6.25E−09	2610	1.500	1.017		71	0.150	28,428	1.78E−04
3/1.24E−05	Gulf-Deep	0.0024		0.2630	7.83E−09	2610	3.000	2.325		71	0.370	160,311	1.25E−03
3/1.24E−05	Pacific-Reg I	0.0147		0.0289	5.27E−09	200	16.000	23.080		152	1.000	701,632	3.70E−03
3/1.24E−05	Pacific–Reg II	0.0147		0.1440	2.62E−08	1200	2.700	2.005		152	1.000	365,712	9.60E−03
3/1.24E−05	Pacific-Reg III	0.0147		0.8270	1.51E−07	2070	0.686	0.327		152	0.254	26,133	3.94E−03
3/1.24E−05	Atlantic-Reg I	0.0196		0.0337	8.19E−09	200	16.000	23.080		203	1.000	937,048	7.67E−03

(Continued)

TABLE 4.2 (*Continued*)

Risk Scenarios													
Initiating Event No./ Frequency/ Year	*Type of Impact (Land-Population Density) (Water-depth/ Region)*	*Fraction of Earth Area Impacted*	*Fraction of Land by Population Density (High, Medium, Low)*	*Impact Fraction by Water Depth/ Region*	*Risk Scenario Frequency/ Year*	*Coastline Impacted (km)*	*Amplitude of Wave at Landfall (Meters)*	*Distance Water Travels Inland (km)*	*Blast Damage Area (km^2)*	*Population Density (per km^2)*	*Fatality Fraction*	*Population at Risk*	*Fatalities/ Year*
3/1.24E−05	Atlantic-Reg II	0.0196		0.1540	3.74E−08	1200	2.700	2.005		203	1.000	488,418	1.83E−02
3/1.24E−05	Atlantic-Reg III	0.0196		0.8120	1.97E−07	3310	0.686	0.327		203	0.254	55,809	1.10E−02
4/2.15E−06	Land-High	0.0150	0.0410		1.32E−09				137,000	46	1.000	6,302,000	8.33E−03
4/2.15E−E−06	Land-Medium	0.0150	0.1620		5.22E−09				137,000	22.	1.000	2,945,500	1.54E−02
4/2.15E−06	Land-Low	0.0150	0.7970		2.57E−08				137,000	9	1.000	1,233,000	3.17E−02
4/2.15E−06	Gulf-Shallow	0.0024		0.5260	2.71E−09	2610	0.210	0.067		71	0.049	610	1.66E−06
4/2.15E−06	Gulf-Medium	0.0024		0.2100	1.08E−09	2610	1.500	1.017		71	0.150	28,428	3.08E−05
4/2.15E−06	Gulf-Deep	0.0024		0.2630	1.36E−09	2610	3.000	2.325		71	0.370	160,311	2.18E−04
4/2.15E−06	Pacific-Reg I	0.0147		0.0289	9.13E−10	200	16.000	23.080		152	1.000	701,632	6.41E−04
4/2.15E−06	Pacific-Reg II	0.0147		0.1440	4.55E−09	1200	2.700	2.005		152	1.000	365,712	1.66E−03
4/2.15E−06	Pacific-Reg III	0.0147		0.8270	2.61E−08	2070	0.686	0.327		152	0.254	26,133	6.83E−04
4/2.15E−04	Atlantic-Reg I	0.0196		0.0337	1.42E−09	200	16.000	13.080		203	1.000	937,048	1.33E−03
4/2.15E−04	Atlantic-Reg II	0.0196		0.1540	6.49E−09	1200	2.700	2.005		203	1.000	488,418	3.17E−03
4/2.15E−04	Atlantic-Reg III	0.0196		0.8120	3.42E−08	3310	0.686	0.327		203	0.254	55,809	1.91E−03
5/2.66E−07	Land-High	0.0150	0.0410		1.64E−10				742,000	15	1.000	11,352,600	1.86E−03
5/2.66E−07	Land-Medium	0.0150	0.1620		6.46E−10				742,000	11	1.000	8,384,600	5.42E−03
5/2.66E−07	Land-Low	0.0150	0.7970		3.18E−09				742,000	9	1.000	6,678,000	2.12E−02
5/2.66E−07	Gulf-Shallow	0.0024		0.5260	3.36E−10	2610	0.210	0.067		71	0.005	610	2.05E−07
5/2.66E−07	Gulf-Medium	0.0024		0.2100	1.34E−10	2610	1.500	1.017		71	0.150	28,428	3.81E−06
5/2.66E−07	Gulf-Deep	0.0024		0.2630	1.68E−10	2610	3.000	1.325		71	0.370	160,311	2.69E−05
5/2.66E−07	Pacific-Reg I	0.0147		0.0289	1.13E−10	200	16.000	23.080		152	1.000	701,632	7.93E−05
5/2.66E−07	Pacific—Reg II	0.0147		0.1440	5.63E−10	1200	2.700	2.005		152	1.000	365,712	2.06E−04
5/2.66E−07	Pacific-Reg III	0.0147		0.8270	3.23E−09	2070	0.686	0.327		152	0.254	26,133	8.45E−05
5/2.66E−07	Atlantic-Reg I	0.0196		0.0337	1.76E−10	200	16.000	23.080		203	1.000	937,048	1.65E−04
5/2.66E−07	Atlantic-Reg II	0.0196		0.1540	8.03E−10	1200	2.700	2.005		203	1.000	488,418	3.92E−04
5/2.66E−07	Atlantic-Reg III	0.0196		0.8120	4.23E−09	3310	0.686	0.327		203	0.254	55,809	2.36E−04
6/2.00E−08	Global											280,000,000	5.60E+00
Total													**2.23E+01**

injuries, environmental effects, and property damage. The most common risk measure is human fatalities and, like the case study on hurricanes, was the choice here. It should be noted that the primary reference for assumptions on the physics and characteristics of asteroid risk is Ref. 18.

Chapter 1 identified a six-step process for performing a quantitative risk assessment, which is the process that follows.

4.2.1 Definition of the System During Normal Conditions (Step 1)

In terms of scoping and the definition of the system for this case study, we have chosen to limit the asteroid impact area to the region and population represented by the 48 states.

Choosing the 48 states as the impact area or the "system" for assessing the risk of asteroids has distinct advantages. One major advantage is the availability of excellent data on the population distribution and topography, including shorelines, of the entire region. Having good data is critical to an accurate representation of the probability and consequences of a wide range of asteroids impacting the Earth. In addition, the 48 states are surrounded by different oceans with different characteristics, which lead to different consequences for the same initiating event in different locations.

One global calculation was made to determine the frequency of very high energy asteroids that could result in the extinction of the human species. No attempt was made to establish a continuum between the 48 state regional consequences and global consequences. Thus, Figure 4.1 representing the regional 48 states was truncated at one million fatalities and the global calculation was a standalone assessment discussed later in this chapter.

4.2.2 Identification and Characterization of System Hazards (Step 2)

The hazard is of course the thousands of asteroids orbiting in space. The circumstances necessary for catastrophic fatalities from a high energy asteroid impacting the 48 states are

- Limited advance warning of the asteroid impact which prevents (1) any deflection of the asteroid away from the Earth while in space or (2) evacuation of a substantial portion of the population from the impact area.
- A high energy asteroid striking either the 48 states directly or striking the bordering oceans causing a major tsunami.

Deflection or Evacuation
This quantitative risk assessment case study takes no credit for any deflection of high energy asteroids away from the Earth or evacuation of people from the asteroid impact area.

High Energy Asteroids
With respect to high energy asteroids impacting the 48 states, they were divided into six initiating events based on impact energy as shown in the table below. No consideration was given to asteroids whose impact energy was less than 0.1 megatons of TNTE. Initiating Events 1 through 5 are considered to have regional impact.

Definition of Initiating Events

Initiating Event	*Impact Energy (Megatons of TNTE)*
1	0.1–10
2	10–10,000
3	10,000–100,000
4	100,000–1,000,000
5	1,000,000–10,000,000
Global[d]	

4.2.3 Structuring of the Risk Scenarios and Consequences (Step 3)

The key scenario parameters are

- Impact energy of the asteroid.
- Whether the asteroid strikes water or land.
- If the asteroid strikes water, what is the depth of the water at the impact site and how far from land is the asteroid impact site?
- If the asteroid strikes water or land, what is the population density in the damage area?

The first parameter (the energy impact of the asteroid) became the basis for the initiating events of the risk scenarios. The other parameters became "top events" (branch points) in the event trees used to define the logic of the intervening events in the scenarios leading to fatalities. The logic for the five regional initiating events is shown

[d] According to Toon, the potential for global impact starts to be possible for asteroids with impact energy greater than 10,000 megatons TNTE and becomes almost certain for asteroids with impact energy greater than 10 million megatons TNTE.

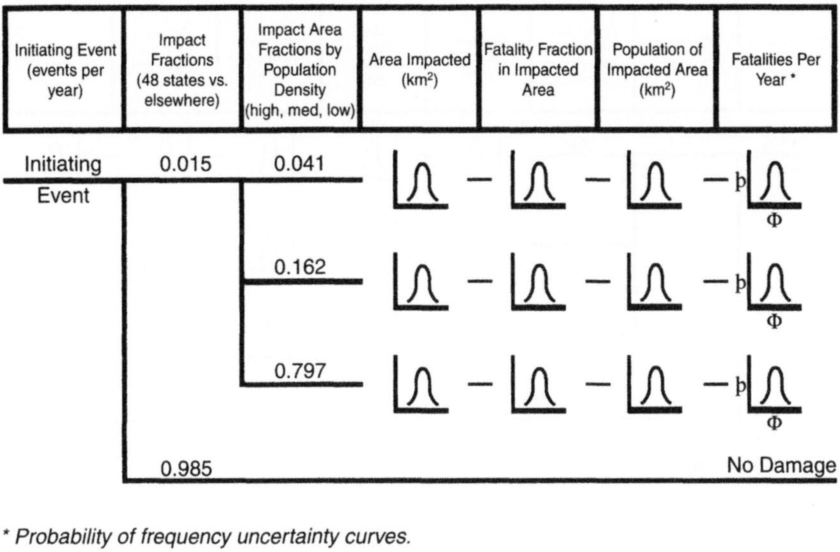

FIGURE 4.2. Asteroid scenarios for land impact.

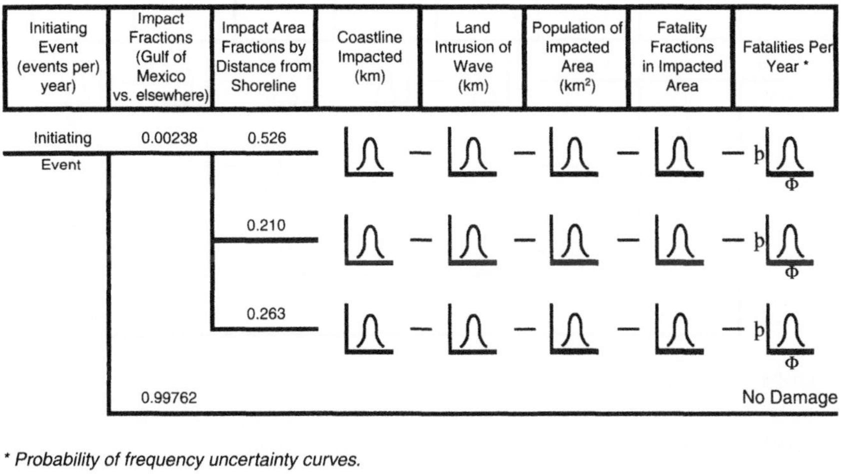

FIGURE 4.3. Asteroid scenarios for the Gulf of Mexico.

in the conceptual event tree diagrams of Figures 4.2–4.5 for the different impact areas considered.

The conceptual event trees show how the parameter uncertainties are propagated through the sequences for asteroids impacting land, the Gulf of Mexico, the Pacific Ocean, and the Atlantic Ocean,

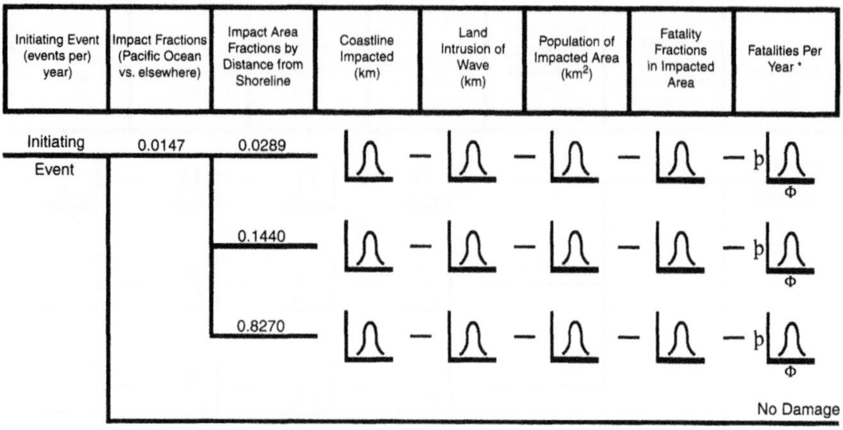

* *Probability of frequency uncertainty curves.*

FIGURE 4.4. Asteroid scenarios for the Pacific Ocean.

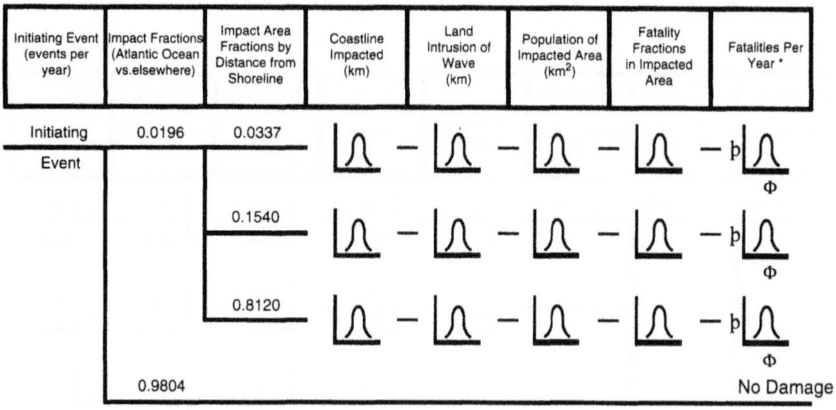

* *Probability of frequency uncertainty curves.*

FIGURE 4.5. Asteroid scenarios for the Atlantic Ocean.

respectively. The parameter uncertainties are accounted for with probability of frequency curves illustrated conceptually in the figures.

Each of the five regional initiating events starts the logic in the regional event tree and each initiating event leads to twelve outcomes except for Initiating Event 1, which has only three outcomes. For asteroids between 0.1 megatons and 10 megatons (Initiating Event 1), the asteroid will most likely burn up in the atmosphere causing a possible air blast on the Earth's surface. If the asteroid (Initiating Event 1) impacts over water, no tsunami is created. There is no accounting for ships in the water impact area. In all, there are 51 regional outcomes. See Table 4.2.

Initiating Events

As described earlier, six initiating events were defined for this case study. The mean frequency and recurrence interval of each initiating event are shown in the following table.

Initiating Event	*Impact Energy (Megatons of TNTE)*	*Recurrence Interval (Years)*	*Frequency (Per Year)*
1	0.1–10	8	1.34×10^{-1}
2	10–10,000	260	3.84×10^{-3}
3	10,000–100,000	81,000	1.24×10^{-5}
4	100,000–1,000,000	470,000	2.15×10^{-6}
5	1,000,000–10,000,000	3,800,000	2.66×10^{-7}
Global		50,000,000	2.00×10^{-8}

The frequencies and energies of the initiating events are based on NASA and Toon data previously referenced. The details of how the NASA and Toon data were processed to account for uncertainty for Initiating Events 1–5 are presented in Appendix C. The logic for the global initiating event is a single sequence, which covers fatalities of the total population at risk (280 million people for the 48 states or 6 billion people for the world). By definition, the outcome of this

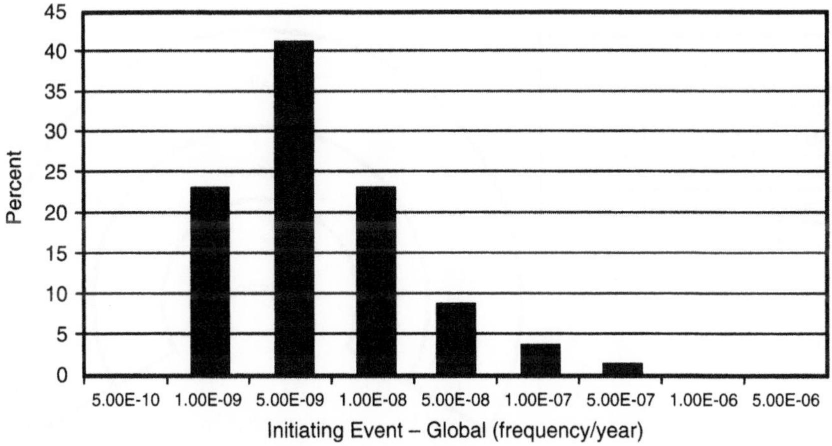

FIGURE 4.6. Uncertainty distribution, global initiating event frequency.

initiating event is certain; it is the frequency of the event that is uncertain. Based on an interpretation of Toon's work, Figure 4.6 was derived as the frequency distribution of the global initiating event with a mean frequency of 2.00×10^{-8} as shown in the above table.

To avoid the distraction of too frequently referring to Appendix C, let it be observed that the details for most of the numerical results of the balance of this chapter are contained in Appendix C. In a few instances, reference to Appendix C is repeated where it is desired to call attention to very specific supporting information.

Impacts—Water or Land

Different regional phenomena occur depending on whether the high energy asteroid impacts the Earth over water or land. For a global impact, location is not a factor.

Water Impact

For a water impact, the phenomenon of interest is a tsunami that could cause shoreline damage. Initiating Event 1 causes an airburst over water but does not cause a tsunami. Initiating Events 2–5 have sequences leading to fatalities caused by the tsunami. Table 4.2 shows the mean value of fatalities per year for each sequence.

The total surface of the Gulf of Mexico included in the risk assessment, 1.2 million km^2, was divided into three water depths: shallow, medium, and deep as illustrated in Figure 4.7.

The total surface of the Pacific Ocean (7.5 million km^2) was divided into three regions: Region I—less than 100 km from land,

Depth	Area (km^2)	Percentage of Area
Shallow	640,000	53
Medium	260,000	21
Deep	320,000	26
Total	1,220,000	

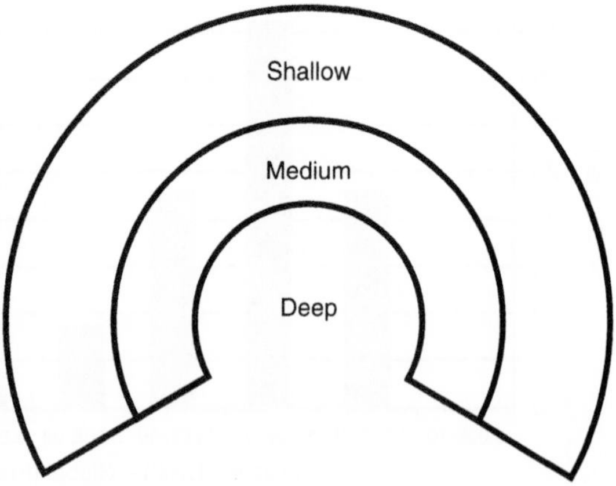

FIGURE 4.7. Tsunami model—Gulf of Mexico.

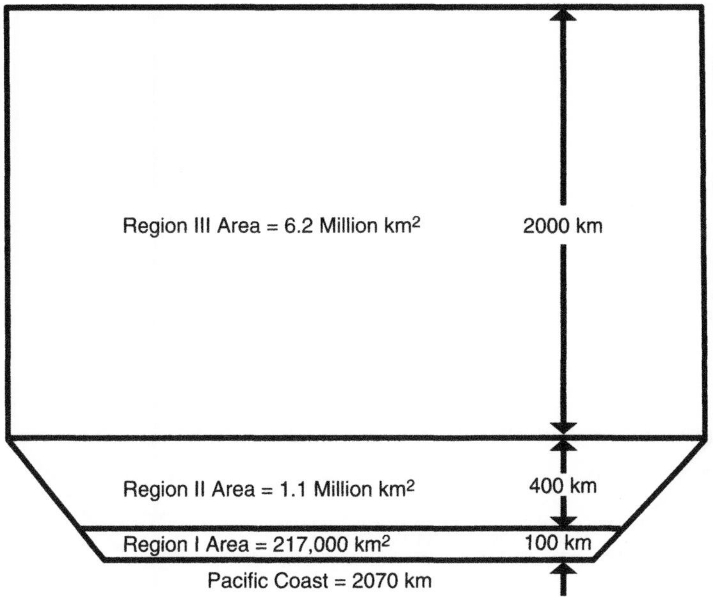

FIGURE 4.8. Tsunami model—Pacific Ocean.

Region II—from 100 to 500 km to land, Region III—from 500 to 2500 km to land (Figure 4.8).

The total surface of the Atlantic Ocean (10.1 million km^2) was divided into three areas: Region I—less than 100 km from land, Region II—from 100 to 500 km to land, Region III—from 500 to 2500 km to land (Figure 4.9).

For a water impact, a key parameter is the amplitude of the tsunami at the edge of the high energy asteroid impact site. This amplitude will depend on the impact energy of the asteroid and the depth of the water. A high energy asteroid impact in shallow water will not result in a large amplitude tsunami at the edge of the asteroid impact site.

Another key parameter for a water impact is the amplitude of the tsunami at the coastline. The amplitude of the tsunami at the coastline will depend on both the amplitude of the tsunami at the edge of the high energy asteroid impact site and the distance from the asteroid impact site to the surrounding coastline.

The damage area from the tsunami will depend on the length of coastline impacted by the tsunami and the distance that the tsunami goes inland. The number of people at risk will be the damage area times the population density. Unless the amplitude of the tsunami at the coastline is high, it is assumed that not everyone in the damage area is a fatality. A fatality fraction parameter was developed to account for different fatality consequences.

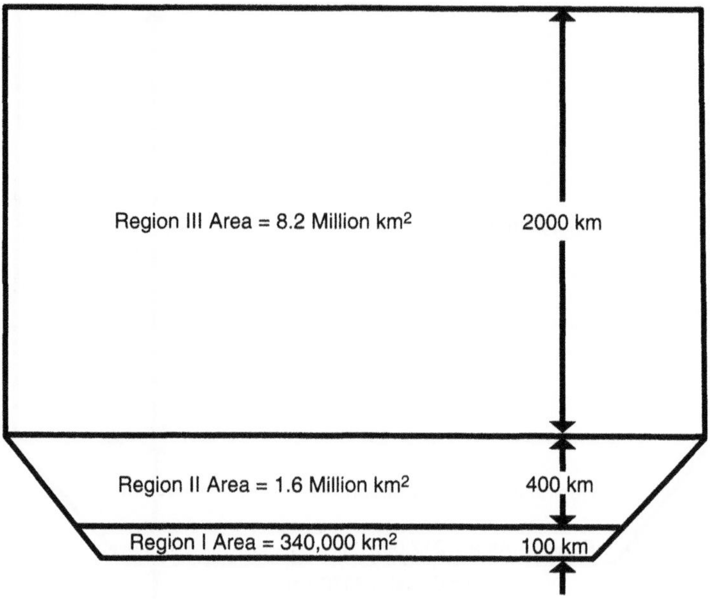

FIGURE 4.9. Tsunami model—Atlantic Ocean.

The detailed calculations of (1) the amplitude of the tsunami at the edge of the high energy asteroid impact site, (2) the amplitude of the tsunami at the coastline, and (3) the distance the tsunami goes inland are contained in Appendix C.

Land Impact

If the high energy asteroid impacts the Earth over land, the primary phenomenon will be the blast effect of the asteroid. There is a secondary phenomenon that arises since the asteroid impact will induce a seismic wave in the ground surrounding the asteroid impact site. Based on Toon, this secondary phenomenon was not included in the risk assessment because the blast impact dominates the consequences. A seismic wave in the blast area is not expected to cause any additional fatalities. When the seismic wave spreads out past the blast area, the magnitude of the seismic wave is not likely to be large enough to cause additional fatalities.

The damage area from the land blast is taken from Toon. The number of people at risk will be the damage area times the population density. For Initiating Events 1 and 2, a fatality fraction is used to account for the fact that some people will survive in the damaged area. Once the asteroid impact energy exceeds 10,000 megatons, the fatality fraction is assumed to be 1.0.

The mean values for the blast areas for the five regional initiating events are shown below.

Initiating Event	*Blast Area (km^2)*
1	176
2	1420
3	26,000
4	137,000
5	742,000

Population Density

Coastal

If the high energy asteroid impacts water, the population at risk is the coastal population for the areas surrounding the asteroid impact site. In general, the tsunami generated by the asteroid will impact large areas of coastline. The coastal population densities (Gulf of Mexico, Pacific, and Atlantic) in this case study are based on spreading the coastal population (year 2000 U.S. census data) into bands (80 or 100 km wide) along the entire coast.

Coastline	*Population Density (km^2)*
Gulf of Mexico	71.4
Pacific Ocean	152
Atlantic Ocean	203

Land

If the high energy asteroid impacts land, the population densities in the risk assessment are based on using year 2000 U.S. census data for metropolitan areas to define high population density sites (metropolitan areas greater than 200 people/km^2) and medium density sites (metropolitan areas less than 200 people/km^2). The balance of the population was spread over the remaining area of the 48 states.

Population Density	*Total Population*	*Total Land Area (km^2)*	*Population Density (km^2)*	*Land Area Fraction*
High	116,157,107	310,828	374	0.041
Medium	108,546,506	1,244,337	87.2	0.162
Low	54,879,824	6,109,891	9	0.797
Total	279,583,437	7,665,056	36.5	1.000

Fatality Fractions

Water

The number of fatalities following a high energy asteroid impact in the ocean will depend on the tsunami amplitude at the shoreline. In this case study, the assumption was made that the larger the tsunami amplitude at the shoreline, the larger the fatality fraction. For the Atlantic and Pacific oceans, a tsunami amplitude at the shoreline calculated to be 2 m or more was assumed to result in a fatality fraction of 1.0. For the Gulf of Mexico, an amplitude at the shoreline calculated to be 6 meters or more was assumed to result in a fatality fraction of 1.0.

Land

For high energy asteroids that land with an impact energy greater than 10,000 megatons TNTE, the assumption is made that everyone in the blast area is a fatality. The uncertainty distributions for the fatality fractions for Initiating Events 1 and 2 are shown in Appendix C, Figures C.36 and C.37.

4.2.4 Quantification of the Likelihoods of the Scenarios (Step 4)

The risk model was quantified by propagating each initiating event through its respective event tree, and compiling the results for all six initiating events.

There are twelve possible sequences for each regional initiating event except for Initiating Event 1, which has only three outcomes. There is only one possible sequence of the global initiating event. As noted earlier, Table 4.2 provides the details of the various parameters used in the risk assessment including the initiating events, the branch points, and the final results. The last column "Fatalities per Year" is the mean number of fatalities per year for each sequence.

Appendix C provides details for the quantification of the high energy asteroid event trees and includes the uncertainty distributions used in this case study. The uncertainties in the initiating event frequencies and the population at risk for each initiating event were propagated through the event tree models to develop composite uncertainties for each sequence. The frequency uncertainty distributions were combined with the corresponding fatality uncertainties, and the results were assembled in the traditional complementary cumulative distribution function format shown in Figure 4.1. See Chapter 2 and Kaplan and Garrick[19] for details of the methods used in producing the complementary cumulative distribution function curves.

4.2.5 Assembly of the Scenarios Into Measures of Risk (Step 5)

The overall results from the risk assessment process are represented by the set of curves shown in Figure 4.1. The computer codes "Crystal Ball" and "EXCEL" were used for the calculations. Figure 4.1 plots the cumulative frequency of exceeding a specific number of fatalities, including an explicit measure of the uncertainty. For example, Figure 4.1 shows that the mean frequency of exceeding 10,000 fatalities is approximately one in 6800 years. This calculation is based on the historical asteroid evidence, the population densities, and the estimated damage area. Figure 4.1 also shows the uncertainty about these results. For example, the 90% confidence interval (from 5% to 95%) for the probability of exceeding 10,000 fatalities spans a factor of approximately 300, from about one in 1900 years to about one in 520,000 years.

As noted in Figure 4.1, the uncertainties are considerable, but this is to be expected for rare events such as large high energy asteroids impacting the Earth, for which there is no recorded history. It is interesting to note that the magnitude of the uncertainty is fairly constant over a wide range of consequences. This may be explained by the fact that while we have more data on the impact of smaller asteroids, we have a better understanding of the number and trajectories of the larger asteroids as they are easier to find and track.

4.2.6 Interpretation of the Results (Step 6)

The information in Table 4.2 can be organized in many different ways. Table 4.3 rearranges Table 4.2 to list the sequences by the mean frequency of each individual sequence in decreasing numerical order. The top sequence is a sequence where an Initiating Event 1 asteroid (asteroid with impact energy less than 10 megatons TNTE) impacts land over a low population density area. The mean frequency of occurrence is approximately one in 600 years and would result in approximately 500 fatalities.

As demonstrated in this chapter, it is possible to perform a quantitative risk assessment of the risk of fatalities from asteroids impacting the Earth using the techniques of Chapter 2. The primary purpose of the case studies is to demonstrate the methodology, not to get results that would necessarily be a basis for decision-making. Had the case studies been more detailed, there would be an opportunity to improve both the database and the models resulting in a reduction in the uncertainties and a more confident basis for decision-making. Database improvements could result in less information uncertainty. Possible improvements in the modeling would be a more accurate representation of the consequences, particularly with respect to the damage areas and their dependency on impact energy. Also needed are improvements in the collateral effects (such as environmental impact) of high energy asteroid impacts.

TABLE 4.3
Mean annual frequencies of asteroid risk scenario by decreasing frequency of sequence for the 48 states

Risk Scenarios														
Initiating Event No./ Frequency Per Year	*Type of Impact (Land-Population Density) (Water-Depth/ Region)*	*Fraction of Earth Area Impacted*	*Fraction of Land by Population Density (High, Medium, Low)*	*Impact Fraction by Water Depth/ Region*	*Risk Scenario Frequency Per Year*	*Coastline Impacted (km)*	*Amplitude of Wave at Landfall (Meters)*	*Distance Water Travels Inland (km)*	*Blast Damage Area (km^2)*	*Population Density (Per km^2)*	*Fatality Fraction*	*Population at Risk*	*Fatalities Per Year*
1/1.34E−01	Land-Low	0.0150	0.7970		1.60E−03				176	9	0.332	526	8.42E−01
1/1.34E−01	Land-Medium	0.0150	0.1620		3.26E−04				176	87	0.332	5095	1.66E+00
1/1.34E−01	Land-High	0.0150	0.0410		8.24E−05				176	374	0.332	21,854	1.80E+00
2/3.82E−03	Atlantic-Reg III	0.0196		0.8120	6.08E−05	3310	0.520	0.231		203	0.151	23,438	1.42E+00
2/3.82E−03	Pacific-Reg III	0.0147		0.8270	4.64E−05	2070	0.520	0.231		152	0.151	10,975	5.10E−01
2/3.82E−03	Land-Low	0.0150	0.7970		4.57E−05				1420	9	0.570	7285	3.33E−01
2/3.82E−03	Atlantic-Reg II	0.0196		0.1540	1.15E−05	1200	2.048	1.398		203	1.000	340,553	3.93E+00
2/3.82E−03	Land-Medium	0.0150	0.1620		9.28E−06				1420	87	0.570	70,580	6.55E−01
2/3.82E−03	Pacific-Reg II	0.0147		0.1440	8.09E−06	1200	2.048	1.398		152	1.000	254,995	2.06E+00
2/3.82E−03	Gulf-Shallow	0.0024		0.5260	4.82E−06	2610	0.210	0.067		71	0.049	610	2.94E−03
2/3.82E−03	Atlantic-Reg I	0.0196		0.0337	2.52E−06	200	12.085	14.800		203	1.000	600,880	1.52E+00
2/3.82E−03	Gulf-Deep	0.0024		0.2630	2.41E−06	2610	2.275	1.592		71	0.280	83,069	2.00E−01
2/3.82E−03	Land-High	0.0150	0.0410		2.35E−06				1420	374	0.570	302,716	7.11E−01
2/3.82E−03	Gulf-Medium	0.0024		0.2100	1.93E−06	2610	1.500	1.017		71	0.150	28,428	5.47E−02
2/3.82E−03	Pacific-Region I	0.0147		0.0289	1.62E−06	200	12.085	14.800		152	1.000	449,920	7.30E−01

3/1.24E−05	Atlantic-Reg III	0.0196		0.8120	1.97E−07	3310	0.686	0.327		203	0.254	55,809	1.10E−02
3/1.24E−05	Pacific-Reg III	0.0147		0.8270	1.51E−07	2070	0.686	0.327		152	0.254	26,133	3.94E−03
3/1.24E−05	Land-Low	0.0150	0.7970		1.48E−07				26,000	9	1.000	234,000	3.47E−02
3/1.24E−05	Atlantic-Reg II	0.0196		0.1540	3.74E−08	1200	2.700	2.005		203	1.000	488,418	1.83E−02
4/2.15E−06	Atlantic-Reg III	0.0196		0.8120	3.42E−08	3310	0.686	0.327		203	0.254	55,809	1.91E−03
3/1.24E−05	Land-Medium	0.0150	0.1620		3.01E−08				26,000	42	1.000	1,112,800	3.35E−02
3/1.24E−05	Pacific-Reg II	0.0147		0.1440	2.62E−08	1200	2.700	2.005		152	1.000	365,712	9.60E−03
4/2.15E−06	Pacific-Reg III	0.0147		0.8270	2.61E−08	2070	0.686	0.327		152	0.254	26,133	6.83E−04
4/2.15E−06	Land-Low	0.0150	0.7970		2.57E−08				137,000	9	1.000	1,233,000	3.17E−02
6/2.00E−08	Global											280,000,000	5.60E+00
3/1.24E−05	Gulf-Shallow	0.0024		0.5260	1.57E−08	2610	0.210	0.067		71	0.049	610	9.56E−06
3/1.24E−05	Atlantic-Reg I	0.0196		0.0337	8.19E−09	200	16.000	23.080		203	1.000	937,048	7.67E−03
3/1.24E−05	Gulf-Deep	0.0024		0.2630	7.83E−09	2610	3.000	2.325		71	0.370	160,311	1.25E−03
3/1.24E−05	Land-High	0.0150	0.0410		7.63E−09				26,000	208	1.000	5,408,000	4.12E−02
4/2.15E−06	Atlantic-Reg II	0.0196		0.1540	6.49E−09	1200	2.700	2.005		203	1.000	488,418	3.17E−03
3/1.24E−05	Gulf-Medium	0.0024		0.2100	6.25E−09	2610	1.500	1.017		71	0.150	28,428	1.78E−04
3/1.24E−05	Pacific-Reg I	0.0147		0.0289	5.27E−09	200	16.000	23.080		152	1.000	701,632	3.70E−03
4/2.15E−06	Land-Medium	0.0150	0.1620		5.22E−09				137,000	21	1.000	2,945,500	1.54E−02
4/2.15E−06	Pacific-Reg II	0.0147		0.1440	4.55E−09	1200	2.700	2.005		152	1.000	365,712	1.66E−06
5/2.66E−07	Atlantic-Reg III	0.0196		0.8120	4.23E−09	3310	0.686	0.327		203	0.254	55,809	2.36E−04
5/2.66E−07	Pacific-Reg III	0.0147		0.8270	3.23E−09	2070	0.686	0.327		152	0.254	26,133	8.45E−05
5/2.66E−07	Land-Low	0.0150	0.7970		3.18E−09				742,000	9	1.000	6,678,000	2.12E−02
4/2.15E−06	Gulf-Shallow	0.0024		0.5260	2.71E−09	2610	0.210	0.067		71	0.049	610	1.66E−06
4/2.15E−06	Atlantic-Reg I	0.0196		0.0337	1.42E−09	200	16.000	23.080		203	1.000	937,048	1.33E−03
4/2.15E−06	Gulf-Deep	0.0024		0.2630	1.36E−09	2610	3.000	2.325		71	0.370	160,311	2.18E−04

(Continued)

TABLE 4.3 (*Continued*)

Risk Scenarios													
Initiating Event No./ Frequency Per Year	*Type of Impact (Land-Population Density) (Water-Depth/ Region)*	*Fraction of Earth Area Impacted*	*Fraction of Land by Population Density (High, Medium, Low)*	*Impact Fraction by Water Depth/ Region*	*Risk Scenario Frequency Per Year*	*Coastline Impacted (km)*	*Amplitude of Wave at Landfall (Meters)*	*Distance Water Travels Inland (km)*	*Blast Damage Area (km^2)*	*Population Density (Per km^2)*	*Fatality Fraction*	*Population at Risk*	*Fatalities Per Year*
4/2.15E−06	Land-High	0.0150	0.0410		1.32E−09				137,000	46	1.000	6,302,000	8.33E−03
4/2.15E−06	Gulf-Medium	0.0024		0.2100	1.08E−09	2610	1.500	1.017		71	0.150	28,428	3.08E−05
4/2.15E−06	Pacific-Reg I	0.0147		0.0289	9.13E−10	200	16.000	23.080		152	1.000	701,632	6.41E−04
5/2.66E−07	Atlantic-Reg II	0.0196		0.1540	8.03E−10	1200	2.700	2.005		203	1.000	488,418	3.92E−04
5/2.66E−07	Land-Medium	0.0150	0.1620		6.46E−10				742,000	11	1.000	8,384,600	5.42E−03
5/2.66E−07	Pacific-Reg II	0.0147		0.1440	5.63E−10	1200	2.700	2.005		152	1.000	365,712	2.06E−04
5/2.66E−07	Gulf-Shallow	0.0024		0.5260	3.36E−10	2610	0.210	0.067		71	0.049	610	2.05E−07
5/2.66E−07	Atlantic-Reg I	0.0196		0.0337	1.76E−10	200	16.000	23.080		203	1.000	937,048	1.65E−04
5/2.66E−07	Gulf-Deep	0.0024		0.2630	1.68E−10	2610	3.000	2.325		71	0.370	160,311	2.69E−05
5/2.66E−07	Land-High	0.0150	0.0410		1.64E−10				742,000	15	1.000	11,352,600	1.86E−03
5/2.66E−07	Gulf-Medium	0.0024		0.2100	1.34E−10	2610	1.500	1.017		71	0.150	28,428	3.81E−06
5/2.66E−07	Pacific-Reg I	0.0147		0.0289	1.13E−10	200	16.000	23.080		152	1.000	701,632	7.93E−05
Total													**2.23E+01**

To the authors, the mean value results of the limited scope risk assessment in this chapter appear to be higher than what has occurred in the 48 states. For example, the most likely high energy asteroid impact has a mean interval of approximately 600 years. Such impacts have not been recorded. This may only be because of the wide range of uncertainty involved. The 90% confidence interval varies over many hundreds of years and the absence of a record that corroborates a central tendency parameter like the "mean" may not be relevant. Nevertheless, an important result of the case study is that the risk of fatalities from asteroids impacting the 48 states is very small on a per year basis. The mean value of 22 fatalities per year among a population of 280 million is relatively insignificant with respect to other phenomena that have the potential to produce fatalities in the U.S. Even at the high end of the confidence interval, approximately 86 fatalities per year, the value is low compared to other phenomena that result in fatalities in the 48 states.

The one characteristic about high energy asteroids that commands attention is that they are a threat that has the potential to lead to species extinction. The need for quantification of such a global threat is obvious.

It is also interesting to note the relationship of the energy of the asteroid (as defined in the initiating events) to its contribution to fatalities per year. Initiating Event 2 (asteroids from 10 to 10,000 megatons TNTE) is the highest contributor (54%) to fatalities per year. Only two other initiating events have a significant contribution to fatalities per year: Initiating Event Global (25%) and Initiating Event 1 (19%). The most likely asteroids to impact Earth do not have the highest risk of fatality.

It is clear that asteroids are not among the high risks facing society. The asteroid risk is unique because it is one of the few threats that can have global consequences and possibly even lead to the extinction of species. The good news is that high energy asteroids that could threaten human civilization are predictable and avoidable if actions are taken prior to impact. This fact alone has prompted speculation on methods for altering the trajectory or changing the speed of a threatening large asteroid to avoid a collision with Earth. There are programs to monitor threatening large asteroids, but to date there has not been a national or international commitment of major funding for intercepting a threatening asteroid.

A point made in this book is the need for more quantitative information on risks to support rational and scientifically based assessments of threats to society. The case studies provide some insights on the relative importance of the risks considered. One insight from this case study is that actions to reduce the regional asteroid risk should be a lower priority for many locations than protecting those locations from other threats such as major hurricanes.

4.3 Future Action

The National Aeronautics and Space Administration is a logical organization to perform assessments of the asteroid risk for the U.S. The literature indicates that NASA has the capability to perform such risk assessments but has not yet integrated quantitative risk assessment into their basic decision making process.

This risk assessment indicates the advisability of concentrating on asteroids that could impact Earth with impact energy of ten megatons TNTE or more.

It is important that individual nations perform assessments of the risks of asteroids, as the more likely asteroids are not of the size to have global impact. The most likely asteroid impact will have a regional impact that is location specific. In general, nations with little or no coastline should have smaller fatality risk because they are not vulnerable to tsunamis caused by asteroid impacts on the oceans.

4.4 The Risk from an Asteroid Impacting New Orleans

To enable a comparison of risks to a specific region, a quantitative risk assessment of a high energy asteroid impacting New Orleans was completed. Results of this analysis were used to compare the quantitative risk assessment of major hurricanes striking New Orleans to the quantitative risk assessment of high energy asteroids impacting New Orleans.

This case study considered a high energy asteroid impacting New Orleans as a result of (1) a direct land impact, (2) impacting the Gulf of Mexico, and (3) an asteroid impact having global consequences. Changes were made to the 48 states model to adapt the model to impact only one metropolitan area with a high population density. Care must be taken when making comparisons of the results of the New Orleans assessment with those results for the 48 states.

A summary of selected results follows.

- The mean frequency is one in 110,000 years of a high energy asteroid impacting New Orleans resulting in 10,000 or more fatalities. See Figure 4.10. There is considerable uncertainty in this result. The 90% confidence interval of the frequency ranges from one in 29,000 years to one in 3,200,000 years.
- The most likely asteroid to cause catastrophic fatalities has impact energy between 10 and 10,000 megatons TNTE and lands in the shallow water of the Gulf of Mexico.
- Spreading the fatalities over a per year basis, the mean number of fatalities from an asteroid hitting New Orleans was approximately three per year. See Table 4.4. The 90% confidence interval ranges from 0.1 fatalities per year to 12 fatalities per year.

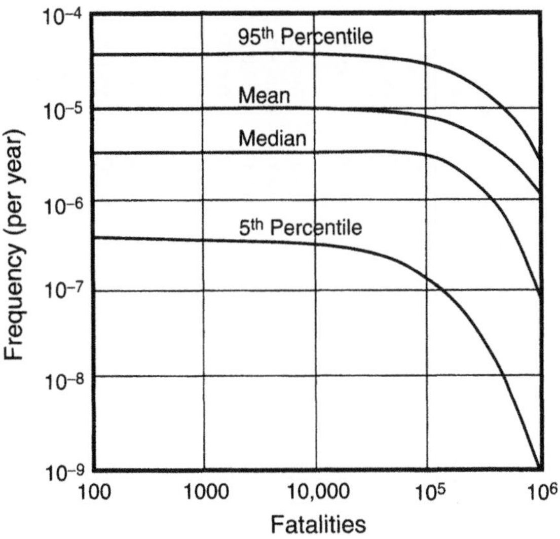

FIGURE 4.10. Asteroid risk for metropolitan New Orleans.

- Approximately 98% of the New Orleans fatality risk comes from asteroids having impact energies between 10 and 10,000 megatons TNTE. Approximately 99% of the New Orleans fatality risk comes from tsunamis.

The approach is to follow the same steps as above and simply indicate the differences in the two models, that is, the 48 state model and New Orleans.

4.4.1 Definition of the System During Normal Conditions (Step 1)

For the case of the vulnerability of New Orleans to a high energy asteroid, the "system" is the city itself, its infrastructure, and its population.

4.4.2 Identification and Characterization of System Hazards (Step 2)

No change from Section 4.2.2.

4.4.3 Structuring of the Risk Scenarios and Consequences (Step 3)

Initiating Events

No changes were made to the regional initiating event frequencies. No changes were made to Figure 4.2 (the land impact event tree) or

TABLE 4.4

Mean annual frequencies of asteroid risk scenarios and split fractions of intervening events by initiating events for New Orleans

Risk Scenarios: Initiating Event No./ Frequency Per Year	Risk Scenarios: Type of Impact (Land-Population Density) (Water-Depth/ Region)	Fraction of New Orleans Area Impacted	Fraction Impacting High Density Population (Land)	Impact Fraction of Gulf by Water Depth	Risk Scenario Frequency Per Year	Coastline Impacted (km)	Amplitude of Wave at Landfall (Meters)	Total Population	Blast Damage Area (km²)	Population Density (Per km²)	Fatality Fraction	Population at Risk	Fatalities Per Year
1/1.34E−01	Land–High	1.17E−05	1.0000		1.568E−07				176	224	0.332	13,089	2.05E−02
2/3.82E−03	Land—High	1.17E−05	1.0000		4.470E−09				1420	224	0.570	181,306	8.10E−03
2/3.82E−03	Gulf—Shallow	0.0024		0.526	4.822E−06	1	0.210	1.3E+06		1	0.100	133,000	6.41E−01
2/3.82E−03	Gulf—Medium	0.0024		0.210	1.925E−06	1	1.500	1.3E+06		1	0.300	399,000	7.68E−01
2/3.82E−03	Gulf—Deep	0.0024		0.263	2.411E−06	1	2.275	1.3E+06		1	0.560	744,800	1.80E+00
3/1.24E−05	Land—High	1.17E−05	1.0000		1.451E−11				5980	224	1.000	1,339,520	1.94E−04
3/1.24E−05	Gulf –Shallow	0.0024		0.526	1.565E−08	1	0.210	1.3E+06		1	0.100	133,000	2.08E−03
3/1.24E−05	Gulf—Medium	0.0024		0.210	6.250E−09	1	1.500	1.3E+06		1	0.300	399,000	2.49E−03
3/1.24E−05	Gulf—Deep	0.0024		0.263	7.827E−09	1	3.000	1.3E+06		1	0.740	984,200	7.70E−03
4/2.15E−06	Land—High	1.17E−05	1.0000		2.516E−12				5980	224	1.000	1,339,520	3.37E−05
4/2.15E−06	Gulf—Shallow	0.0024		0.526	2.714E−09	1	0.210	1.3E+06		1	0.100	133,000	3.61E−04
4/2.15E−06	Gulf—Medium	0.0024		0.210	1.084E−09	1	1.500	1.3E+06		1	0.300	399,000	4.32E−04
4/2.15E−06	Gulf—Deep	0.0024		0.263	1.357E−09	1	3.000	1.3E+06		1	0.740	984,200	1.34E−03
5/2.66E−07	Land—High	1.17E−05	1.0000		3.112E−13				5980	224	1.000	1,339,520	4.17E−06
5/2.66E−07	Gulf—Shallow	0.0024		0.526	3.358E−10	1	0.210	1.3E+06		1	0.100	133,000	4.47E−05
5/2.66E−07	Gulf—Medium	0.0024		0.210	1.341E−10	1	1.500	1.3E+06		1	0.300	399,000	5.35E−05
5/2.66E−07	Gulf—Deep	0.0024		0.263	1.679E−10	1	3.000	1.3E+06		1	0.740	984,200	1.65E−04
6/2.00E−08	Global				2.000E−08							1,339,520	2.68E−02
Total					**9.38E−06**								**3.25E+00**

to Figure 4.3 (the Gulf of Mexico impact event tree). Figure 4.4 (the Pacific Ocean impact event tree) and Figure 4.5 (the Atlantic Ocean impact event tree) do not apply to New Orleans. No changes were made to the global impact frequency but appropriate changes were made to the population at risk.

Impacts Water or Land

No changes were made to structuring the scenarios with respect to whether the high energy asteroid impacts water or land except to delete the sequences resulting from asteroid impact in the Atlantic and the Pacific oceans.

Population Density

Population data from the year 1990 and year 2000 U.S. census was used for the values of population and land area. The total population of the New Orleans metropolitan area was taken to be 1.34 million in a land area of 5980 km^2, which resulted in a population density of 224 people/km^2.

A large portion of the New Orleans metropolitan area is below sea level. This unique characteristic of New Orleans affected the impact of a tsunami. For New Orleans, the asteroid that lands in the Gulf of Mexico has an impact somewhat analogous to a hurricane that strikes New Orleans but is different in critical areas. For asteroids, there is no evacuation assumed. People inside the New Orleans levee system will be protected from the tsunami as long as the levee system remains intact. However, it is likely that the levee would be breached by the tsunami. People outside the levee system would be subjected to the direct impact of the tsunami.

The New Orleans risk model did not consider the distance the tsunami would travel inland. Because of the low-lying nature of the New Orleans area, the tsunami that reached the coast of the Gulf of Mexico was assumed to impact the entire population of New Orleans regardless of the amplitude of the tsunami. The tsunami would impact New Orleans by going up the Mississippi River, by entering Lake Pontchartrain, and by traversing over the low-lying land (mostly wetlands) separating New Orleans from the Gulf of Mexico.

As noted above, for land impact New Orleans was treated as a location with a high population density. All sequences with medium and low population densities were deleted from the New Orleans risk assessment.

Fatality Fractions

The fatality fraction for New Orleans following a high energy asteroid impact in the Gulf of Mexico was assumed to be twice the fatality fraction used in the 48 states model due to the unique characteristics of New Orleans as elaborated on in Appendix C. No changes were made to the fatality fractions for those asteroids that directly impacted New Orleans.

4.4.4 Quantification of the Likelihood of the Scenarios (Step 4)

The New Orleans risk model was quantified by propagating each initiating event through its respective event tree.

The 52 possible sequences in the 48 states model were reduced to 18 sequences as shown in Table 4.4. There are 17 regional sequences and 1 global sequence. Of the 17 regional sequences, 5 of the sequences represent asteroid impact on land and 12 of the sequences represent asteroids impacting the Gulf of Mexico. Table 4.4 provides the details of the various parameters used in the risk assessment including the initiating events, the branch points, and the final results. The last column "Fatalities per Year" is the mean number of fatalities per year for each sequence.

Appendix C provides details for the quantification of the asteroid event trees for New Orleans. No changes were made to the uncertainty distributions except for the fatality fraction following an asteroid impact in the Gulf of Mexico.

4.4.5 Assembly of the Scenarios into Measures of Risk (Step 5)

The uncertainties in the initiating event frequencies and the people at risk for each initiating event were propagated through the event tree models to develop composite uncertainties for each sequence. The frequency uncertainty distributions were then combined with the corresponding fatality uncertainties, and the results were assembled in the traditional complementary cumulative distribution function format shown in Figure 4.10.

The overall results from the quantitative risk assessment for New Orleans are represented by the set of curves shown in Figure 4.10. The computer codes "Crystal Ball" and "EXCEL" were used for the calculations. Figure 4.10 plots the cumulative frequency, with uncertainty, of exceeding a specific number of fatalities. Figure 4.10 shows that the mean frequency of exceeding 10,000 fatalities is approximately one in 110,000 years. The 90% confidence interval ranges from one in 29,000 years to about one in 3,200,000 years.

4.4.6 Interpretation of the Results (Step 6)

The information in Table 4.4 can be organized in many ways to obtain different perspectives on the results.

Table 4.5 rearranges Table 4.4 to list the sequences by the frequency of each individual sequence in decreasing numerical order

TABLE 4.5
Mean annual frequencies of asteroid risk scenarios and split fractions by decreasing frequency for New Orleans

Risk Scenarios													
Initiating Event No./ Frequency Per Year	*Type of Impact (Land-Population Density) (Water-Depth/ Region)*	*Fraction of New Orleans Area Impacted*	*Fraction Impacting High Density Population (land)*	*Impact Fraction of Gulf by Water Depth*	*Risk Scenario Frequency Per Year*	*Coastline Impacted (km)*	*Amplitude of Wave at Landfall (meters)*	*Total Population*	*Blast Damage Area (km^2)*	*Population Density (Per km^2)*	*Fatality Fraction*	*Population at Risk*	*Fatalities Per Year*
2/3.82E−03	Gulf—Shallow	0.0024		0.526	4.822E−06	1	0.210	1.3E+06		1	0.100	133,000	6.41E−01
2/3.82E−03	Gulf—Deep	0.0024		0.263	2.411E−06	1	2.275	1.3E+06		1	0.560	744,800	1.80E+00
2/3.82E−03	Gulf—Medium	0.0024		0.210	1.925E−06	1	1.500	1.3E+06		1	0.300	399,000	7.68E−01
1/1.34E−01	Land-High	1.17E−05	1.0000		1.568E−07				176	224	0.332	13,089	2.05E−02
3/1.24E−05	Gulf—Shallow	0.0024		0.526	1.565E−08	1	0.210	1.3E+06		1	0.100	133,000	2.08E−03
3/1.24E−05	Gulf—Deep	0.0024		0.263	7.827E−09	1	3.000	1.3E+06		1	0.740	984,200	7.70E−03
3/1.24E−05	Gulf—Medium	0.0024		0.210	6.240E−09	1	1.500	1.3E+06		1	0.300	399,000	2.49E−03
2/3.82E−03	Land—High	1.17E−05	1.0000		4.469E−09				1420	224	0.570	181,306	8.10E−03
4/2.15E−06	Gulf—Shallow	0.0024		0.526	2.714E−09	1	0.210	1.3E+-06		1	0.100	133,000	3.61E−04
4/2.15E−06	Gulf—Deep	0.0024		0.263	1.357E−09	1	3.000	1.3E+06		1	0.740	984,200	1.34E−03
4/2.15E−06	Gulf—Medium	0.0024		0.210	1.084E−09	1	1.500	1.3E+06		1	0.300	399,000	4.32E−04
5/2.66E−07	Gulf—Shallow	0.0024		0.526	3.358E−10	1	0.210	1.3E+06		1	0.100	133,000	4.47E−05
5/2.66E−07	Gulf—Deep	0.0024		0.263	1.680E−10	1	3.000	1.3E+06		1	0.740	984,200	1.65E−04
5/2.66E−07	Gulf—Medium	0.0024		0.210	1.341E−10	1	1.500	1.3E+06		1	0.300	399,000	5.35E−05
3/1.24E−05	Land—High	1.17E−05	1.0000		1.451E−11				5980	224	1.000	1,399,520	1.94E−04
4/2.15E−06	Land—High	1.17E−05	1.0000		2.516E−12				5980	224	1.000	1,339,520	3.37E−05
5/2.66E−07	Land—High	1.17E−05	1.0000		3.112E−13				5980	224	1.000	1,339,520	4.17E−06
6/2.00E−08	Global				2.000E−08							1,339,520	2.68E−02
Total					**9.376E−06**								**3.25E+00**

(mean values). The top sequence is a sequence involving an asteroid with impact energy between 10 and 10,000 megatons TNTE impacting the Gulf of Mexico over shallow water. The mean frequency of occurrence is approximately one every 200,000 years and would result in approximately 133,000 fatalities.

The mean value of approximately three fatalities per year is low compared to other phenomena that result in fatalities in New Orleans. Even at the high end of the uncertainty interval, approximately 12 fatalities per year, the value is low compared to other phenomena that result in fatalities for New Orleans.

References

[1] Conway, B. A., *Optimal Interception and Deflection of Earth-Approaching Asteroids Using Low-Thrust Electric Propulsion*, Mitigation of Hazardous Comets and Asteroids, Cambridge, 2004, pp. 292–312.

[2] NASA (National Aeronautics and Space Administration), *Near-Earth Object Science Definition Team, Study to Determine the Feasibility of Extending the Search for Near-Earth Objects to Smaller Limiting Diameters*, NASA Office of Space Science, Solar System Exploration Div., Washington, DC, 2003, http://neo.jpl.nasa.gov/neo/neoreport030825.pdf.

[3] Lewis, J. S., *Comet and Asteroid Impact Hazards on a Populated Earth*, Academic Press, San Diego, CA, 2000.

[4] Morrison, D., "Asteroid and Comet Impacts: the Ultimate Environmental Catastrophe," *Philos. Trans. R. Soc.* 2006, *364*, 2041–2054.

[5] Chapman, C. R., Impact lethality and risks in today's world: Lessons for interpreting earth history, In: Koeberl, C., MacLeod, K. G. (Eds.), *Catastrophic Events and Mass Extinctions: Impacts and Beyond*, Geological Society of America, Boulder, CO, 2002, Special Paper, Vol. 356, pp. 7–19.

[6] Chapman, C. R., *Earth Planet. Sci. Lett.* 2004, *222*, 1–15.

[7] Alvarez, L. W., Asaro, E., Michel, H. V., "Extraterrestrial Cause for the Cretaceous—Tertiary Extinction," *Science* 1980, *208*, 1095–1099.

[8] Shoemaker, E. M., "Asteroid and Comet Bombardment of the Earth," *Annu. Rev. Earth Planet Sci.* 1983, *11*, 461–494.

[9] Belton, M. J. S., Morgan, T. H., Samarasinha, N., Yeomans, D. K. (Eds.), *Mitigation of Hazardous Comets and Asteroids*, Cambridge University Press, Cambridge, MA, 2004.

[10] Morrison, D. (Ed.), *The Spaceguard Survey Report of the NASA International Near-Earth-Object Detection Workshop*, NASA Publication, 1992, http://impact.arc.nasa.gov.

[11] Burrows, W. E. 2006. The Survival Imperative, Appendix B. White Paper Summarizing Findings and Recommendations from the 2004 Planetary Defense Conference, Protecting Earth from Asteroids, Forge, New York.

[12] Sommer, G., *Astronomical Odds: A Policy Framework for the Cosmic Impact Hazard*. Rand Corporation, Santa Monica, CA, 2005, (http://www.rand.org/publications/RGSD/RGSD184).

[13] In Stokes, G. (Ed.), *Study to Determine the Feasibility of Extending the Search for Near Earth Objects to Smaller Limiting Diameters*, Report of the NASA NEO Science Definition Team. 2003, http://neo.jpl.nasa.gov/neo/report.html.

[14] NRC (National Research Council), *New Frontiers in the Solar System: An Integrated Exploration Strategy.* National Academy Press, Washington, DC, 2002, http://nap.edu.

[15] NRC (National Research Council), *Astronomy and Astrophysics in the New Millennium*, National Academy Press, Washington, DC, 2001, http://nap.edu.

[16] Chapman, C. R., The asteroid/comet impact hazard: Homo Sapiens as dinosaur? In: Sarewitz, D., Pielke, R. A., Byerly, R., (Eds.), *Prediction: Science, Decision Making, and the Future of Nature*, Island Press, Washington, DC, 2000, 107–134.

[17] Chesley, R. S., and Ward, S. N., "A Quantitative Assessment of the Human and Economic Hazard from Impact-Generated Tsunami," *Nat. Hazards* 2006, *38*, 355–374.

[18] Toon, O. B., Zahnle, K., Morrison, D., "Environmental Perturbations Caused by the Impacts of Asteroids and Comets," *Rev. Geophys.* 1997, *35*(1), 41–78.

[19] Kaplan, S., and Garrick, B. J., "On the Quantitative Definition of Risk," *Risk Anal.* 1981, *1*(1).

CHAPTER 5

Case Study 3: Terrorist Attack on the National Electrical Grid

This case study is an extension of a study that the same authors performed for a report prepared by a committee of the National Academy of Engineering of the United States that was published not by the Academy but as a special issue of the international journal, *Reliability Engineering & System Safety*.[1] This case study is not a full scope quantitative risk assessment, but a scoping analysis to examine the need and value added of such an assessment. The goal is to obtain insights on the risk of a terrorist attack of a specific region of the electrical grid serving several large metropolitan areas.

The infrastructure for the reliable supply of electricity is critical to the nation's well being. This case study presents a risk assessment of a hypothetical terrorist attack on a portion of the national electrical grid. Vulnerabilities to terrorist attacks have been addressed in other studies.[2,3] The energy sector is currently responding to the threat of possible attacks through risk management practices.[4] Risk management is an integral part of the electricity sector's program of "critical infrastructure protection." As such, these efforts are defined as "safeguarding the essential components of the electric infrastructure against physical and electronic threats in a manner consistent with appropriate risk management, with both industry and industry–government partnerships, while sustaining public confidence in the electricity sector."

This case study involves a risk assessment of a combined cyber[5] and physical attack on a hypothetical electric power grid. The threat assessment part of the case study is more limited than the vulnerability assessment; this is primarily due to security considerations and a lack of resources to examine classified threat information (see Chapter 2, Section 2.4.2, and Figure 2.4 for the distinctions between "threat" and "vulnerability" assessments). The vulnerability assessment develops the consequences to the point of inflicting various degrees of damage to the electrical grid, but not to the point of the human fatality risk. When the assessment is specialized to a specific location, the risk measure will be in terms of consequences to the population at risk as was done for the previous case studies on hurricanes and asteroids.

The authors have attempted to provide enough detail to convey the ideas of the risk assessment methodology without resorting to overly technical jargon. For some readers (managers, policy makers, etc.), there may be too much detail. For the technical community, there may not be enough detail. The authors suggest that policy makers concentrate on the first few and last few pages of the chapter, which cover the essentials. More technically inclined readers may want to more closely examine the detailed analytical steps of the risk assessment process.

This case study shows how vital infrastructure systems can be analyzed to expose their vulnerabilities and to provide a basis for taking corrective actions either to avert or to mitigate the consequences of a terrorist attack. The risk assessment leads to specific recommendations, derived from the supporting evidence that could not have been easily deduced, or supported, without this formal approach.

In the previous case studies on hurricanes and asteroids, a summary of results was presented in the introductory material to facilitate not having to wade through a lot of technical jargon to get to some bottom line information. For scoping analyses such as presented here on a terrorist attack and in the next case study on abrupt climate change the results take on a different meaning. That meaning is what can be learned from risk assessment scoping analyses that is relevant to the real world. Experience indicates that much can be learned especially if care is taken to have the scoping models reflect typical, if not actual, conditions to be expected in the real world. Of course, the further compromise in the scoping analyses is that the measures of risk are precursors to the desired measure of consequences to the population at risk. In particular, the scoping analyses truncate the damage states prior to health effects on the population. Quantifying health effects to humans is a much greater challenge and is usually not undertaken until the scoping analysis indicates the need to do so. Thus, the summary of results that follows must be interpreted in the context of how well the hypothetical system being analyzed represents reality. Past practice has indicated, in fact, that such scoping analyses most often do provide important insights into the risk of the actual system, although with much greater uncertainty than would be the case for a comprehensive risk assessment.

5.1 Summary of Insights from the Scoping Analysis

Besides serving the purpose of guiding the modeling approach in a quantitative risk assessment, a scoping analysis provides insights into the magnitude of the risk of the system being analyzed. Of course, such results are suspect until they are backed up with a quantitative assessment of the specific system at its specific location. Meanwhile, the scoping results usually bound the risk of the intended application and provide information for making a decision to proceed or not to proceed with the analysis.

As a general indication of the risk of a terrorist attack of a regional electrical grid having the properties and conditions chosen for this scoping analysis, we are 90% confident that the frequency of a long-term regional power outage[a] from a terrorist attack is one event in 5250 to 67 years. The mean frequency is approximately one in 250 years. Whether such an event meets the criteria of a catastrophic event (10,000 fatalities or greater) is dependent on many things yet to be considered such as (1) how much longer than 24 h might the outage be, (2) exactly how dependent the metropolitan area(s) are on electric power, (3) what emergency systems exist for such an event, and (4) the alternatives for evacuation and outside support.

This scoping analysis indicates that the most likely consequence of a terrorist attack is not a long-term outage, but rather a shorter and more localized power outage such as the outage of a single vulnerable network within a regional grid that possibly serves a large metropolitan area. For this case we are 90% confident that the frequency of a power disruption is one event in 225 to 20 years. The mean frequency is one event in 53 years. The result would most likely be a short-term outage or possibly only a power transient.

Other insights provided by the scoping analysis follow.

- The analysis clearly shows that the attackers' chance of success in causing a long-term outage is very network[b] dependent within a regional grid. Different networks in a grid have different vulnerabilities that are not obvious until actual attack scenarios are detailed in a structured manner typical of risk modeling. Networks and systems considered important for grid stability and normal operation, and thus most often best protected, may not be the best targets for creating extended outages.

[a] A long-term power outage is defined in this study as a complete loss of power for 24 h or greater.

[b] A network is an interconnected system of generators, substations, and major transmission lines.

- The likelihood of a successful attack is much greater for a physical attack than for a cyber attack. This is probably because it is easier to build in software barriers than it is to "harden" diverse and expansive physical plant systems that are highly interconnected.
- The relative high degree of success of physical attacks is attributed primarily to the fact that many critical systems such as substations, extra high-voltage transmission lines, and switching facilities are often located in remote areas with minimal security or regular surveillance.
- Coordinated physical attacks on local substations in vulnerable networks and a cyber attack on the regional supervisory control and data acquisition (SCADA) systems is most important to the overall risk of combined network and regional power outages.
- The analysis indicated that there would be a very low likelihood of successful intrusion into the regional SCADA systems, although there is a great deal of uncertainty in the estimates.

Despite a plethora of systems involved in a regional electrical grid, the scoping analysis suggests that a limited number of systems stand out as vulnerable targets for terrorists. Examples are strategically located substations that may not be critical to the network's electrical stability, but very critical in triggering a long-term outage within a network.

5.2 A Scoping Analysis to Support a Quantitative Risk Assessment of a Specific Terrorist Attack

This scoping analysis follows the pattern of the six-step process to quantitative risk assessment introduced in Chapter 1 and employed for Case Studies 1 and 2. In the absence of this being a complete risk assessment, the steps are still employed to note what scoping analysis effort relates to what risk assessment step.

5.2.1 Definition of the System During Normal Conditions (Step 1)

In an increasingly interconnected world, technology-based systems and networks are becoming more and more interdependent. An attack on one system can have far-reaching, cascading effects on other systems and on society as a whole. The system in this example is a hypothetical portion of a national electric power grid, which is tightly linked to other vital systems and, therefore, is an attractive target for terrorists. The consequences of an attack on a national electric power grid that leads to long term-outages, say greater than 48 h, could cascade into major disruptions in transportation, communications, sanitation, food supplies, water supplies, and other vital systems.

The first step of the analysis process is to understand how the electricity supply system works, so that departures from normal, successful operation can be examined. Once the normal system status is clearly understood, vulnerabilities that require special analysis can be identified.

The hypothetical system for this case study includes a portion of the national electrical grid that supplies power to a region of the country. Within that region are several interconnected networks that contain electrical generators, transmission lines, and distribution facilities that deliver power to major metropolitan areas and individual consumers.

The Region

Figure 5.1 represents a major region in the national electric power grid. Each network corresponds to a large metropolitan area, such as New York City, Philadelphia, Boston, etc. Networks are interconnected to form a regional grid (such as the northeast corridor or the western states). In Figure 5.1, Network 1 is interconnected with four neighboring networks through ties T12, T13, T14, and T15. These "interties" form the transmission system and are typically extra-high voltage (EHV) transmission lines that provide the major pathways for power flow throughout the region and between cities. Regional grid operations are typically coordinated through established protocols that are designed to ensure economical transfers of power through the interties and to prevent failures from cascading and causing widespread disruptions in power (such as the August 2003 blackout in the northeast U.S.).

Figure 5.1 shows that external power can be routed to Network 1 through several parallel interties. In some parts of the country, the available interconnections are limited. Well known examples include the north-south ties through the Western Interconnection, ties from the southern power pools to the Electric Reliability Council of Texas Interconnection, and limited ties to Florida through the Eastern Interconnection. Because these regional-specific features must be carefully taken into account, realistic risk assessments cannot be performed generically.

In addition, the U.S. Department of Energy has identified several transmission "bottlenecks" at various interties throughout the U.S. electrical grid.[6] Bottlenecks occur at points where major tie lines are frequently loaded to a large fraction of their available capacity and

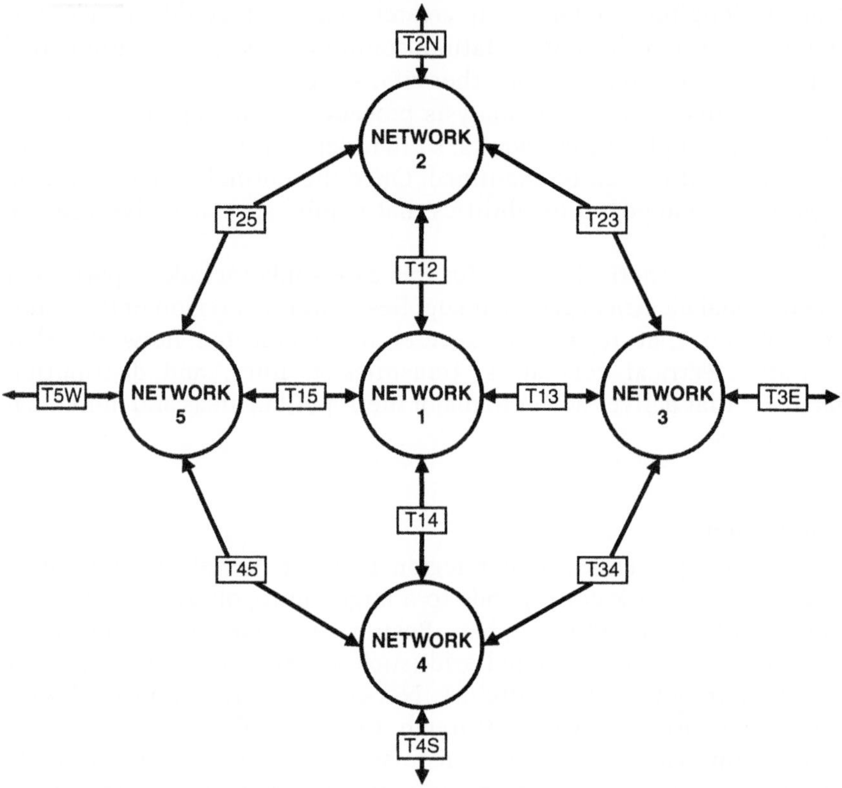

FIGURE 5.1. Example regional grid.

thus have limited reserve capacity for additional power flows during emergency situations. These bottlenecks represent critical chokepoints in the transfer of power between interconnected networks.

The Network

Figure 5.2 shows an expanded view of Network 1. The network contains an interconnected system of generators, substations, and major transmission lines. Network 1 has five major generating stations (G1 through G5) that are responsible for generating power, and four major transmission substations (S1 through S4) that distribute power. Generation, transmission, and power flows in each network are typically coordinated through a centralized operations and control facility. Network control centers have the primary responsibility of scheduling power purchases from generating units, allocating generation and

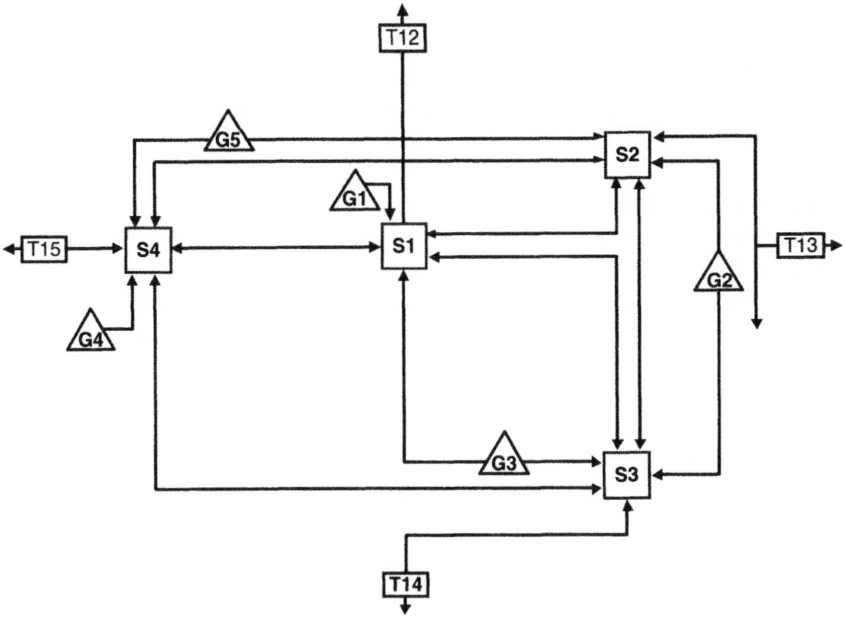

FIGURE 5.2. Generating stations and substations in Network 1.

loads to available transmission lines, ensuring network stability and reliability, and responding to emergencies.

Elements of the Network and Region

This analysis focuses on four elements of the electrical grid: substations, transmission lines, SCADA systems, and energy management systems (EMSs). Each represents a potential point of vulnerability and therefore must be defined.

Substations

Substations are the transfer points for energy flows within the electrical grid. Each substation contains transmission line termination points, as well as circuit breakers and bus bars that interconnect the transmission lines with various circuits. Major substations also contain transformers that reduce intertie transmission line voltages to network transmission levels. Each substation contains metering equipment, protection relays, and switching circuits that control the operation of the connected generation, transmission, and distribution supplies.

In Figure 5.2, substation S1 contains monitoring, control, and protection circuits for all power output from generating station G1; part of the power output from generating station G3; network transmission line connections to substations S2, S3, and S4; and regional transmission line interconnection T12.

Transmission Lines

Transmission lines are the conduits that transfer energy throughout the grid. Because of their importance to system operation, this assessment focuses primarily on EHV transmission lines that transmit energy from individual generators to the major substations in each network.

Substation S1 contains the following six transmission line connections: line G1-S1 connects the output from generating station G1; line G3-S1 connects the output from generating station G3; lines S1–S2, S1–S3, and S1–S4 connect to the other substations in the network; and line T12 is the regional intertie to Network 2.

In practice, each transmission line typically contains two or more parallel circuits, either mounted on overhead towers or routed underground. Because land space for EHV transmission corridors is often limited, several transmission lines may be routed through the same right-of-way. For example, transmission line T12 is the long distance intertie line to Network 2. However, lines G1 and T12 are located in a common right-of-way for part of their route to substation S1. Similarly, lines S1–S2 and S1–S3 leave substation S1 together before they split.

Supervisory Control and Data Acquisition Systems

Each network SCADA system provides integrated parameter monitoring, data processing, and automatic control of circuit switching, load smoothing, and regulation of voltage and frequency throughout the network. The SCADA system also provides status displays for all major equipment and transmission lines, parametric trends, alarms, and a manual control interface for the load-control center operators.

The SCADA system "oversees" the network and responds to changing conditions. For example, if generating station G1 trips off line, the consequential voltage and frequency fluctuations may require rapid, active circuit switching to route additional power to substation S1. The SCADA system automatically controls energy transfers by switching the appropriate circuit breakers, and it increases output from the remaining generators to compensate for the lost generating capacity. If the fluctuations cannot be stabilized, the SCADA system implements preprogrammed automatic protection protocols to open the connections to selected substations and restores stable conditions throughout the remainder of the network.

Similar supervisory and control functions are also performed by SCADA systems at the regional level.

Energy Management Systems

An EMS can be loosely thought of as providing input to the SCADA control system. The EMS determines the most cost-effective configuration of power production, transmission, and distribution throughout the network, considering the required criteria for system stability, safety, and reliability. An EMS typically provides the fundamental information and computation capability to perform real-time network analyses, to provide strategies for controlling system energy flows, and to determine the most economical mix of power generation, power purchases, and sales.

For example, if generating station G4 trips off line, the EMS will determine if it is more cost-effective to increase output from generating station G5, to start local peaking units (auxiliary units to supplement high network demands), to increase energy flow through interconnection T15, or to implement other options. The selected strategy depends on the system status at the time of the transient and preprogrammed protocols for rapid recovery of stable load flows at the lowest available cost for emergency replacement power.

Specific Details of the Example System

The following system characteristics are defined.

- The total available generating capacity in Network 1 is not sufficient to meet load demands during periods of peak energy usage (e.g., summer weekdays).
- Most of the customers in Network 1 are supplied through connections to substations S1, S2, and S3.
- Substation S4 serves primarily as an EHV transmission intertie, and it carries only a small fraction of the total network distribution load.
- Interconnection T12, the primary intertie between Network 1 and the region, is a potential transmission bottleneck.

5.2.2 Identification and Characterization of System Hazards (Step 2)

Once a system is defined, the hazards associated with it can be identified and characterized. In the risk sciences, the word "hazard" is usually defined as "a potential source of danger or damage," but does not necessarily imply the infliction of damage. A risk scenario is a sequence of events that links the hazard to the final damage state. For example, a chemical plant with an inventory of toxic chemicals can contain a variety of hazards, but only through risk scenarios (i.e., accidents or malicious acts) can the hazards be manifested as an actual damage state.

For this particular case study, the source of danger is defined as a potential terrorist action. Specific scenarios that will be developed further in this case study are (1) physical attacks on critical substations within Network 1, (2) physical attacks on transmission lines and switching facilities in the regional grid, and (3) a coordinated simultaneous cyber attack on the regional SCADA systems.

5.2.3 Structuring of the Risk Scenarios and Consequences (Step 3)

Scenario development, the fundamental building block of every risk assessment, follows a structured format that answers two of the risk triplet questions.

- What can go wrong?
- What are the consequences?

A variety of logic models and analytical tools are used to develop risk scenarios. These include event–sequence diagrams (ESDs) that display important elements of the evolving scenario, failure modes and effects analyses (FMEAs) that tabulate possible contributing causes, and event trees or fault trees that display functional and logical relationships among threats, targets, vulnerabilities, and consequences.

Two common methods are used for scenario development. One involves going forward from an initial disturbance of the system; the other works backward from the undesirable end state.

1. Given a set of initiating events, the structuring of scenarios is done so that the end state (the damage state or undesired event) of each scenario is the condition that terminates the scenario. This approach is used for full-scope risk assessments that trace an upset from its initiation to its final impact on the system. Scenarios constructed in this way form what is called an event tree.
2. Given an end state (the undesired event), a systematic logical process is used to project backwards and to determine the potential scenarios that could cause the end state. This approach yields what is called a fault tree.

These methods are often used together to construct an encompassing set of risk scenarios. Obviously, a comprehensive examination of the electrical grid vulnerabilities might identify a great number of possible threat scenarios for a particular set of consequences or damage levels. It is impractical in this example to demonstrate a complete risk assessment of all possible damage conditions. Therefore, we define a small number of possible scenarios and link them to defined damage levels.

Definition of Damage Levels and Consequences

In Section 5.2.2, the source of danger (the terrorist action) was defined, and potential threats were identified. In this step of the analysis, six potential end states are defined and are linked to initiating events through the scenario development process.

- Damage Level 0 (no damage)—no significant network or regional power outages
- Damage Level 1—transient outage to Network 1
- Damage Level 2—transient outage to the region (and Network 1)
- Damage Level 3—long-term outage to Network 1
- Damage Level 4—long-term outage to Network 1 and transient outage to the region
- Damage Level 5—long-term outage to the region (and Network 1)

Damage Level 0 (included for analysis completeness) accounts for the possibility that the terrorists may fail to cause any significant damage. Actions that prevent or effectively mitigate an attack scenario result in successful termination of the event, and these scenarios are assigned to Damage Level 0.

For this case study, transient damage means a complete loss of power for a period of 4–24 h. Long-term damage means a complete loss of power for more than 24 h. For example, Damage Level 1 means that Network 1 (and only Network 1) experiences a power outage of up to 24 h. Damage Level 4 means that Network 1 experiences a power outage of more than 24 h, and the entire region experiences an outage of up to 24 h. The damage levels are used primarily to focus the scenario construction process and to show that a clear definition of the undesired consequence is critical to a structured risk assessment.

In this case study, the damage levels do not explicitly account for public health and safety consequences. However, a sustained outage of electric power would clearly cause chaos and helplessness, especially in an urban environment. Depending on the duration of the outage and the infrastructure interdependencies, the consequences could be catastrophic. Cascading events could lead to the loss of communications and transportation systems, clean water, sanitation, health care, security, and food supplies. In this general context, the six damage levels are defined with an expected increasing degree of severity with respect to their consequences on the local and regional populations. Of course, based on the extent and duration of the power outages considered in this example, it would be possible to assess more detailed health and safety consequences for a specific urban setting using the same analysis techniques.

Derivation of the Attack Scenarios

The risk scenarios show how specific damage levels can result from physical attacks on the system hardware, cyber attacks on system controls, and combinations of these attacks. First, a potential physical attack is discussed to illustrate how an event is generated. Second, a cyber attack is presented. The cyber attack may either initiate additional failures or further compound the effects of the physical damage.

Physical Attack on Network 1

Numerous physical methods could be used to damage equipment at each substation in Network 1, with varying degrees of damage to the network and the region. For example, carbon fibers, Mylar strips, or other contaminants could be sprayed over buses and transformers to cause severe short circuits. Explosives could be used to destroy key transformers, circuit breakers, and bus sections. Attackers could also damage circuit breaker controls at substation operating panels.

To help model these scenarios, it is assumed that a threat assessment revealed a high likelihood that detailed information about the electrical grid has been made available to terrorists. To illustrate the analysis method, substation S1 is examined first because it controls the full output from generating station G1, part of the output from generating station G3, and, most important, the termination of the key regional interconnection T12.

To generate the attack scenarios, five sequential questions are asked.

1. Does the attack succeed? Success means that substation S1 is physically attacked, and it is completely disabled.
2. Do all of the other generating stations in Network 1 fail? Electrical grids are typically designed so that one substation can trip off line without destabilizing the entire network. However, it is conceivable that a fault-initiated clearance of all circuits at substation S1 could cause a sufficient drop in voltage and frequency to initiate automatic load shedding and circuit isolation at all other

substations, thereby causing all remaining generators to trip off line. Therefore, our model must account for this possibility.

3. How long does the Network 1 outage last? This will depend on how quickly contingency plans can be implemented to restore power to the network.
4. Does the transient propagate throughout the region? In the event of a physical attack that destabilizes Network 1, it is very likely that the regional protection signals would automatically open the remaining interconnections (T13, T14, and T15) and prevent the transient from propagating to the adjacent networks. However, the possibility of failures that cause cascading damage at the regional level must be considered for a comprehensive analysis of transient and long-term outages.
5. How long does the regional outage last? This will depend on how quickly contingency plans can be implemented to enable the system to recover. It is assumed that if there is a transient outage of Network 1, then the maximum time for a regional outage is also transient (i.e., less than 24 h).

Figure 5.3 shows the systematic thought process that is used to develop the attack scenarios and to assign their consequences to the damage levels. Branches may be added to account for other protective barriers in each system. The primary purpose of this exercise is to create a comprehensive framework for identifying vulnerabilities and to ensure a balanced consideration of all possible consequences. Figure 5.3 also illustrates how the scenario development process can be used to represent complex combinations of failures and cascading effects. A full-scale risk assessment would further examine the effects from attacks on each substation, as well as multiple substations at once.

Cyber Attack in Conjunction with a Physical Attack

If a physical attack has destabilized Network 1, how might a terrorist prevent isolation of the network and allow the transient to propagate into the regional grid? One method would be to coordinate the physical attack with a cyber attack that keeps the circuit breakers closed at the network-region interties.

The growing complexity of the electric power grid, coupled with economic incentives for trading energy across regions, has significantly increased reliance on computerized control systems and data communication networks to control the components of the electrical grid. This leads to a potential vulnerability that can be exploited by terrorists. A cyber attack may be attractive to terrorists for many reasons.

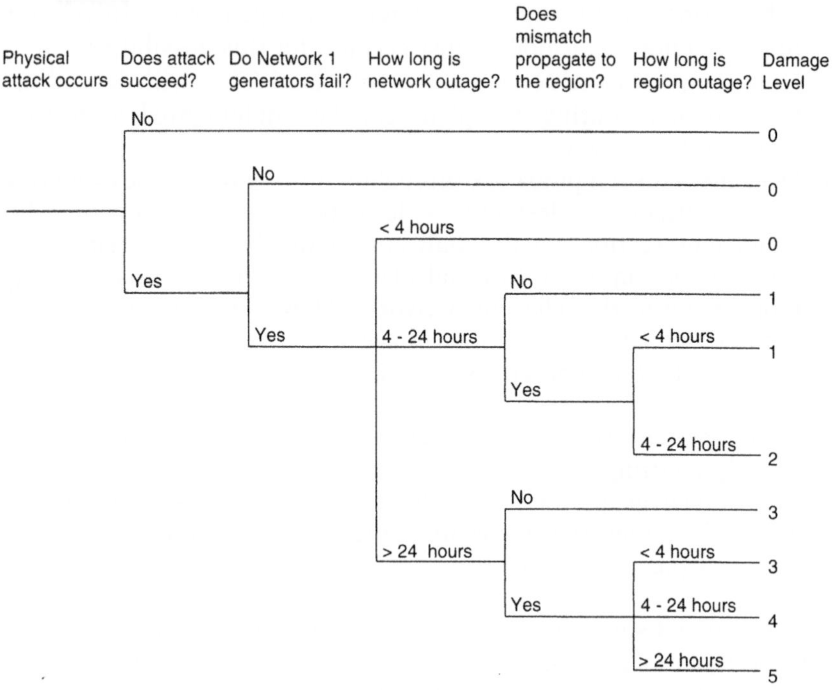

FIGURE 5.3. Thought process for attack scenarios.

- A cyber attack on the electric power grid would not require a physical presence in the United States. The attack could be planned, coordinated, and carried out from almost anywhere in the world where there is a connection to the Internet, thus eliminating the security risks and expenses of infiltrating human agents into the U.S. where they and their plans might be discovered.
- Significant damage could be done with minimal investment. This same logic has been hypothesized as a motivating factor behind the September 11 attacks. The attacks were planned to produce the most damage with the least investment (e.g., a high return on investment for terrorist monetary and personnel resources).
- The terrorists would use our own resources to attack us. This is also compatible with one of the hypothesized characteristics of the current threat; the major resources for an attack are supplied by the target nation. The open, unregulated nature of the Internet in the United States provides a wide-open pathway to targets. This Internet pathway provides not only a way to reach SCADA systems, but is also an invaluable resource for identifying potential targets and providing technical information critical to the success of an attack.

As Figure 5.3 shows, power outages in Network 1 may propagate into the regional grid if the regional SCADA emergency protection and control functions are disabled. Thus, one possible way for terrorists to cause a regional outage is to ensure that two successive events occur: (1) a very large power mismatch in Network 1 must be created (e.g., by a physical attack) and (2) the initiating transient created in Network 1 must propagate through and disable the region (e.g., by a cyber attack).

One possible way to achieve Damage Level 4 or Damage Level 5 is to ensure that the power mismatch created by a physical attack cannot be quickly corrected by combinations of available generation and automatic load shedding in Network 1 or by automatic supplies from the interconnected regional grid. After Network 1 is brought down, additional attack strategies would be necessary to ensure that the Network 1 failures cascade throughout the regional grid. Thus, intruders must override or block the regional SCADA protection and control systems that contain the frequency stabilization, load shedding, and isolation protocols. If the major regional interties remain connected to the faults in Network 1, the entire grid will quickly collapse. Individual network protection and control systems will attempt to maintain stable power flows within each of the other networks. However, if a network depends heavily on bulk power flows from the regional grid, it is very likely that the internal network control systems will not stabilize voltage or frequency. Widespread automatic shedding of loads and generation will then cause additional outages and exacerbate instabilities in other networks along the line. Of course, causing this level of regional damage would typically require more resources and coordination than an attack that affects only Network 1.

Development of the Risk Model

The triplet definition of risk is the fundamental framework for risk assessment. Risk is measured in terms of scenarios (what may happen), likelihood (how likely is it to happen), and consequences (what are the results). Risk is not a number, but a collection of numbers, or more precisely a collection of curves that display scenarios, likelihoods, and consequences. The so-called "risk parameter" is usually expressed as the frequency with which an undesired event occurs. Since this frequency is never known exactly, the state of knowledge about the numerical value of this frequency is expressed as a probability curve against the possible numerical values of the frequency. This probability curve is used in the Bayesian sense, and it expresses the state of knowledge about the frequency, based on all the relevant available evidence. Probability interpreted in this way embodies the notion of uncertainty. The undesired event(s) can be a fixed level of damage, such as the total destruction of a building, or a varying parameter, such as the number of fatalities or injuries with probability as a parameter. Dollars are also a widely used parameter for measuring risk. In many situations, combinations of risk measures are used.

In this section we will examine how the model for assessing the risk of power failures at the network and regional levels is constructed and, separately, how each type of attack is modeled and quantified. The quantification of the attack scenarios follows the process described in Chapter 2. The first step of the process is to develop a model that provides a framework for systematically evaluating the causes, frequencies, and consequences of each undesired condition. Experience has shown that the "top down" perspective employed here is the best way to ensure that the analyses are complete. The scope of the model must be broad enough to account for all possible causes and all possible consequences. The model must also be sufficiently detailed to support realistic engineering evaluations of various threats and vulnerabilities and to provide clear information about the contributors to each undesired event. The model must support quantitative analysis of each potential contributor, including rigorous treatment of uncertainties throughout the analysis process.

The parameter selected for measuring risk is based on the success rate of different levels of damage to the electrical grid. The probability of frequency concept developed in Chapter 2 is a convenient method for calculating risk because it not only represents the frequency with which a specific consequence may occur, but it also communicates the analyst's uncertainty in that frequency and, therefore, in the risk. In this case study, the success rate for different levels of damage (a form of frequency) was chosen as a convenient parameter. Thus, the probability of the success rate for achieving different consequences, or damage levels, is the basis for measuring risk.

Top-level Event Tree

Figure 5.4 shows a simplified top-level event tree that may be used to quantify the levels of damage in this case study. The following items briefly summarize the scope and definition of each top event listed in the figure.

Network 1 damage. The Network 1 Damage top event represents the success rate for attackers damaging sufficient equipment in

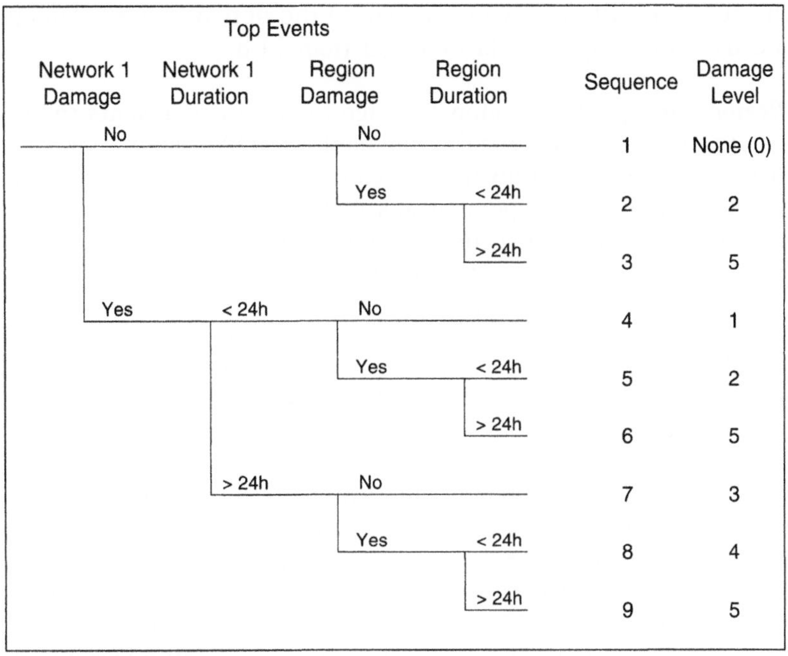

Damage Levels:
0 = No damage to Network 1 or the regional grid
1 = Transient damage to Network 1, no damage to the regional grid
2 = Transient damage to Network 1, transient damage to the regional grid
3 = Long-term damage to Network 1, no damage to the regional grid
4 = Long-term damage to Network 1, transient damage to the regional grid
5 = Long-term damage to Network 1, long-term damage to the regional grid

FIGURE 5.4. Top-level event tree for grid damage.

Network 1 to cause a power outage throughout the network. The horizontal path from the Network 1 Damage top event occurs if the attackers do not disable enough equipment to cause a network power outage. The failure path from the Network 1 Damage top event (the vertical path in the event tree) occurs if the attack results in damage to the network power supplies.

Network 1 duration. The Network 1 Duration top event is questioned only if the attack causes a power failure in Network 1. It evaluates the duration of the network power outage. The horizontal path from the Network 1 Duration top event occurs if the attack causes only enough damage to disable power for less than 24 h (i.e., a transient power outage for this case study). The failure path from the Network 1 Duration top event (the vertical path in the event tree) occurs if the attack results in severe damage to the network power supplies and causes an outage that lasts longer than 24 h.

Region damage. The Region Damage top event represents the success rate for attackers damaging regional SCADA system controls or transmission system hardware that causes power failures throughout the regional grid. This type of attack may be launched independently of an attack on Network 1, or the attacks may be coordinated. Thus, the Region Damage top event is questioned after both success and failure of the Network 1 Damage top event. The horizontal path from the Region Damage top event occurs if the intruders do not disable the regional grid. The failure path from the Region Damage top event (the vertical path in the event tree) occurs if the intruders cause a regional power outage.

Region duration. The Region Duration top event is questioned only if the attack causes a power failure in the regional grid. It evaluates the duration of the regional power outage. The horizontal path from the Region Duration top event occurs if the attack causes only enough damage to disable power for less than 24 h (i.e., a transient power outage for this case study). The failure path from the Region Duration top event (the vertical path in the event tree) occurs if the attack results in severe damage to the regional grid power supplies and causes an outage that lasts longer than 24 h.

Possible Outcomes

Sequence 1 in Figure 5.4 occurs if the attackers do not achieve any of their objectives. Even if there are some localized power outages in Network 1 or in portions of the regional grid, the outages are not severe enough or of long enough duration to satisfy the damage criteria of concern for this analysis. Sequence 1 terminates in a condition that is considered to be functional success of the regional and network power supplies, and it is assigned to Damage Level 0.

Sequence 2 may occur if the intruders successfully initiate a cyber attack on the regional SCADA systems, causing them to send out anomalous protection and control signals that result in widespread, short-term power outages throughout the grid. These regional outages also affect Network 1. Sequence 2 may also occur if the attackers damage the regional grid hardware, but the damage is not severe enough to prevent restoration of power within 24 h. In these scenarios, the local attackers are not able to cause sufficient damage to prolong the outages in Network 1 beyond 24 h. Therefore, Sequence 2 terminates in a condition that is equivalent to Damage Level 2.

Sequence 3 occurs if the attackers cause severe damage throughout the regional grid, which cannot be repaired or circumvented within 24 h. These conditions result in a long-term regional power outage, including Network 1, and Sequence 3 is assigned to Damage Level 5.

Sequence 4 occurs if the attackers achieve sufficient damage in Network 1 to cause transient power outages throughout a large portion of the network, but no disruption in the regional power supplies. Sequence 4 terminates in a condition that is equivalent to Damage Level 1.

Sequence 5 is functionally similar to Sequence 2. A coordinated attack damages enough equipment in Network 1 and the regional grid to cause short-term power outages. However, the regional attacks are not severe enough to prolong the outages beyond 24 h. Therefore, Sequence 5 terminates in a condition that is equivalent to Damage Level 2.

Sequence 6 occurs if the attackers achieve sufficient local damage in Network 1 to cause only a transient power outage in the network. However, the attackers also disable the regional SCADA system controls or cause severe damage throughout the regional grid, which cannot be repaired or circumvented within 24 h. These conditions result in a long-term regional power outage, including Network 1, and Sequence 6 is assigned to Damage Level 5.

Sequence 7 occurs if the attackers cause sufficient damage to Network 1 to cause widespread, long-term power outages throughout a large portion of the network, but no disruption in regional power supplies. Sequence 7 terminates in a condition that is equivalent to Damage Level 3.

Sequence 8 occurs if the attackers cause sufficient damage to Network 1 to cause widespread, long-term power outages throughout a large portion of the network. The intruders also successfully initiate a cyber attack on the regional SCADA systems causing them to send out anomalous protection and control signals resulting in widespread, short-term power outages throughout the grid, but the damage is not severe enough to prevent power restoration within 24 h. In these scenarios, power is restored to the region within 24 h, but Network 1 remains deenergized for an extended period of time. Therefore, Sequence 8 terminates in a condition that is equivalent to Damage Level 4.

Sequence 9 occurs if the attackers achieve all of their objectives. The local attackers cause sufficient damage in Network 1 to cause widespread, long-term power outages throughout a large portion of the network. The attackers also disable the regional SCADA system controls or cause severe damage throughout the regional grid, which cannot be repaired or circumvented within 24 h. These conditions result in a long-term regional power outage, including Network 1, and Sequence 9 is assigned to Damage Level 5.

Detailed Model of the Network Attack Scenarios

The top-level event tree in Figure 5.4 is logically complete, and it provides a framework for evaluating the success rate of each potential level of damage. In practice, however, it is often necessary to increase the level of detail in the supporting analyses to examine the threats, vulnerabilities, and causes that may contribute to each undesired condition. The increased detail facilitates a more systematic evaluation of each potential cause of failure, and it provides a logical framework for assessing the effectiveness of specific mitigation measures. The detailed evaluations also often reduce the inherent uncertainties in approximate high-level estimates, or identify the most important sources of uncertainty in each estimate.

Event Tree Logic Structure

The event tree in Figure 5.5 develops a more detailed analysis of the Network 1 Damage and Network 1 Duration top events. The expanded logic includes more details about attacks on the three critical substations in Network 1, and the corresponding likelihoods of short-term and long-term network power outages. The scope and definition of each top event listed in the figure are summarized below.

SUB S1. The SUB S1 top event evaluates whether the attackers destroy enough equipment in substation S1 to functionally disable its power generation and transmission interconnections. The horizontal path from the SUB S1 top event occurs if the attackers do not achieve their goal. The substation may be partially damaged, or the impacts may temporarily disrupt power. However, the damage is not sufficient to incapacitate the major interconnections for more than 4 h. The failure path from the SUB S1 top event (the vertical path in the event tree) occurs if the attackers cause enough damage to completely disable substation S1 for at least 4 h.

SUB S2. The SUB S2 top event is similar to the SUB S1 top event. It evaluates whether the attackers destroy enough equipment in substation S2 to functionally disable its power generation and transmission interconnections.

SUB S3. The SUB S3 top event is similar to the SUB S1 top event. It evaluates whether the attackers destroy enough equipment in substation S3 to functionally disable its power generation and transmission interconnections.

NET ST. The NET ST top event evaluates the conditional likelihood that each level of substation damage causes at least a transient power outage throughout Network 1. The likelihood of a network

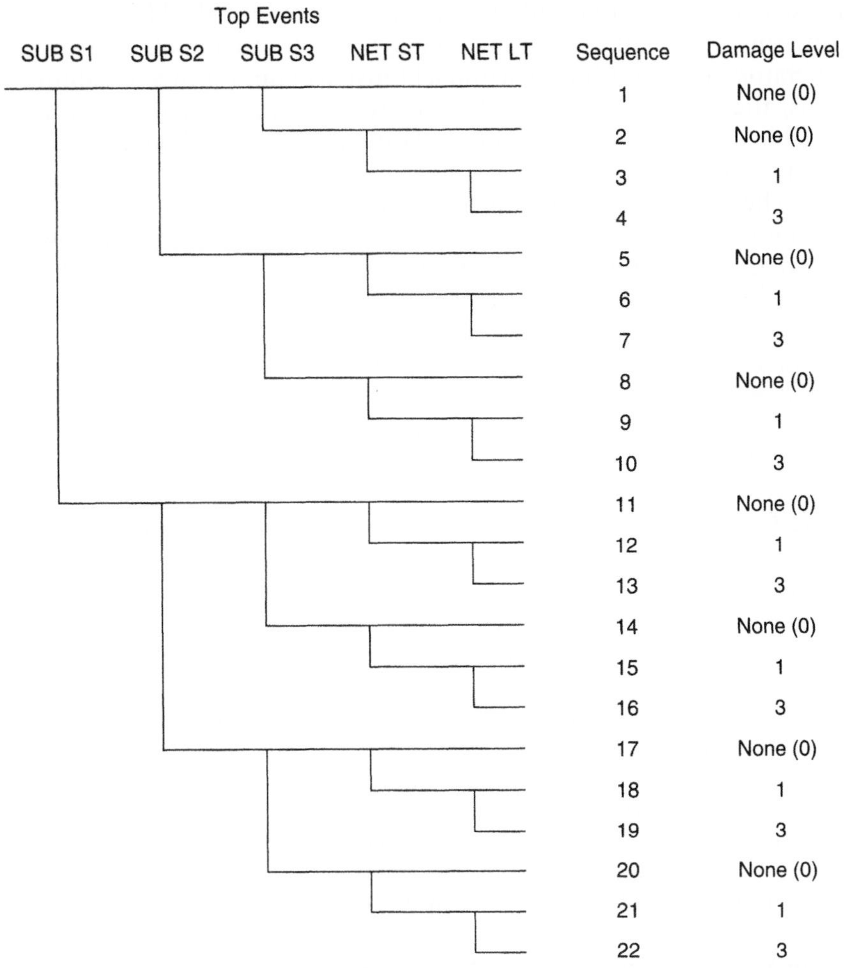

FIGURE 5.5. Event tree with increased detail for Network 1 damage and duration.

outage depends on the specific combination of substations that are damaged, their generation and transmission interconnections, and the network loading conditions at the time of the attack. Thus, the assigned value for consequential failure of the NET ST top event is different for each combination of substation damage conditions.

Electrical grids are typically designed so that one substation can trip off line without destabilizing the entire network. However, it is conceivable that a fault-initiated clearance of all circuits at one substation could cause severe fluctuations in voltage and frequency throughout the network, especially during conditions of high loading.

These cascading faults could initiate automatic load shedding and circuit isolation at all other substations, thereby causing all remaining generators to trip off line. The model must account for this possibility. Thus, the status of the NET ST top event is questioned after every possible combination of substation failures, including damage to only one substation.

The horizontal path from the NET ST top event occurs if the substation damage is not severe enough to cause a widespread power outage throughout Network 1, or if the resulting network outage duration is less than 4 h. These conditions are considered to be functional success of the network power supplies. The failure path from the NET ST top event (the vertical path in the event tree) occurs if the substation damage is severe enough to cause a network power outage that lasts at least 4 h (i.e., at least a transient network outage for this analysis).

NET LT. The NET LT top event evaluates the conditional likelihood that each level of substation damage causes a long-term power outage throughout Network 1 with duration longer than 24 h. The assigned value for consequential failure of the NET LT top event depends on the same combinations of substation damage and network loading conditions that affect the NET ST top event.

The horizontal path from the NET LT top event occurs if the substation damage is not severe enough to cause a network outage with duration longer than 24 h. These conditions are functionally equivalent to a transient network outage for this analysis. The failure path from the NET LT top event (the vertical path in the event tree) occurs if the substation damage is severe enough to cause a network power outage that lasts longer than 24 h.

Event Sequences and Possible Outcomes

Sequence 1 in Figure 5.5 occurs if the attackers do not cause enough damage to incapacitate any of the three critical substations. Short-term, localized power disruptions may occur in some areas, but the outages are not of sufficient severity or duration to satisfy the damage criteria of concern for this analysis. The success path from the NET ST top event also occurs if the attacks do not inflict enough damage to cause widespread outages throughout the network, or if the outage duration is less than 4 h. Thus, Sequences 1, 2, 5, 8, 11, 14, 17, and 20 end in a condition that is considered to be functional success of the network power supplies.

Sequence 3 in the event tree occurs if the damage to substation S3 is severe enough to cause a transient power outage throughout a large portion of Network 1. The failure path from the NET ST top event occurs whenever the achieved level of damage is severe enough to cause a network outage that lasts longer than 4 h. The success path from the NET LT top event occurs if the outage duration is limited to less than 24 h. This transient outage condition occurs in Sequences 3, 6, 9, 12, 15, 18, and 21, and it is equivalent to Damage Level 1.

Sequence 4 in the event tree occurs if the damage to substation S3 is severe enough to cause a long-term power outage throughout a large portion of Network 1. The failure path from the NET LT top event occurs whenever the achieved level of damage is severe enough to cause a network outage that lasts longer than 24 h. This outage condition occurs in Sequences 4, 7, 10, 13, 16, 19, and 22, and it is equivalent to Damage Level 3.

In practice, for a more detailed evaluation of the possible contributions to network power outages, the Network 1 Damage and Network 1 Duration top events in Figure 5.4 can be replaced by the entire event tree in Figure 5.5. Of course, other types of logic models can be used to accomplish the same goal (e.g., a fault tree that is logically equivalent to Figure 5.5). More detailed models may be developed to further subdivide and evaluate the various threats and vulnerabilities that contribute to each top event. For example, numerous potential attack scenarios with specific requirements for attacker resources and corresponding likelihoods of success may be examined for substation S1. Coordination strategies for attacks on multiple targets may also be examined, which may introduce important dependencies among the analyses for each substation. For the purposes of this example, the level of detail is developed only as far as shown for the integration of Figures 5.4 and 5.5.

Complete Model for the Risk Assessment

Figure 5.6 shows the complete event tree that is used to quantify the various attack scenarios and possible damage levels for this case study.

Event Tree Logic Structure

The final event tree is logically equivalent to the top-level event tree that is shown in Figure 5.4. The order of the regional and network top events is reversed to improve computation efficiency in the final model. However, the possible attack scenarios, damage contributions, and consequences remain the same. Figure 5.6 combines the detailed event tree for Network 1 damage from Figure 5.5 with the following two additional top events that evaluate possible damage at the regional level.

GRID H. The case study includes the simplifying assumption that long-term regional power outages require substantial physical damage to critical EHV intertie transmission lines, transformers, or buswork

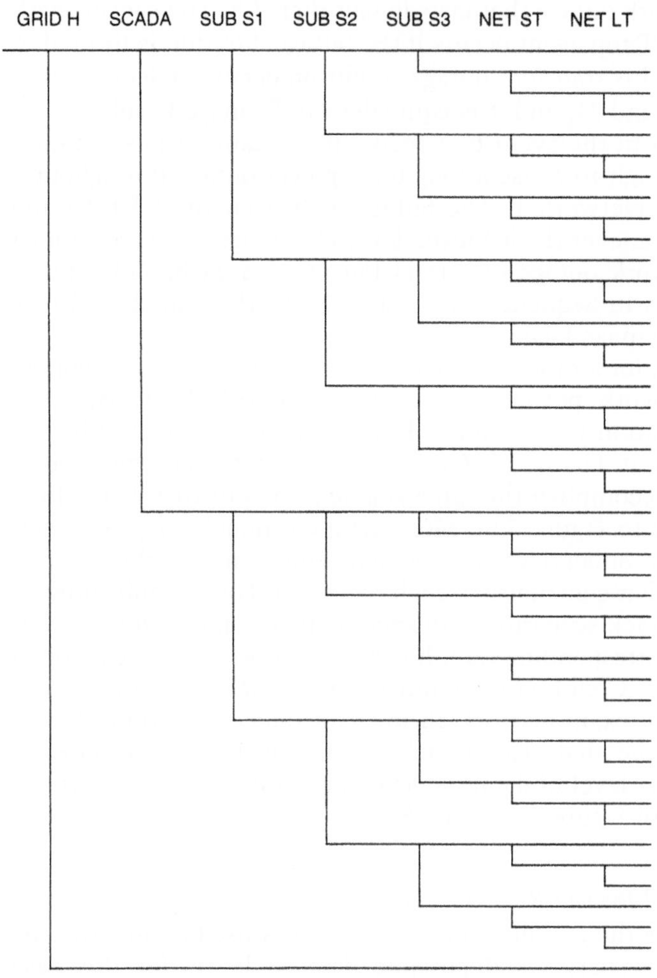

Sequence	Damage Level
1	0
2	0
3	1
4	3
5	0
6	1
7	3
8	0
9	1
10	3
11	0
12	1
13	3
14	0
15	1
16	3
17	0
18	1
19	3
20	0
21	1
22	3
23	2
24	2
25	2
26	4
27	2
28	2
29	4
30	2
31	2
32	4
33	2
34	2
35	4
36	2
37	2
38	4
39	2
40	2
41	4
42	2
43	2
44	4
45	5

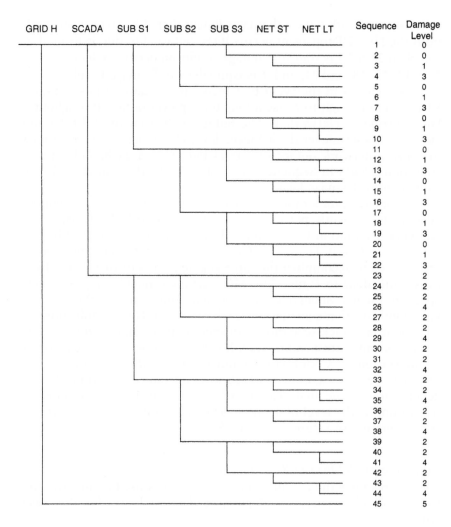

FIGURE 5.6. Complete event tree for case study.

at major regional switching stations. If the transmission system hardware remains intact, it is assumed that regional control personnel can restore power to major portions of the grid within 24 h. Some individual networks may remain deenergized for longer than 24 h, but the key regional interties can be reconnected and grid integrity can be reestablished. Thus, it is assumed that only widespread physical damage may cause a long-term outage throughout the entire region.

The GRID H top event evaluates whether the attackers physically destroy enough transmission system hardware to cause a severe long-term power outage at the regional level. The horizontal path from the GRID H top event occurs if the attackers do not achieve their goal. The grid hardware may be partially damaged, or the impacts may temporarily disrupt power. However, the damage is not sufficient to incapacitate the major regional interconnections for more than 4 h. The failure path from the GRID H top event (the vertical path in the event tree) occurs if the attackers cause enough damage to completely disable the regional grid for more than 24 h.

SCADA. A cyber attack on the regional SCADA control systems may cause them to send out anomalous protection and control signals causing widespread, short-term power outages throughout the grid. If the transmission system hardware is intact, it is assumed that regional control personnel can override or disable the spurious SCADA signals, restart generating stations, reclose transmission interties, and manually restore power to large portions of the grid within 24 h. Thus, the case study includes the assumption that attacks on the SCADA control systems may cause only a transient power outage at the regional level.

The SCADA top event evaluates whether the intruders successfully launch a cyber attack that functionally disables the regional SCADA control systems and disrupts power throughout the region. The horizontal path from the SCADA top event occurs if the attackers do not achieve their goal. The regional controls may be partially disabled, or the impacts may only temporarily disrupt power. However, the damage is not sufficient to incapacitate the major regional interconnections for more than 4 h. The failure path from the SCADA top event (the vertical path in the event tree) occurs if the cyber attack causes enough damage to completely disable the regional grid for longer than 4 h.

Event Sequences and Possible Outcomes

The event tree in Figure 5.6 contains 45 event sequences that account for all possible combinations of damage to the regional grid hardware, the regional SCADA controls, and the three critical substations in Network 1. Each sequence is assigned to the corresponding damage level that characterizes the extent and duration of the resulting power outage. Thus, the event tree in Figure 5.6 provides a comprehensive logical framework to quantify the frequency and the consequences from various possible terrorist threat scenarios.

5.2.4 Quantification of the Likelihood of the Scenarios (Step 4)

The most important function of a risk model is to organize the problem logically and provide a structured format for the systematic examination and evaluation of contributing threats and vulnerabilities. Figure 5.6 provides a logical framework with enough detail to perform a top-level evaluation of the risk associated with each level of damage considered in this case study. The scope and logical structure of the model fulfills two of the fundamental elements of the risk triplet (i.e., what can go wrong and what are the consequences).

Completion of the third element of the triplet (i.e., how likely is it) requires careful examination of the available supporting evidence to derive meaningful numerical estimates for quantification of the risk. This is often the most difficult part of the risk assessment process. Thus, this case study requires the development of realistic, quantitative estimates for the frequency of each potential terrorist threat scenario and the corresponding likelihood that a particular attack may disrupt local or regional power, including consistent evaluations of the uncertainties in each estimate.

Evaluation of the Terrorist Threat

Risk is most often measured in terms of the frequency of an undesired consequence, including a consistent evaluation of the corresponding uncertainty. In this context, the case study requires quantitative estimates for the frequency of various possible terrorist threats that involve attacks on elements of the local and regional electrical grids. The scope of the analyses is simplified by considering seven discrete types of potential threats.

Scenario 1: Physical attacks on the local substations in Network 1.
Scenario 2: Physical attacks on the regional transmission system hardware (e.g., major EHV transmission lines, transformers, switching facilities, etc.).
Scenario 3: Coordinated physical attacks on the local substations in Network 1, and physical attacks on the regional transmission system hardware.
Scenario 4: A cyber attack on the regional SCADA systems.
Scenario 5: Coordinated physical attacks on the local substations in Network 1, and a cyber attack on the regional SCADA systems.
Scenario 6: Coordinated physical attacks on the regional transmission system hardware, and a cyber attack on the regional SCADA systems.
Scenario 7: Coordinated physical attacks on the local substations in Network 1, and physical attacks on the regional transmission system hardware, and a cyber attack on the regional SCADA systems.

Figure 5.7 is a simple decision tree logic structure that shows how these attack scenarios are related. In principle, this or similar

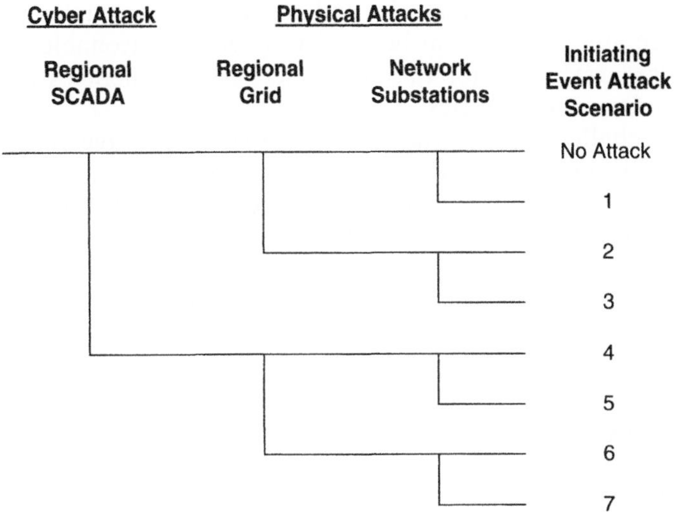

FIGURE 5.7. Decision tree for initiating event attack scenarios.

types of decision models may be used to assess qualitative or quantitative estimates for the scenario frequencies, as suggested by some contemporary researchers.[7] This case study does not use Figure 5.7 as a quantification tool per se, but rather as a simple representation to show how the seven attack scenarios cover the range of possible threats that are evaluated in the analysis.

Estimation of the attack scenario frequencies (the "initiating events" for this case study) is the most challenging element of the analysis process. For the most part, we do not have documented empirical data to directly support these estimates. Discussions with security experts have indicated that contemporary databases compile evidence of terrorist activities in a variety of qualitative and quantitative formats. This evidence is updated continuously through numerous intelligence gathering sources. The data are evaluated through diverse and complementary methods to identify specific characteristics and trends that signal both intent and potential implementation of terrorist acts.

Security experts and federal agencies are understandably very reluctant to share information about the format, contents, or methods that are used to process these data. The authors of this case study also have a responsibility to protect the confidence of specific information that may compromise sensitive intelligence. Therefore, this section of the analysis summarizes only the first stage of a fundamental Bayesian methodology for systematic evaluation of experiential evidence. Here, we show how the available evidence may be structured to support a quantitative analysis, and we show how our engineering experience and insights can be used to develop reasonable "state of knowledge" estimates for the attack scenario frequencies and their corresponding uncertainties.

The challenge is to transform limited qualitative information from intelligence reports and quantitative information (if any) from historical experience into consistent uncertainty distributions about the frequency of each potential attack scenario. We must also keep in mind that this case study applies to terrorist attacks against a particular target (i.e., a specific metropolitan electric power network and the surrounding regional grid). In other words, our estimates will apply to the annual frequencies of attacks on a particular segment of the electric power supply grid, not the annual frequency of attacks that may occur anywhere in the entire U.S. national grid.

Attack Scenario 4

We start the analysis with Scenario 4 because we have the most readily available experience and evidence for these types of attacks. Cyber attacks represent the lowest risk to the intruders. Abundant resources are also available for cyber attacks. Initial probing incursions may be launched by agents who are only loosely connected with a coordinating organization. Rather extensive industrial security experience has shown that intrusions into a variety of commercial, governmental, and financial computer systems (most often motivated by monetary gain) can be initiated from remote Internet terminals in foreign countries using inexpensive hardware and automation software, with little risk of identifying the perpetrators. Anecdotal information from security experts indicates that these types of challenges occur very frequently, perhaps on a daily basis for some particularly attractive targets. They are most often thwarted by automated firewall protection systems without the need for direct human intervention. In this case study, we are specifically interested in the frequency of cyber attacks that are motivated by an intrusion into the SCADA system for a particular regional electric power grid. The "initiating event" for these scenarios is simply a focused challenge to the automated security functions and human oversight that should protect against intrusion. The vulnerability assessment then evaluates the effectiveness of those barriers. It is apparent from the anecdotal evidence that these types of focused attacks for the specific purpose of gaining SCADA system entry occur less frequently than other commercial computer intrusions with criminal, malicious, financial, or nuisance intent. However, the evidence also indicates that challenges to electrical grid computer networks do occur with some regularity, although the specific motivations may not be known or publicly divulged.

The frequency distribution for Scenario 4 is derived from these insights and experience. It assesses an upper bound of approximately once in 6 months for this type of attack, and a lower bound of approximately once in 5 years. These upper and lower bounds are assigned as the 90% confidence interval of a probability distribution for the frequency of attack Scenario 4. In other words, we are 90% confident that the actual frequency of these attacks on a particular SCADA system is somewhere between twice per year and once in 5 years. There is a 5% probability that the attacks may occur more frequently than twice per year, and a 5% probability that they may occur less frequently than once in 5 years. For analytical convenience, we use a lognormal distribution to represent the shape of our probability function. (Experience from many similar analyses has shown that the lognormal function generally provides a good physical representation of observed distributions of equipment and human performance data.) Based on this assessment, the median estimate for the frequency of Scenario 4 is approximately one attack in 19 months, and the mean frequency is approximately one attack in 15 months.

Attack Scenario 1

Physical attacks on the local network substations require substantially more coordination and resources than cyber attacks. These attacks probably require a team of approximately 6–10 intruders, if the targets are limited to three critical substations. Research is also required to identify the specific targets and the loading conditions that will maximize the power disruption. Thus, planning and preparations for these attacks would also likely involve individuals with some education and experience in the design and operation of large electric power systems. The materials needed to accomplish these attacks are not sophisticated, and they are readily available without need for special fabrication or transportation across international borders. The attackers face moderate risk of discovery, if the substations are located in an urban environment.

The assessment of the frequency for Scenario 1 makes use of these general observations and the lack of documented evidence that these types of coordinated physical attacks have been attempted to date. In particular, it is estimated that approximately four or five major metropolitan areas in the United States might satisfy the attackers' motives for a well publicized disruption of an urban electric power supply infrastructure. Anecdotal estimates by security experts and electric power industry personnel indicate that a terrorist attack on some target in the U.S. electrical grid may be likely in the time frame of the next 5–10 years. Many experts also believe that the most likely attack method will involve small groups of intruders who are intent on physical destruction of the power supply network. Thus, based on these assessments, it seems reasonable to estimate that the total frequency of attacks on any particular one of the four or five key metropolitan targets may be in the range of one in 20 years to one in 50 years. We also note that there have been no reported attacks of this nature in 4 years since September 11, 2001 (a total of 16–20 years of experience for the four or five key metropolitan areas). Of course, there is substantial uncertainty about these estimates. There are also many other technical, physical, financial, and organizational factors that may affect the attackers' selection of one particular target over another.

Our frequency distribution for Scenario 1 assesses an upper bound of approximately one in 20 years for this type of localized attack, and a lower bound of approximately one in 500 years. These upper and lower bounds are assigned as the 90% confidence interval of a lognormal probability distribution. In other words, we are 90% confident that the actual frequency of these attacks is somewhere between one in 500 and one in 20 years. There is a 5% probability that the attacks may occur more frequently than one in 20 years, and a 5% probability that they may occur less frequently than one in 500 years. This uncertainty range may seem quite broad to many security experts who are trained to assess the "worst case" frequencies and consequences from potential threats. However, it represents only a factor of 25 in the range of frequencies between our upper and lower estimates. Risk assessments of many diverse industries and activities have often shown that this degree of uncertainty is actually rather small, when compared with the range of actual experience for "rare event" hazards.

Based on this assessment, our median estimate for the frequency of Scenario 1 is approximately one attack in 100 years, and our mean frequency is approximately one attack in 62 years. It is again noted that these estimates apply for a particular metropolitan electric power network, and not the entire U.S. electrical grid.

Attack Scenario 2

Physical attacks on the regional transmission grid require even more coordination and resources than attacks on the local substations. These attacks also probably require a team of approximately 6–10 intruders, depending on the grid configuration. However, the targets are typically more dispersed, and more options are available to re-route power through undamaged interconnections. Rather extensive training, research, and system monitoring are required to identify the specific targets and the loading conditions that will maximize the power disruption. Again, the materials needed to accomplish these attacks are not sophisticated, and they are readily available. The individual attack forces face low risk of discovery at each target.

However, they must coordinate the attack locations over a relatively large geographic area, and they must synchronize their timing to assure that the grid will collapse as desired. These organization and communications requirements expose the attackers to possible discovery during the planning phases or shortly prior to the final decision to launch the attack.

The frequency distribution for Scenario 2 is based on the judgment that the more complex planning and coordination, and the increased chance of discovery for these attacks make them somewhat less attractive than the localized attacks in Scenario 1. However, the uncertainty range is also larger to account for the possibility that some terrorist groups may prefer to accept these risks for the potential payback of a wider disruption of the electrical grid. The frequency distribution assesses an upper bound of approximately one in 25 years for this type of distributed attack, and a lower bound of approximately one in 2500 years. These upper and lower bounds are assigned as the 90% confidence interval of a lognormal probability distribution. Thus, we are 90% confident that the actual frequency of these attacks is somewhere between one in 2500 and one in 25 years.

Based on this assessment, the median estimate for the frequency of Scenario 2 is approximately one attack in 250 years, or a factor of 2.5 times lower than the median frequency of Scenario 1. The mean frequency for Scenario 2 is approximately one attack in 94 years, or approximately 66% of the mean frequency of Scenario 1. The variations between the median value and mean value estimates are due to the shape of the lognormal probability distribution and the fact that the uncertainty for Scenario 2 spans a factor of 100 in the range of frequencies between the upper and lower estimates, while the uncertainty for Scenario 1 spans only a factor of 25. As before, it is noted that these estimates apply for attacks on the regional grid surrounding a particular metropolitan area, and not the entire U.S. electrical grid.

Attack Scenario 3

Coordinated physical attacks on the local network substations and the regional grid probably require a team of at least 10, or as many as 20, well trained intruders. The individual attackers face their greatest risk of discovery in the urban environment of the local substations. However, the relatively large size of the team imposes even more requirements for planning, coordination, and pre-attack communications, compared with Scenario 2. It is expected that these considerations may deter many potential intruders, who may favor strategies that require fewer personnel and less central coordination.

The frequency distribution for Scenario 3 is based on the judgment that the increased team size, complexity, and planning for these coordinated regional and local attacks make them less likely than the attacks in Scenario 2. However, the uncertainty range is also somewhat larger to again account for the possibility that some terrorist groups may prefer to accept these risks. The frequency distribution assesses an upper bound of approximately one in 82 years for this type of coordinated attack, and a lower bound of approximately one in 12,250 years. (These somewhat unusual upper and lower bounds were derived from the assigned broader uncertainty range and the scaled median frequency from Scenario 2. Thus, they are not "round" numbers, like the corresponding estimates for Scenarios 1 and 2.) Based on these estimates, we are 90% confident that the actual frequency of these coordinated attacks is somewhere between one in 12,250 and one in 82 years.

The median estimate for the frequency of Scenario 3 is approximately one attack in 1000 years, or a factor of 4 times lower than the median frequency of Scenario 2. The mean frequency for Scenario 3 is approximately one attack in 314 years, or approximately 3 times lower than the mean frequency of Scenario 2. Again, these variations between the median value and mean value estimates are due to the shape of the lognormal probability distribution and the fact that the uncertainty for Scenario 3 is somewhat broader than the uncertainty for Scenario 2.

Attack Scenarios 5, 6, and 7

Scenarios 5, 6, and 7 evaluate physical attacks on the electrical grid that are coordinated with a cyber attack on the regional SCADA systems. Based on the available evidence and intelligence, it is not clear whether terrorists would conclude that these types of coordinated activities are more or less attractive than independent physical attacks or cyber intrusions. It is apparent that cyber attacks represent the lowest risk to the intruders, and they may be launched with a relatively small allocation of resources. Thus, from the perspective of a fully integrated global organization, it would seem that coordination of cyber and physical attacks would add little cost, with the potential of greater disruption to the electrical grid. However, it is also clear that these coordinated activities require substantially more planning and communications than the focused physical attacks. In practice, the additional logistics and increased chances of discovery during the preparation phases may make a coordinated strike less attractive, compared with physical attacks that may be planned and implemented by a relatively isolated cell of a few individuals.

Discussions with some security experts indicate that disparities between the ideological motivations of specific terrorist groups or the personal motivations of their leaders may also weigh against these integrated attack scenarios. For example, one group may favor relatively widespread and sophisticated attacks on the U.S. financial, political, and public services infrastructure. Another group may favor more opportunistic and isolated physical destruction and violence. If these assessments have some validity, then an integrated attack may require coordination from two or more groups with somewhat different objectives. Thus, from this perspective, it would also seem that this type of combined attack may occur less often than isolated physical attacks or cyber intrusions.

Based on these assessments, we estimate that the median frequency of a combined cyber and physical attack is somewhat lower than the median frequency of the corresponding independent physical attack. Thus, we assign median estimates for the frequencies of Scenarios 5, 6, and 7 that are each a factor of 4 times lower than the respective median frequencies of Scenarios 1, 2, and 3.

It is apparent that the uncertainties about these combined attacks are even larger than the uncertainties about the physical attacks. Therefore, we also extended the range of each probability distribution to account for these higher uncertainties. For example, the frequency distribution for Scenario 5 assesses an upper bound of approximately one in 27 years for this type of coordinated attack, and a lower bound of approximately one in 6000 years. We have highest uncertainty about a fully integrated physical and cyber attack (Scenario 7), with an upper bound of approximately one in 200 years, and a lower bound of approximately one in 80,000 years.

Summary of Attack Scenario Frequencies

Table 5.1 summarizes our uncertainty distributions for the frequency of each terrorist attack scenario.

At the completion of this type of assessment, it is often useful to perform some simple checks to ensure that our numerical estimates

TABLE 5.1
Probability distributions for the frequency of terrorist attacks

	Annual Frequency of Attack (Events Per Year)				
Attack Scenario	*5th Percentile*	*Median*	*95th Percentile*	*Mean*	*Error Factor*
1	2.00E–03*	1.00E–02	5.00E–02	1.61E–02	5
2	4.00E–04	4.00E–03	4.00E–02	1.07E–02	10
3	8.16E–05	1.00E–03	1.22E–02	3.19E–03	12.2
4	2.00E–01	6.32E–01	2.00	8.08E–01	3.2
5	1.67E–04	2.50E–03	3.75E–02	9.69E–03	15
6	6.67E–05	1.00E–03	1.50E–02	3.88E–03	15
7	1.25E–05	2.50E–04	5.00E–03	1.31E–03	20

*2.00 = 0.002 and is notation used for very small or very large numbers.

are not unreasonable, compared with our current insights and experience. In the discussion of Scenario 1 in the "Attack Scenario 1" section under Section 5.2.4, we noted that it seems reasonable to estimate that the total frequency of physical attacks on any particular one of four or five key U.S. metropolitan electrical grids may be in the range of one in 20 years to one in 50 years. All of our attack scenarios, except Scenario 4, involve some degree of physical threat against either the local network substations or the regional transmission grid. From Table 5.1 the sum of the median frequencies of these six physical attack scenarios is approximately 0.019 events per year, or approximately one attack in 53 years. The sum of the mean scenario frequencies is approximately 0.045 events per year, or approximately one attack in 22 years. Thus, our overall numerical estimates are quite consistent with the expert opinion "state of knowledge" regarding these potential terrorist threats. The preceding sections have described our thought process and our rationale for the relative frequencies of the seven attack scenarios, based on our current understanding of the available intelligence, and the motivations and goals of various terrorist groups. Of course, we also have substantial uncertainty about these assessments, which is fully explained and displayed in our quantitative estimates.

Use of Bayesian Methods

The frequency estimates in Table 5.1 represent the most fundamental element of a Bayesian methodology (see Chapter 2) for systematic evaluation of experiential evidence. They are a consistent quantitative expression of our current "state of knowledge," considering the available experience and insights from our experts. The estimates also explicitly display and quantify the rather large uncertainties about these attack scenarios. These uncertainties are an integral element of the risk assessment process. In a progressive Bayesian approach, successive additions to the available qualitative information and data allow us to systematically refine the assessments and uncertainties, and to derive improved quantitative estimates of the risk.

For example, one application of this process is a formal elicitation of security and intelligence experts to assess their respective best estimates and their uncertainties about each threat scenario, in the same manner as summarized in the preceding sections. These expert estimates are then combined probabilistically to develop a composite frequency for each scenario that consistently accounts for each expert's assessment, each expert's personal uncertainty, and the variability among the polled experts.

As additional intelligence is received, it is then added to the existing information base through a Bayesian update of the experts' state of knowledge. For example, specific Internet "chatter" or reported challenges to electrical grid SCADA computer system firewalls may alert security monitors to the increased likelihood of a cyber attack. This information may then be used to update the frequency estimates for cyber-related attack scenarios. If this cyber intelligence is received in conjunction with information about mobilization of personnel and weapons for a terrorist cell in the same region, this evidence may strongly reinforce the estimated frequency of a specific coordinated attack scenario.

Thus, a formal Bayesian evaluation of the available expert knowledge and the evolving intelligence would provide both an improved understanding of the absolute and relative frequencies of each threat scenario and a consistent methodology to update these estimates as more data are received.

Evaluation of System Vulnerabilities

Quantification of the risk model also requires consistent numerical assessments of the vulnerability of each element in the local and regional grids to damage during a terrorist attack. For the case study, these estimates are necessarily simplified and are not derived from detailed analyses of any particular electrical network or regional control system. However, they illustrate the types of analyses, thought processes, and inputs that are typically developed to support the risk assessment process. In their simplest form, as in this example, inputs may be based on the experience and judgment of experts. Even though these high-level screening analyses are typically only approximate and often include large uncertainties, they are useful to quickly focus attention on specific elements of the problem or parts of the analysis that merit more careful, or more detailed evaluation. Additional detail may then be added to the models for those elements, and their supporting analyses refined to identify the most important contributing causes or to reduce the initial uncertainties.

Physical Attacks on Network 1 Substations

Top events SUB S1, SUB S2, and SUB S3 in Figure 5.6 represent the likelihood that attackers destroy enough equipment in each substation to disable its generating supplies and transmission interconnections. In this model, each critical substation is assigned a different vulnerability to attack.

Substation S1. It is assumed that substation S1 is located in an urban environment and is the most heavily protected of the three substations. It may be surrounded by protective walls, may be continually manned by utility personnel, and may be checked by local police during their normal neighborhood surveillance patrols.

Substation S2. It is assumed that substation S2 is located in a suburban or partially rural environment and is the least protected of the three substations. It may be surrounded by a chain link fence, may not be manned, and may not be subject to routine surveillance by local police.

Substation S3. It is assumed that substation S3 is located in an urban environment but is only partially protected. For example, it may be surrounded by protective walls and checked by local police during their normal neighborhood surveillance patrols, but it may not be continually manned.

A simple probability distribution can be developed to assess the likelihood that attackers could successfully enter each substation and cause extensive damage to critical transformers, circuit breakers, buses, and controls. The histogram in Figure 5.8 applies to substation S1.

This simple histogram does not rigorously display the discrete probability boundaries over the full range of the cumulative probability distribution function. Nevertheless, it is useful for demonstrating the fundamental concepts that would be used for a more numerically rigorous representation of the uncertainties. The sample histogram shows the following information.

- There is a 5% probability that the attackers would succeed in 5% of their attacks on substation S1 (i.e., that 1 of 20 attacks would be successful).

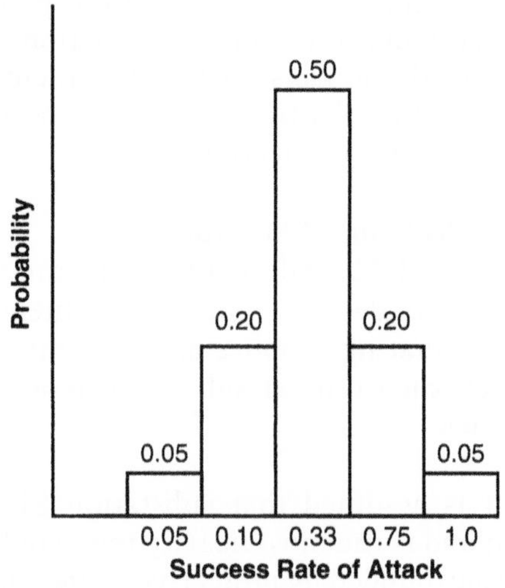

FIGURE 5.8. Histogram showing success rate of an attack on Substation S1.

- There is a 20% probability that the attackers would succeed in 10% of their attacks on substation S1 (i.e., that 1 of 10 attacks would be successful).
- There is a 50% probability that the attackers would succeed in 33% of their attacks on substation S1 (i.e., that 1 of 3 attacks would be successful).
- There is a 20% probability that the attackers would succeed in 75% of their attacks on substation S1 (i.e., that 3 of 4 attacks would be successful).
- There is a 5% probability that the attackers would always succeed in their attacks on substation S1 (i.e., that every attack would be successful).

According to these estimates, the mean likelihood of a successful attack on substation S1 is approximately 0.39 (i.e., approximately 10 of 26 attacks would be successful). These estimates are obviously not derived from detailed models of specific attack scenarios or from a detailed evaluation of the specific substation vulnerability to each attack. However, these types of estimates can be developed relatively easily, based on information from experts familiar with potential attack strategies, resources, and specific vulnerabilities of the target. If the case study results show that attacks on substation S1 are potentially important to one of the undesired damage levels, more extensive analyses would be justified.

Table 5.2 summarizes the probability distributions for a successful attack on each substation, considering its specific vulnerabilities. These estimates account for the conditional likelihood of success after an attack is launched, but they do not explicitly account for pre-attack planning to identify key targets, evaluate critical network-loading conditions, develop logistics for the attack teams, etc. These factors would obviously also influence the overall likelihood of a successful attack, especially a coordinated offensive on multiple targets. In a more detailed analysis, these factors could be included as additional inputs to the models for each substation, or they could be

TABLE 5.2
Estimated success rates for physical attacks on Network 1 substations

	Probability					
Substation	*0.05*	*0.20*	*0.50*	*0.20*	*0.05*	*Mean Success Rate*
S1	0.05	0.10	0.33	0.75	1.0	0.39
S2	0.75	0.85	0.90	0.95	1.0	0.90
S3	0.10	0.30	0.50	0.80	1.0	0.53

evaluated in a separate part of the risk model that specifically examines the planning, resources, and logistics of the attack.

Network 1 Outage Vulnerability
The NET ST and NET LT top events in Figure 5.6 account for the conditional likelihood that each possible substation damage scenario may cause a transient power outage or an extended outage throughout Network 1. The description of the network indicates that substation S1 is most important because it controls the full output from generating station G1, part of the output from generating station G3, and the termination of key regional interconnection T12. Substation S2 is next in importance because it contains the connections from generating stations G2 and G5, which are not directly connected to substation S1, and regional interconnection T13. Substation S3 is the least important of the three critical substations.

The network is designed to withstand the complete loss of any one substation under normal loading conditions. However, under severe loading conditions, attack-initiated faults might cascade to other substations and generating units. Therefore, the models for the NET ST and NET LT top events must assign a likelihood of network failure after any combination of substations is damaged. The models and the quantification of these top events are correlated. For example, the NET ST top event evaluates the likelihood that a particular substation damage condition will cause a network power outage that lasts at least 4 h. The NET LT top event then evaluates the conditional likelihood that the damage condition will cause a long-term outage that lasts longer than 24 h.

Table 5.3 summarizes estimates for the conditional likelihood of each type of outage, depending on the specific combination of damaged substations. These estimates are derived from the general

TABLE 5.3
Estimated conditional likelihood for Network 1 power outages

	Damage to Network		*Duration of Outage*	
Damaged Substation(s)	*No Outage*	*Outage (Longer than 4 h)*	*Transient (4 to 24 h)*	*Long-term (>24 h)*
S1	0.500	0.500	0.300	0.200
S2	0.667	0.333	0.208	0.125
S3	0.800	0.200	0.125	0.075
S1 and S2	0.050	0.950	0.283	0.667
S1 and S3	0.200	0.800	0.300	0.500
S2 and S3	0.333	0.667	0.367	0.300
S1, S2, and S3	0.000	1.000	0.050	0.950

information about each substation and its relative importance to overall power generation and distribution within the network. For example, if only substation S1 is damaged, Table 5.3 indicates that there is a 50% likelihood that the damage may cascade into a network outage that lasts longer than 4 h. In other words, approximately one-half of the attacks that damage only substation S1 will result in a network power outage, due to unexpected cascading failures or unusual system loading conditions. Of these failures, it is estimated that approximately 60% will result in transient outages of duration less than 24 h, and 40% will result in outages that last longer than 24 h. Of course, in a more detailed analysis, additional supporting information for these estimates could be derived from dynamic load-flow simulations, models of system response, interviews with network operations personnel, etc.

The conditional likelihoods in Table 5.3 are expressed only as point-estimate values to illustrate the relative importance of each substation damage condition and its contribution to transient or long-term power outages. The corresponding numerical values that are assigned to top events NET ST and NET LT must also account for the logical relationships between these top events. In particular, the numerical value for top event NET LT accounts for the fraction of all network outages that are long-term. Thus, if only substation S1 is damaged, Table 5.3 shows that the point-estimate value for top event NET ST would be 0.500 (the likelihood of a network outage longer than 4 h). The corresponding conditional value for top event NET LT would be 0.400. These values would then ensure that 30% of the substation S1 damage scenarios result in a transient outage, and 20% result in a long-term outage.

Of course, there is substantial uncertainty about these estimates, which must also be consistently represented in the risk model quantification process. Table 5.4 summarizes simplified probability histograms for the conditional likelihood that each substation damage condition causes an outage with duration longer than 4 h. The "outage mean likelihood" values in Table 5.4 correspond to the point-estimates in Table 5.3 for the fraction of substation damage scenarios that cause a network outage that is longer than 4 h. These histograms are used directly for quantification of top event NET ST. Thus, it is evident that seven different numerical values apply for top event NET ST, depending on the specific combination of damaged substations. These damage conditions are shown graphically by the branching logic in Figure 5.6. The risk models contain internal logic rules to ensure that the correct numerical values are used when each damage condition is quantified.

Table 5.5 shows the corresponding probability histograms for the conditional likelihood that each substation damage condition causes a

TABLE 5.4
Estimated conditional likelihood top event NET ST outage

Contribution to Failure of Top Event NET ST						
	Probability					
Damaged Substation(s)	*0.05*	*0.20*	*0.50*	*0.20*	*0.05*	*Outage Mean Likelihood*
S1	0.10	0.25	0.50	0.75	0.90	0.50
S2	0.05	0.15	0.33	0.50	0.75	0.34
S3	0.01	0.05	0.20	0.30	0.60	0.20
S1 and S2	0.90	0.92	0.95	0.98	1.00	0.95
S1 and S3	0.60	0.70	0.80	0.90	1.00	0.80
S2 and S3	0.33	0.50	0.67	0.85	1.00	0.67
S1, S2, and S3	1.00	1.00	1.00	1.00	1.00	1.00

TABLE 5.5
Estimated conditional likelihood for top event NET LT outage

Contribution to Failure of Top Event NET LT						
	Probability					
Damaged Substation(s)	*0.05*	*0.20*	*0.50*	*0.20*	*0.05*	*Outage Mean Likelihood*
S1	0.05	0.10	0.40	0.67	0.90	0.40
S2	0.05	0.10	0.35	0.60	0.80	0.36
S3	0.02	0.05	0.35	0.60	0.80	0.35
S1 and S2	0.33	0.50	0.75	0.95	1.00	0.73
S1 and S3	0.20	0.40	0.67	0.80	0.95	0.63
S2 and S3	0.10	0.25	0.50	0.67	0.80	0.48
S1, S2, and S3	0.90	0.92	0.95	0.98	1.00	0.95

long-term outage of longer than 24 h. These histograms are used for quantification of top event NET LT.

Physical Attacks on Regional Grid Hardware

Top event GRID H in Figure 5.6 evaluates the likelihood that physical attacks damage a sufficient number of EHV intertie transmission lines, major transformers, or other equipment in critical switching facilities to disrupt power throughout the region for longer than 24 h. To accomplish this level of damage, the intruders must carefully coordinate their attacks on several diverse facilities that are typically distributed over a wide geographic area. The attacks must occur closely in time to avoid discovery or effective mobilization of security forces. The attacks must also destroy a large number of transmission lines or switching stations to ensure that the regional interties are disabled and that power cannot be rerouted through the remaining interconnections. However, in many cases, the EHV transmission lines and switching facilities are often located in remote areas with minimal security or regular surveillance.

For this case study, a simple probability distribution is also developed to assess the likelihood that the attackers succeed in these goals. The probability histogram is shown in Figure 5.9. It is interpreted in the same manner as the histogram shown in Figure 5.8 for damage to substation S1. According to these estimates, the mean likelihood of a successful physical attack on the regional grid hardware is approximately 0.21 (i.e., approximately 21 of 100 attacks would be successful). These estimates are obviously not derived from detailed models of specific attack scenarios or from a detailed evaluation of a specific regional grid configuration. If the case study results show that physical attacks on the grid hardware are potentially important to long-term regional power outages, more extensive analyses would certainly be justified.

Cyber Attacks on Regional SCADA Systems

Top event SCADA in Figure 5.6 evaluates whether the intruders successfully launch a cyber attack that functionally disables the regional SCADA systems, causing widespread, short-term power outages throughout the grid. The cyber attack scenario for this case study

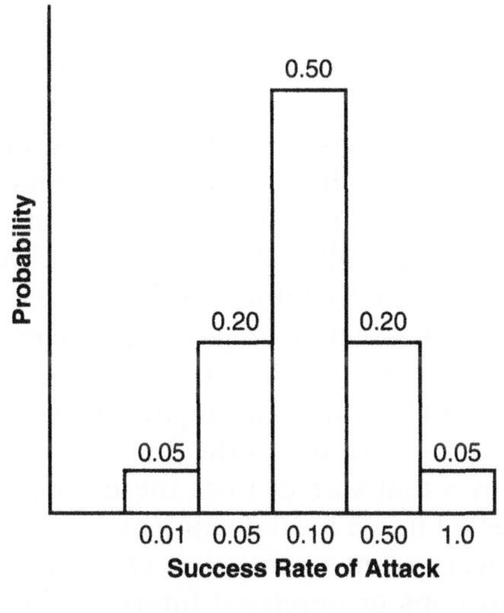

FIGURE 5.9. Histogram showing success rate of an attack on regional grid hardware.

takes place over a 3-week period. Although it is possible the attack could be orchestrated in much less time, it is assumed that the characteristics of a September 11 type attack (e.g., cautious and careful planning) would be in operation. Thus, a 3-week timeline might be more typical. The cyber attack is divided into five phases: (1) discovery, (2) launch platform acquisition, (3) target selection, (4) target reconnaissance and compromise, and (5) initiation of an actual attack on the electric power grid.

Discovery phase. The discovery phase of the operation begins with the identification of potential targets and the assembly of critical information about them. Actors, a term for terrorists used in the intelligence community, with very little computer knowledge could carry out this phase of the attack, and there is a good chance that the activities during this phase would be carried out by individuals other than those who would be responsible for the final attack. This would compartmentalize resources and protect higher level technical operatives from possible exposure and loss.

The first step would be to identify potential targets via the Internet. This could be done using one of hundreds of search engines by typing in keywords, such as "power company," "electric power," "power and light," or other common phrases associated with electric utilities. In just a few hours, a large number of U.S. electric utility companies could be identified. Alternatively, the names of every private and municipal electric utility in the U.S. could be collected in electronic format in less than 5 min from a publicly available government website.

The next step in the discovery phase would be to find the computer systems of the electric utility companies that are connected to the Internet. Like most institutional entities with a presence on the Internet, electric utility companies have registered and reserved large ranges of IP addresses. Registered IP addresses are unique to the registered entity; they are the "electronic address" by which they can be reached from anywhere else on the Internet.

One efficient way to collect these addresses would be to access one of thousands of "whois" engines on the Internet. In just seconds, these publicly available search engines can search millions of IP address registration records and identify the addresses associated with keywords, such as "XYZ Power and Light" or other specific electric utility company names collected in the first part of the discovery process. The IP addresses that surface from these "whois" searches could then be cut and pasted into a local document, such as an Excel spreadsheet on the discovery team's computer. Once this has been accomplished, tens of millions of unrelated Internet addresses would have been eliminated, and a database of potential electric utility computer systems would have been assembled. It is likely at this point that the discovery team would encrypt their electric utility system database, burn it on to a compact disc, and hand it off to a courier who would physically carry it to the attack team. This would prevent it from being intercepted by the National Security Agency or another intelligence-gathering organization.

Launch platform acquisition. For security reasons, the actual attack team would most likely be located in a country other than the one in which the discovery team resides. The attack team would probably include several intermediate-level computer users and one expert computer hacker. Their first task would be to compromise a series of computers from which to launch the attack. Computer attacks are typically carried out through a series of computers, which makes it very difficult to trace the source of the attack, if it is even discovered. The attack team would prowl computer networks in countries where computer security is poor or nonexistent. Using autorooters, port scanners, and other tools that are readily available on the Internet, they would scan computer networks in these vulnerable countries looking for computer systems with vulnerabilities that could be exploited. Once found, the computers would be compromised. The attackers would arrange administrative privileges on these machines and then go dormant, covering their tracks by deleting log entries and using other stealth techniques. In this manner, they would build a set of computer systems from which they could launch their cyber attacks remotely.

Target selection. The actual portion of the electrical grid selected as a target might depend on an a priori selection of targets by higher-level operatives in the terrorist organization to coordinate with a physical attack on the power grid or even on another interdependent infrastructure target. However, the terrorist organization might also settle for a target of convenience and leave the decision up to the attack team.

In any event, once an electric utility had been selected as a target, the attack group would activate some of the computers exploited in the platform acquisition phase, transferring the autorooter and port scanning tools to the compromised computers. Next, those tools would be used against the utility's range of IP addresses in the discovery team database. Many autorooters are sophisticated and automated, that is, they can try multiple attack strategies against a large range of machines. When they are successful, they can install a number of surveillance/reconnaissance tools that would automatically cover up any sign that the utility computer had been compromised.

Several classes of commercially available products are designed to protect against these kinds of attacks. A number of computer firewall products are designed to recognize and deflect attacks like the one described above by restricting all incoming and outgoing network traffic unless the administrator of the firewall designates it. A second class of security devices, intrusion-detection systems, monitors incoming and outgoing network traffic for digital signatures of known cyber attack tools and ploys. Although these security techniques are often effective, they are not 100% effective. In fact, they are often compromised by misconfigurations by the administrator. An acute shortage of well trained computer security professionals is a contributing factor to the problem of computer security.

Target reconnaissance/compromise. The initial electric utility computer system that was compromised in the previous stage would most likely be an administrative server, web server, or other computer not directly involved in the SCADA system, and, therefore, not the final target of the cyber attack. Cyber attacks with preplanned goals or objectives, such as the one in the terrorist scenario, usually use "attacks by increment" strategies.

In this phase, the computer system compromised in the previous stage would be used as the home base for the cyber attackers. They would next attempt to find out the purpose of the compromised computer and then assess the number of other computers in the network that "trust" the compromised computer, and to what extent. They could then use these trust relationships to inspect other computer systems on the network, as well as to discover other local networks. The cyber attackers might also install packet sniffers to listen in on network traffic for packets destined for ports specific to a particular SCADA software system. Once they found SCADA port traffic, they could identify the computer systems being used as SCADA systems.

If the compromised computer does not provide a pathway to the SCADA network, the attackers would go back to the previous phase and attempt to compromise another externally visible computer system in the utility company's IP range. Another possible outcome might be that another vulnerable computer system (but not a SCADA controller) on a connected, but different network in the utility system would be identified; this computer would also be compromised, and reconnaissance could then be initiated from a newly compromised machine.

Initiation of attack. The final step would involve compromising one or more of the computer systems that run the SCADA system. These systems would be attacked using the same autoroot and exploit tools that gained access to the initial computer in the electric utility. Once the SCADA system was compromised, the amount of damage inflicted on the components of the power grid reachable by the compromised SCADA system would depend on the attack team's knowledge of electric power systems.

Ideally, one member of the attack team would be a power engineer trained in the basics of power generation and distribution systems. The damage inflicted could be significantly increased by knowledge of the specific power system and components that would be under the control of the terrorist group.

Table 5.6 provides a timeline of the SCADA system intrusion.

The overall intruder success rate for the SCADA top event in Figure 5.6 can be estimated by evaluating each step in the intrusion process and then combining the various event probabilities. The composite success rate of an intrusion that gains full control over the SCADA system is estimated to be approximately 0.000013 success per attempt (i.e., approximately 1 success in 75,000 attempts). This estimate is based on a combination of the estimated success probabilities in Table 5.6, and it includes a very large uncertainty. For this case study, the composite expert estimate was used as the median value of a lognormal uncertainty distribution with an assigned error factor of 10. This means that the experts were 90% confident that the likelihood of success would be within a factor of ± 10 of the estimated value. The parameters of this uncertainty distribution are shown in Table 5.7.

Finally, the event tree shown in Figure 5.6 provides the logical framework to complete the quantification of the frequency of the scenarios and each grid damage level that is examined in this case study. The quantification process first uses the information in Table 5.1 for the frequency of each terrorist attack scenario. The scenarios are reproduced below to summarize the primary focus of each attack.

Scenario 1: Physical attacks on the local substations in Network 1.

Scenario 2: Physical attacks on the regional transmission system hardware (e.g., major EHV transmission lines, transformers, switching facilities, etc.).

Scenario 3: Coordinated physical attacks on the local substations in Network 1, and physical attacks on the regional transmission system hardware.

Scenario 4: A cyber attack on the regional SCADA systems.

Scenario 5: Coordinated physical attacks on the local substations in Network 1, and a cyber attack on the regional SCADA systems.

Scenario 6: Coordinated physical attacks on the regional transmission system hardware, and a cyber attack on the regional SCADA systems.

Scenario 7: Coordinated physical attacks on the local substations in Network 1, and physical attacks on the regional transmission system hardware, and a cyber attack on the regional SCADA systems.

TABLE 5.6
Stages of SCADA system intrusion

Day	*Event*	*Objective*	*Actors*	*Probability of Success*	*Probability Parameters*	*Chokepoint*	*Probability of Successful Chokepoint Intervention*
1	Use Internet search engine to find U.S. power companies.	Identify potential targets and power generation sites.	Low-level operatives.	Near 1.0	Presence of power company and generation site details on Internet.	No	Near 0
1	Search "whois" engine for names of power companies discovered above.	"Who is" records will contain IP address blocks assigned to the company, thereby drastically reducing the search space for exploit targets.	Low-level operatives.	0.8	Some power companies are listed but have blocked IP addresses; others have large blocks of IP addresses registered.	No	Small
3	Deploy autorooter exploit at foreign	Create a network of exploited computers in countries with	Actors with modest to intermediate	1.0	The number of vulnerable computers on the Internet,	No	

	networks in Korea, India, etc. Identify vulnerable systems. Plan exploit to take control of N systems.	many poorly protected networks. Typical tactic is to gain root access to a number of machines and then connect to the target machine through multiple IP connections to hide the true IP address of the attacker.	computer skills.		especially in certain parts of the world, for all purposes makes this an almost certain element of any attack.		
6	Deploy autorooter and portscan exploits on networks and computers captured on Day 3 against IP ranges of power	Look for vulnerable computer systems at U.S. power companies. These systems may be poorly protected web servers,	Actors with intermediate to expert computer skills.	Near 1.0	Some vulnerable computers, especially in the administrative and web server classes, would be somewhere in the search space.	Unlikely	This should not be considered a chokepoint because there are too many entry points to ensure that all of them have been

(*Continued*)

TABLE 5.6 (*Continued*)

Day	*Event*	*Objective*	*Actors*	*Probability of Success*	*Probability Parameters*	*Chokepoint*	*Probability of Successful Chokepoint Intervention*
	company networks discovered on Day 1.	administrative computers, or (if you're really lucky) a computer with direct SCADA duties.					adequately protected.
8	Evaluate most likely targets from list of vulnerable power company computers found on Day 6. Pick top 3 or 4 targets in terms of attractiveness and deploy the appropriate exploit tool to	Gain covert control over the power company computer. The first objective is to determine what purpose the compromised machine serves to determine its potential as a launching pad toward the	Actors with intermediate to expert computer skills.	0.2	The actual probability of success depends on the security posture of the particular power company and the type of computer (web server, administrative machine, etc.) exploited,	Yes	0.9 plus With the proper firewall/ operating system software and system security, the probability of a breach can be kept to a minimum.

	gain covert control over the computer.	SCADA system. Exploit code may be encrypted or packet fragmentation may be used to avoid detection by firewall software.			patches to firewall, and operating systems.		
8	Execute routines to subvert logs that would tip off systems administrator of intrusion.	Intrusion detection avoidance either by a deployed intrusion detection system or a sharp-eyed computer systems administrator.	Actors with intermediate to expert computer skills.	0.4	Once a firewall has been defeated, it is likely the exploit used is one for which the intrusion detection system does not yet have a digital signature. Therefore, once past the firewall, an intrusion is	Yes	0.9 plus A good intrusion detection system can make it difficult for terrorist groups without sophisticated computer knowledge to go undetected.

(Continued)

TABLE 5.6 (*Continued*)

Day	*Event*	*Objective*	*Actors*	*Probability of Success*	*Probability Parameters*	*Chokepoint*	*Probability of Successful Chokepoint Intervention*
					more likely to go unnoticed.		
10	Examine the list of hosts, trusted hosts. "Sniff" packets of traffic going through the compromised machine.	Understand the role of the currently compromised computer in the power company's computer network. The next step is to explore the network the compromised computer is on to find other computers on the network as well as other networks to which it is	Actors with expert computer skills to avoid detection during exploration of other networks on power company's computer infrastructure.	0.1	The probability here refers to the chance that the attacker will be detected during his exploration activities, which are risky because they involve multiple machines and may generate unusual traffic on other networks (e.g., IP addresses normally not	No	

		connected. Note that this activity often occurs one or more days after the successful intrusion. Some initial research shows that attackers often lie low for a day or so after an intrusion and then return to see if their exploit is still present and viable (i.e., undiscovered).			seen passing traffic on a particular network).		
10 to 15	Explore the power company's computer network looking for evidence of	Some of SCADA systems knowledge is required either directly or by following a "cookbook"	Actors with expert computer skills to avoid detection; direct or indirect	0.01	The opportunity to find a SCADA machine depends greatly on the level of	Yes	0.9 plus Probably the best protection is air, that is, keeping the SCADA

(Continued)

TABLE 5.6 (*Continued*)

Day	*Event*	*Objective*	*Actors*	*Probability of Success*	*Probability Parameters*	*Chokepoint*	*Probability of Successful Chokepoint Intervention*
	SCADA activity.	set of SCADA indicators. These indicators would include looking for specific files in certain directories, certain processes that could be identified by doing something as simple as looking at the threads currently being executed on the machine, looking for	expertise with power systems to identify markers that identify a SCADA system.		security and connectivity within the company's computer networks.		systems disconnected from other networks. Of course, the Internet will significantly reduce the odds of an attack.

		traffic on certain ports, looking for particular hardware drivers for devices associated with SCADA hardware/ software interfaces.					
16 to 17	Identify a SCADA computer and a careful process of attacking it with an exploit to gain covert control.	Gain control of a machine inside the SCADA system for intelligence gathering and for use as an exploit launching pad.	Actors with expert computer skills and at least some SCADA experience.	0.3	Once inside the SCADA network, it is likely that the attacker will find machines relatively less protected because of their "center perimeter" location, as well as because many of the characteristics	Maybe	Intelligent security command and control agents that can isolate potential "bad guy" traffic with as little disruption to the power grid as possible may be able to make this a chokepoint.

(*Continued*)

TABLE 5.6 (*Continued*)

Day	*Event*	*Objective*	*Actors*	*Probability of Success*	*Probability Parameters*	*Chokepoint*	*Probability of Successful Chokepoint Intervention*
					of SCADA systems, such as mandatory fast response times to events, preclude extensive use of time-consuming encryption, packet inspection, or authentication.		
17 to 21	Identify additional SCADA computers on the network and run exploits to	Gain control of as many SCADA control systems and devices as necessary to increase the	Actors with expert computer skills and at least some SCADA experience.	0.7		Maybe	Intelligent security command and control agents that can isolate potential "bad guy" traffic

gain control of them as well.	amount of damage that could be done, as well as to reduce the probability that intervention by a power control systems operator would limit or prevent damage to the power grid.		with as little disruption to the power grid as possible could make this a chokepoint.

TABLE 5.7
Probability distribution for a successful SCADA intrusion

Likelihood of Success Per Attempted SCADA Intrusion			
5th Percentile	*Median*	*95th Percentile*	*Mean*
1.3E–06*	1.3E–05	1.3E–04	3.5E–05

*1.3E–06=0.0000013 and is notation used for very small or very large numbers.

The event tree is quantified for each attack scenario, using the corresponding information about the vulnerabilities of the network substations, the grid hardware, the regional SCADA systems, and the conditional likelihood of short-term or long-term network power outages. Of course, the possible attack scenarios do not always challenge every element of the local network and the regional grid. Therefore, the risk model also contains the following conditions that account for the specific characteristics of each attack.

- Top event GRID H is successfully bypassed for attack Scenarios 1, 4, and 5. These scenarios do not involve physical attacks on the regional transmission system hardware.
- Top event SCADA is successfully bypassed for attack Scenarios 1, 2, and 3. These scenarios do not involve cyber attacks on the regional SCADA systems.
- Top events SUB S1, SUB S2, and SUB S3 are successfully bypassed for attack Scenarios 2, 4, and 6. These scenarios do not involve physical attacks on the local network substations.

5.2.5 Assembly of the Scenarios into Measures of Risk (Step 5)

We must now ask if the methodology described in Chapters 1 and 2 meets our expectations. To focus our answer, we must first revisit the questions that were the basis for proposing a quantitative risk-based methodology. The methodology is intended to answer such questions as what are the threats and vulnerabilities; what are the contributing factors, and how do they rank in importance; and what actions will have the biggest payoff in terms of risk reduction for the amount of resources invested.

Once the individual attack scenarios have been quantified, they can be assembled into measures of the risk. This is a matter of first combining all scenarios that terminate in each specific damage category. If the risk measure is a variable, such as fatalities, injuries, or dollars, then the process also involves arranging the scenarios in order of increasing damage, and accumulating the frequencies from bottom to top.

The modes of terrorist attacks examined in this case study were physical damage and a cyber attack. The elements of the electrical grid considered in the vulnerability assessment were local network substations, regional transmission lines and switching facilities, and the regional SCADA systems. The interface between the threat assessment and the vulnerability assessment is the actual attack on the grid itself. This initiating event (i.e., the nature of the attack) is the output of the threat assessment and the input for the vulnerability assessment. Seven different attack scenarios were examined, considering various possible combinations of physical attacks and coordinated cyber attacks.

The risk model shown in Figure 5.6 was quantified using the supporting data summarized in Section 5.2.4. The combined risk results for all damage levels are shown graphically in Figures 5.10–5.13. Figure 5.10 displays the uncertainty distributions for the annual frequency of each specific damage level, and the combined annual frequency of an attack that results in any level of damage to the local network or the region. Figure 5.11 expands the frequency scale to show more details of the uncertainty distributions for the top three damage levels and the total. Figure 5.12 compares the two least likely damage levels (Damage Levels 2 and 4). Figure 5.13 shows the

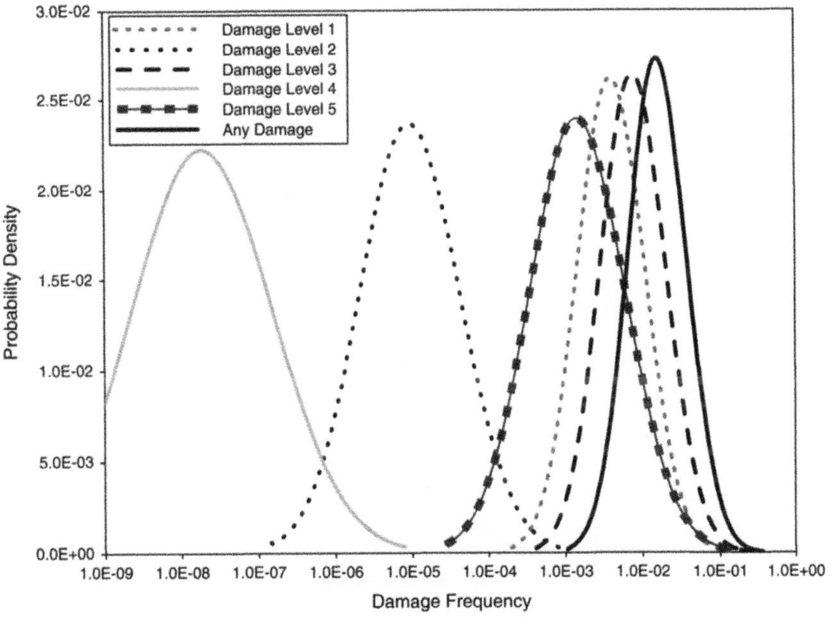

FIGURE 5.10. Combined results for all damage levels.

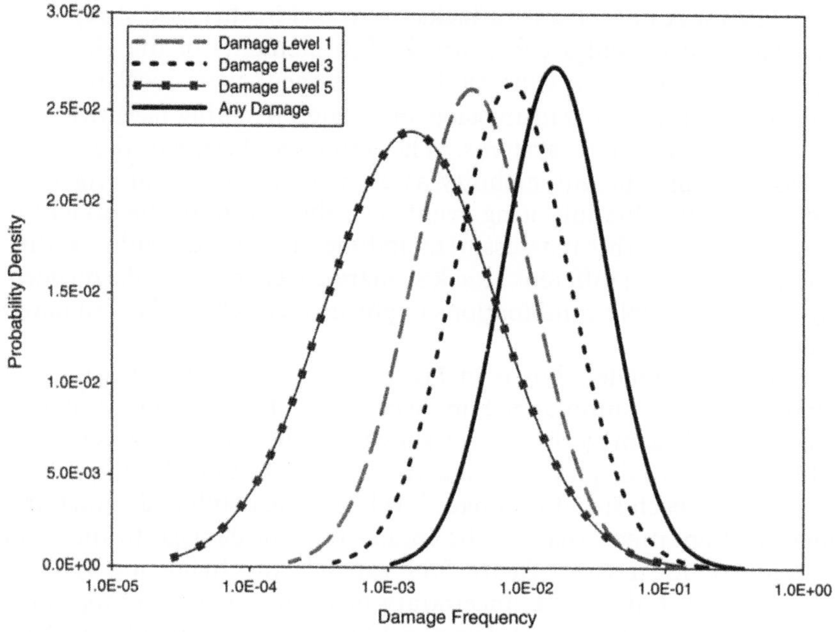

FIGURE 5.11. Results for top damage levels.

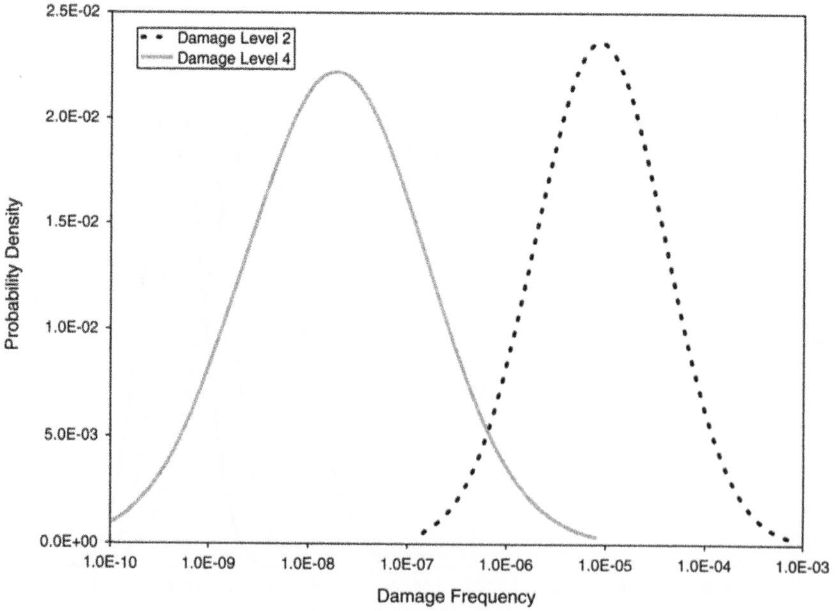

FIGURE 5.12. Results for Damage Levels 2 and 4.

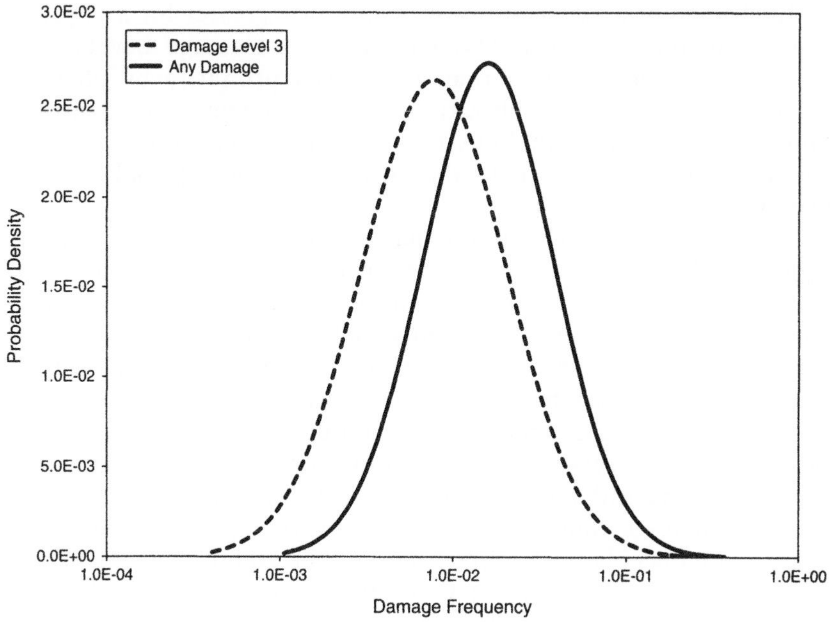

FIGURE 5.13. Results for Damage Level 3 and any damage.

relationship between the most likely damage level (Damage Level 3) and the total frequency of any damage.

5.2.6 Interpretation of the Results (Step 6)

Table 5.8 summarizes selected parameters of the uncertainty distributions for each level of damage. It is evident from Figures 5.10–5.13 and from Table 5.8 that, for the model in this case study,

TABLE 5.8
Selected parameters of uncertainty distribution for each level of damage

Attack Success Rate (Outage Events Per Year)				
Damage Level	*5th Percentile*	*Median*	*95th Percentile*	*Mean*
Any Damage	4.6E–03*	1.6E–02	6.4E–02	2.3E–02
1	8.0E–04	4.0E–03	2.0E–02	6.6E–03
2	7.3E–07	9.5E–06	1.0E–04	2.9E–05
3	1.7E–03	7.7E–03	3.7E–02	1.2E–02
4	6.9E–10	1.9E–08	5.7E–07	1.7E–07
5	1.9E–04	1.5E–03	1.5E–02	4.0E–03

*4.6E–03 = 0.0046 and is notation used for very small or very large numbers.

the likelihood of a successful attack is much greater for a physical attack than for a cyber attack. For example, Damage Levels 1, 3, and 5 account for outages that are caused primarily by physical attacks on the network substations or the regional grid hardware. Damage Levels 2 and 4 include transient outages of the regional grid that result from coordinated cyber attacks on the regional SCADA systems.

The analysis clearly shows that the attackers would have the highest likelihood of causing long-term power outages in only Network 1 (Damage Level 3). The next most likely consequence is a transient power outage that affects only Network 1 (Damage Level 1). Long-term regional power outages (Damage Level 5) occur approximately five times less frequently than outages that affect only Network 1. Thus, the case study application demonstrates that even an abbreviated risk assessment can yield meaningful results on the vulnerability of the grid and the threat of a cyber attack.

The analyses confirm that there are very large uncertainties in the assessments of both the threat of an attack and the system vulnerabilities. The results indicate that the grid would be much more vulnerable to physical attacks than to cyber attacks. However, the methodology also shows that the uncertainties associated with cyber attacks are much greater. For example, Figure 5.12 shows that the uncertainties for Damage Levels 2 and 4 span more than five orders of magnitude on the frequency scale. In contrast, Figure 5.11 shows that the uncertainties for damage levels that result from physical attacks are much smaller, typically spanning less than three orders of magnitude on the frequency scale.

The results of this case study conclude that the most likely consequence of a terrorist attack is a power outage that is localized to only Network 1. For example, there is 90% confidence that the rate of a power disruption in only Network 1 (the sum of Damage Levels 1 and 3) would be between 0.0044 and 0.049 events per year, or approximately one event in 227 to 20 years. The expected mean frequency of a power outage in Network 1 is approximately 0.019 events per year, or one event in 53 years. The threat analysis concluded that the mean frequency of physical attacks on Network 1 is approximately 0.03 attempts per year or one attempt every 33 years, considering all relevant threat scenarios. Thus, the analysis shows that the intruders have an expected success rate of approximately 63% for causing at least a transient power outage in Network 1.

The overall vulnerability of Network 1 is most strongly determined by the relatively high vulnerability of substation S2, in spite of the fact that this substation is not as important to the network's electrical stability as the other more secure substations. Table 5.9 shows that successful attacks on substation S1 contribute to

TABLE 5.9
Substation damage contribution to Network 1 outages

Damage to Substations	*Outage Frequency (Event Per Year)**	*Fraction of Total Damage Levels 1 and 3*
S2 and S3	5.63E–03**	29.4%
S1 and S2 and S3	5.31E–03	27.7%
S1 and S2	4.64E–03	24.3%
S2	2.59E–03	13.5%
S1 and S3	4.95E–04	2.6%
S1	2.86E–04	1.5%
S3	1.97E–04	1.0%

*Combined frequency of a local physical attack, damage to the affected substations, and failure of Network 1 as a consequence of the substation damage.
**5.63E–03 = 0.00563 and is notation used for very small or very large numbers.

approximately 56% of the total frequency of combined Damage Levels 1 and 3. Successful attacks on substation S2 contribute to approximately 95% of the total Network 1 outage frequency, and successful attacks on substation S3 contribute to approximately 61% of the total. (These so-called fractional-importance measures are simply the sum of all scenarios that involve damage to each substation, divided by the total number of scenarios.) Therefore, the overall vulnerability of Network 1 is most strongly determined by the relatively high vulnerability of substation S2, even though this substation is not individually as important to the network power generation and transmission interties as the more secure substation S1.

Damage Level 5 accounts for long-term regional power outages that are caused by physical damage to the EHV transmission lines, transformers, and equipment at critical intertie switching facilities. Table 5.8 shows that there is 90% confidence that the rate of long-term regional power outages would be between 1.9E–04[c] and 1.5E–02 events per year, or approximately one event in 5250 to 67 years. The expected mean frequency of a long-term regional power outage is approximately 4.0E–03 event per year, or one event in 250 years. The threat analysis concluded that the mean frequency of physical attacks on the grid hardware is approximately 1.9E–02 attempts per year, or one attempt every 53 years, considering all relevant threat scenarios. Thus, the analysis shows that the intruders have an expected success rate of approximately 21% for causing severe damage to the regional grid. As noted in the vulnerability assessment, this rather high degree of success is attributed primarily to the fact that the

[c] 1.9E–04 is the same as 0.000194 and is the notation used for very small or very large numbers.

EHV transmission lines and switching facilities are often located in remote areas with minimal security or regular surveillance.

According to the case study analysis of a cyber attack, there would be a very low likelihood of successful intrusion into the regional SCADA systems, although there is a great deal of uncertainty in the estimates. For example, there is 90% confidence that the rate of a transient disruption of regional power (the sum of Damage Levels 2 and 4) would be between 9.0E–07 and 1.1E–04 events per year, or approximately one event in 1.1 million to 9000 years. The expected mean frequency of a transient regional outage is approximately 2.9E–05 events per year, or one event in 34,500 years. The threat analysis concluded that the mean frequency of a cyber attack is approximately 0.82 attempts per year, or one attempt every 15 months, considering all relevant threat scenarios. Thus, the analysis shows that the expected rate of successful cyber attacks is approximately one in every 28,000 attempts. However, considering the very large uncertainties in the threat and vulnerability assessments, there is 90% confidence that the cyber intruders will be successful once in every 910,000 to 7500 attempts.

In this example, even though cyber-initiated events did not constitute a major threat, they could not be ignored. First, there was a wide range of uncertainty in the assessment. Second, unlike a physical attack in which the risks of repeated attempts to the terrorists would be high, a cyber-initiated interruption of power could be attempted many times with very little investment and very little risk to the terrorists. The analysis revealed a need to develop more information to reduce the uncertainty and to explore ways of discouraging repeated attempts.

The case study examined seven possible terrorist attack scenarios that are described in the "Evaluation of the terrorist threat" section under Section 5.2.4. Table 5.10 summarizes the fraction of each grid damage level that is caused by each threat. Some entries in the table are blank, because the particular attack does not challenge elements of the local or regional grid that are included in the damage level. For example, attack Scenario 4 represents a cyber attack on the regional SCADA systems. Within the context of the risk models, this type of attack can result only in a transient regional power outage, or Damage Level 2. Similarly, attack Scenario 2 represents a physical attack on the regional transmission system hardware. This type of attack can result only in a long-term regional power outage, or Damage Level 5.

It is evident from Table 5.10 that attack Scenario 1 is most important to the overall risk of power outages. The consequences from this attack scenario are limited to outages that affect only Network 1, as represented by Damage Level 1 and Damage Level 3. It accounts for slightly more than 45% of the total frequency of any

TABLE 5.10
Attack scenario contribution to each level of damage

	Fractional Contribution to Damage					
Attack Scenario	*Any Damage* %	*Damage Level 1* %	*Damage Level 2* %	*Damage Level 3* %	*Damage Level 4* %	*Damage Level 5* %
1	45.3	54.8	–	54.8	–	–
2	9.7	–	–	–	–	56.1
3	9.9	8.6	–	8.6	–	16.7
4	0.1	–	98.9	–	–	–
5	27.3	33.1	0.7	33.1	90.3	–
6	3.5	–	0.4	–	–	20.4
7	4.1	3.6	0.1	3.6	9.7	6.9

level of grid damage and approximately 55% of the total frequency of Network 1 outages. This attack scenario involves physical attacks on the local substations in Network 1. As discussed above, the overall vulnerability to these attacks is most strongly determined by the relatively high vulnerability of substation S2.

Attack Scenario 5 is most important to the overall risk of combined network and regional power outages. It accounts for slightly more than 27% of the total frequency of any level of grid damage. Table 5.10 also shows that Scenario 5 accounts for approximately 33% of the total frequency of outages that affect only Network 1, and slightly more than 90% of the frequency of combined transient regional outages and long-term damage to Network 1. This attack scenario involves coordinated physical attacks on the local substations in Network 1, and a cyber attack on the regional SCADA systems. The importance of this threat is derived from the following observations.

- The threat assessment process concluded that the intruders may attempt these coordinated attacks at a mean frequency of approximately one in 103 years. However, the threat assessment uncertainty is quite large, with 90% confidence that the attack frequency is between one in 6000 and one in 27 years. These scenarios may be quite attractive to a well organized group, because attacks on the network substations require only a small group of local infiltrators, and the regional SCADA cyber attacks can be initiated with relatively small organizational investment and risk.
- The vulnerability assessment process concluded that the attackers have a relatively high likelihood of successfully disrupting power in the local network. There is a much lower likelihood that the

cyber attackers will be successful. However, when the combined frequency and consequences are evaluated consistently in the risk model, it is evident that these coordinated attack scenarios have the highest payback for the infiltrators.

After the localized attacks of Scenario 1 and the coordinated attacks of Scenario 5, Table 5.10 shows that the next most important scenarios involve coordinated physical attacks throughout the region (attack Scenario 3) and physical attacks on only the regional transmission system hardware (attack Scenario 2).

Measures of Cumulative Risk

The scope of this case study is limited because it does not extend the consequence analyses to include the final step of a full-scope risk assessment. That step would include an integrated evaluation of the public health and safety impacts from each level of damage; economic effects from disruptions to communications, transportation, and commerce; societal disorder from the knowledge that terrorists have damaged a critical element of the public infrastructure, etc.

The process for evaluating these societal risk measures is demonstrated by organizing the case study results in the format of cumulative risk curves, as described in Chapter 2. The five damage levels correspond to increasingly more severe local and regional power outages, from a transient outage that affects only Network 1 (Damage Level 1) to a long-term regional outage (Damage Level 5). If it is assumed that the societal consequences are directly proportional to these damage levels, then the analysis results can be represented by the risk curves that are shown in Figure 5.14. These curves are derived by accumulating the results that are discussed in Section 5.2.5 according to the severity of damage. They show the total frequency for exceeding a particular severity level and the corresponding uncertainties in that frequency. Thus, the curves show that the total mean frequency of exceeding any level of damage for this case study is approximately 0.023 events per year, or one in 43 years. The uncertainty analysis indicates that there is 90% confidence that the actual frequency lies within a factor of approximately 10 around the median frequency.

The annual exceedence frequency decreases as the damage severity increases. Thus, the mean frequency of exceeding Damage Level 3, or worse, is approximately 0.016 events per year, or once in 62 years. The risk curves also show that our uncertainty increases as we evaluate less frequent, more severe conditions. For example, the analysis indicates that there is 90% confidence that the actual frequency lies

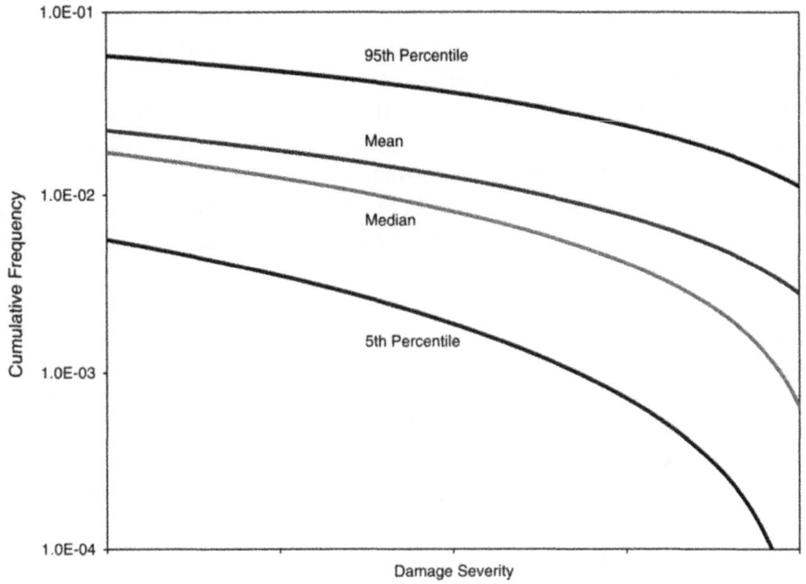

FIGURE 5.14. Cumulative risk curves.

within a factor of approximately 15 around the median frequency for Damage Level 3, or worse. The risk curves terminate at Damage Level 5, the most severe consequence that is evaluated for this case study. The mean frequency for that degree of severity is approximately 0.004 events per year, or once in 250 years. However, the uncertainty in that estimate is very large. There is 90% confidence that the actual frequency lies within a factor of approximately 100 around the median frequency.

Conclusions and Recommendations

To avert cyber-initiated attacks, steps could be taken to reduce the uncertainties in the analysis and to find ways to discourage repeated attempts. For coordinated physical attacks, one very clear action to consider would be to improve the security of substation S2, which was identified as the principal contributor to power outages in Network 1. This priority might not have been evident without an integrated assessment of the vulnerabilities and the potential consequences from the failure of each substation. It is also very clear from Table 5.9 that attacks on multiple substations would greatly increase the likelihood of Network 1 failure. Thus, substation security in general would be an important consideration for improving local network power security. Improved security and surveillance of remote facilities would also reduce the grid vulnerability to severe regional power outages. The relative importance of the substations, regional hardware, and SCADA systems to overall grid vulnerability would not be readily apparent without an integrated model that systematically evaluates each contribution to damage.

Once developed, these models become key elements in a systematic risk management process to evaluate the effectiveness of proposed improvements. As the models and supporting evidence are refined, the updated analysis results display the corresponding changes to the overall grid risk profile, and the contributors to each damage level may be reordered. Based on these risk-based insights, systematic examination of successive improvements would continue until an acceptable level of overall risk was achieved.

This case study example is intended to illustrate how quantitative risk assessment can be used to "turn up the microscope" to expose the risk of an event that is either catastrophic or could become catastrophic. Extensive analysis may not be necessary in situations where the risks are apparent (i.e., when the threats and vulnerabilities can be easily identified). Obvious steps can be taken to reduce the vulnerability to a terrorist attack on many important assets in conventional facilities and buildings. Risk reduction in those situations may include improving ventilation systems, emergency action training, improving security, providing rapid escape systems, identifying protective staging locations, and upgrading emergency response capabilities. However, this case study shows that threats to complex elements of our societal infrastructure may involve multi-faceted attack strategies with uncertain degrees of success and widely different potential consequences. In these situations, the systematic discipline of quantitative risk assessment provides the necessary framework to support effective risk management decisions.

Conclusion

Quantitative risk assessment is an effective method of exposing the risks of complex systems to events that could lead to catastrophic consequences. The hallmark of a quantitative risk assessment is the quantification of uncertainty; uncertainty is the risk of greatest concern.

Recommendation

Quantitative risk assessment should be applied in cases where the consequences can be catastrophic and where there is great uncertainty about the risk scenarios and contributing factors. Meanwhile, the government and private sector should act quickly to reduce the risk to those assets where the payoff can be readily determined.

References

[1] Garrick, B. J., et al, "Confronting the Risks of Terrorism: Making the Right Decisions," *Reliab. Eng. Syst. Saf.* 2000, *86*(Special Issue).

[2] Amin, M, "Security Challenges for the Electricity Infrastructure," *Security and Privacy* 2002, *35*(4), 8–10.

[3] EPRI (Electric Power Research Institute), *Electricity Infrastructure Security Assessment: EPRI Security Overview*, EPRI, Palo Alto, CA, 2002, Electronic Power Research Institute (distribution limited).

[4] NAERC (North American Electric Reliability Council) *The Electricity Sector Response to the Critical Infrastructure Protection Challenge*, North American Electric Reliability Council, Princeton, NJ, 2002.

[5] Alvey, J., "Digital Terrorism: Holes in the Firewall?" *Public Util. Fortnightly* 2002, *140*(6).

[6] DOE (Department of Energy), National Transmission Grid Study, DOE, Washington, DC, 2002.

[7] Eisenhawer, S., Bott, T., Rao, D. V., *Assessing the Risk of Nuclear Terrorism Using Logic Evolved Decision Analysis*, Los Alamos National Laboratory (LA-UR-03-3467), 2003.

CHAPTER 6

Case Study 4: Abrupt Climate Change

This case study, like the previous one on terrorism, has been included to indicate the insights that can be obtained from a risk assessment that is in its earliest stage of development. It too is classified as a "scoping analysis" of a proposed quantitative risk assessment. The context is a scoping analysis to a risk assessment of the north Atlantic coastal states of the United States of an abrupt climate change; a climate change that could possibly be triggered by global warming. The driver considered for a possible abrupt climate change is a disruption of the Atlantic thermohaline circulation (THC). This assessment has not progressed to the point of the case studies of Chapters 3 and 4 in terms of supporting evidence, a fundamental understanding of the system, and the development of definitive scenarios. The assessment is essentially a first pass at modeling a very complex feedback system. The analysis represents the type of modeling that is often done early in a risk assessment to establish boundary conditions and a fundamental understanding of the system. No attempt is made at this time to carry the analysis to the point of human health and safety consequences. The focus is on precursor events to human health effects. In terms of the six-step process of quantitative risk assessment, only Steps 1 and 2 are definitive. Nevertheless, the modeling that has been done is segregated into the remaining four steps of the process to telegraph what modeling activity relates to the risk assessment steps. As the modeling progresses, specific scenarios will be defined and extended to include the impact on the population along the coastline of the northeast United States.

This chapter is a scoping analysis that will eventually lead to a quantitative risk assessment of the human fatality risk of an abrupt climate change in the north Atlantic coastal states. The phenomenon considered for the abrupt climate change is the possible shutdown of the THC in the Atlantic Ocean triggered by global warming.

Human activities have the potential of perturbing regional and global climate by affecting many of the complex interactions among atmosphere, land surface, and oceans. A major component of human induced climate change is related to increases in greenhouse gases (GHGs) such as methane and carbon dioxide (CO_2). Although most attention has focused on the modern industrial era, Ruddiman[1] has argued that anthropogenic climate change started thousands of years ago. But there is no doubt that the rate of change of increases in GHGs in the atmosphere has been most marked over the past century. Combustion of fossil fuels has led to increases in CO_2 from about 280 ppm in the pre-industrial atmosphere to over 350 ppm presently. This level of CO_2 is higher than it has been for at least several hundreds of thousands of years. Furthermore, projections indicate that over the next century, concentrations will reach 800–1000 ppm under a "business-as-usual" scenario.[2]

Assessments of the potential impact of GHG increases have been a part of the research agenda in the climate sciences for many years. For example, the U.S. Global Change Research Program has invested ~$20 billion since 1990 on research into climate variability and change. Because assessments of future scenarios require mathematical models to make projections, much work has been aimed at improving representations of a host of complex physical and biophysical processes within global models. Most impact assessments are based on model projections that show gradual warming of the surface temperatures over the next several centuries.[2]

The coupling between the atmosphere and the oceans is a critical aspect of global climate dynamics. Relative to the atmosphere, the ocean responds to perturbations slowly, and stores and transports large amounts of energy. The transport of large amounts of heat by the oceans feeds back to the atmosphere and is responsible for many aspects of climate; for example, the Gulf Stream substantially moderates the climate of Greenland and northern Europe. The large-scale ocean circulation (Figure 6.1) that is responsible for the transport of heat from the equator toward the poles in the surface of the Atlantic and the return flow of cold, deep water in the reverse direction is part of the global THC. The circulation is driven by density changes between freshwater and saline water and between warm and cold water; hence the term "thermohaline" circulation is used. The circulation is called "overturning" because colder and more saline water at the surface sinks as it is heavier and displaces the less dense water at depth. In this chapter we considered risk indicators associated with a disruption in the THC. In particular, we will consider changes in the rate of Atlantic Ocean overturning (i.e., the flow rate of the

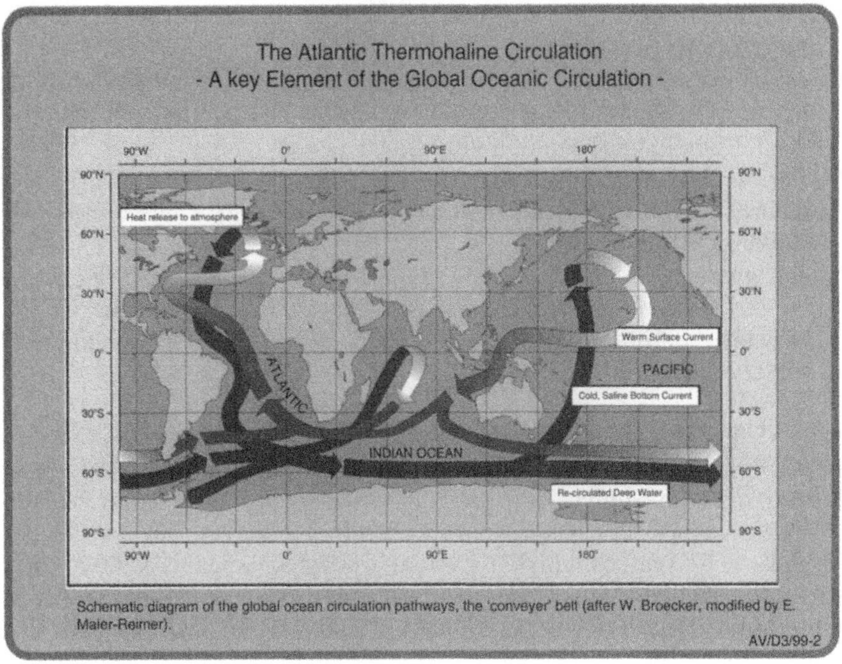

FIGURE 6.1. From CLIVAR: http://www.clivar.org/publications/other_pubs/clivar_transp/d3_transp.htm.

circulation expressed as a volume of water per time), which is a major driver of the THC, but first we jump ahead and present some results.

6.1 Summary of Insights from the Scoping Analysis

While this was only a scoping analysis of the risk of reducing or shutting down the THC and the consequent possibility of an abrupt climate change, some interesting risk insights were developed. Major contributors to the uncertainties in the analysis are the use of very simplified models, the assumed warming rates, and the limited understanding of the relationship between flow reduction and actual climate change. These contributors require much more investigation and analysis to complete a quantitative risk assessment. However, based on the assumptions and evidence presented in the scoping analysis the following results were obtained:

- The overall risk of a severe reduction in the Atlantic overturning is most strongly influenced by the uncertainty in the rate of global warming over the next 150 years.

- The probability that a shutdown will occur in this century is on the order of one in a million.
- At 100 years in the future, we estimate that the expected reduction in the Atlantic overturning will be in the vicinity of 30% with a 90% confidence range of about 12–50%.
- The scoping analysis indicated that there is greater than a 5% chance (a 1 in 20 chance) that the circulation will shut down in the next 300–500 years.
- The analysis further indicated that 300 years or more into the future we are 90% confident that the reduction in the THC flow will be between approximately 17% and 100%. The mean value is approximately 48%.

These results may be significant as there is some evidence that a complete shutdown of the circulation is not necessary to result in an abrupt climate change. In particular, some scientists speculate that as little as a 25% reduction in flow might trigger an abrupt climate change. The real conclusion from the scoping analysis is that there is strong evidence that abrupt climate change due to global warming should be a high-priority candidate for a quantitative risk assessment.

6.2 Scoping Analysis to Support a Quantitative Risk Assessment of Abrupt Climate Change

This scoping analysis is organized along the lines of the six-step process used in the previous case studies. The first two steps (6.2.1 and 6.2.2) are definitive, but the remaining steps (6.2.3–6.2.6) are very much in their early stages of development.

6.2.1 Definition of the System Under Normal Conditions (Step 1)

The system is the interaction of the atmosphere and the north Atlantic Ocean currents under conditions that could possibly lead to an abrupt climate change in the northeast coastal states of the United States.

6.2.2 Identification and Characterization of System Hazards (Step 2)

The hazards are phenomena such as global warming and the release of GHGs, phenomena that could possibly disrupt the coupling between the atmosphere and the oceans in a manner that affects the movement of stored energy in the oceans and the consequent possibility of an abrupt climate change.

6.2.3 Structuring of the Risk Scenarios and Consequences (Step 3)

Unlike the previous case studies where the basic modeling framework was an event tree that provided a scenario (sequence of events) roadmap between initiating events and consequences, because of its early stage of development the approach in this scoping analysis is to utilize the modeling approaches well established in the climate sciences. As the study advances and more modeling results become available, the results will be recast into the event tree and scenario format to be consistent with the practices of scenario based quantitative risk assessment. Among the modeling approaches extensively used in climate studies are so-called "box models." The boxes represent different ocean regions and are linked with input/output parameters characterizing flow properties of the ocean currents. The initiating events for the scenarios using this model are discrete temperature increases in global mean temperature expected over the next several hundred years. The output is information on the overturning rate of the circulation.

More than 40 years ago, Stommel[3] showed that a simple two-box model of the world oceans exhibited two distinct equilibrium solutions and speculated that the real ocean system might be bistable. Broecker[4] used this argument to warn that an alternate stable state for the THC was a shutdown of the circulation with possible large consequences for global climate. The change from an "on" to an "off" state for the THC is thought of as a bifurcation due to nonlinearities in the system and thus would occur suddenly and not gradually. Such a switch would lead to an abrupt climate change, for example, changes of mean temperature of several degrees Centigrade (°C) over several decades. This has led to suggestions that the THC is the "Achilles' heel" of the climate system and that increases in GHGs could be a trigger.[5] Evidence for abrupt climate change in the past is seen in paleoclimate records,[6] and changes in the THC are linked to at least some of these events.

The climate change that attends global warming may have a variety of impacts. Analysis of potential impacts on ecosystems and on human populations of a gradual rise in temperature and associated changes in precipitation suggests that humans will be able to adapt reasonably well but that costs may be quite significant.[7]

The issue that is the focus of this case study, however, is associated with *abrupt* climate change. The question is how human populations might adapt in the face of changes that occur over a decade or several decades and what the ecological and economic consequences will be. Thought has been given to how analyses of consequences of a weakening or a shutdown of the THC might be made,[6,8,9] but the quantification of such analyses is a topic of active research. We will consider a risk assessment that stops short of the actual damage states of primary interest, that is environmental, ecological, economical, and human health impacts, and consider changes in the rate of the Atlantic overturning itself as the end state. The rate of overturning of the ocean currents is believed to be an important precursor event to the above risk measures.

The model results presented in the IPCC report[2] for projected changes in the Atlantic overturning range from no effect to a reduction of ~50% from current conditions. The conclusion drawn in the IPCC report[2] is that the most likely result of GHG warming over the next century will be a gradual weakening of the THC with little likelihood of a collapse in the twenty-first century. It has been suggested, however, that the THC might be moved closer to a threshold due to warming[10] and that a reduction of about 25% in the rate of the Atlantic overturning might precipitate a shutdown of the THC. It is not at all certain that changes in the THC can result in abrupt climate changes with significant impact.

Estimates of various levels of weakening of the Atlantic overturning are needed to produce a sensible risk assessment. Here we consider seven different end states to characterize the risk. We denote by the symbol m the rate of Atlantic overturning and by m_0 the current rate.

Damage State 1—$0.8 \leq m/m_0 < 0.9$
Damage State 2—$0.8 \leq m/m_0 < 0.7$
Damage State 3—$0.7 \leq m/m_0 < 0.6$
Damage State 4—$0.6 \leq m/m_0 < 0$
Damage State 5—shutdown $(m = 0)$ occurs *after* the first 200 years
Damage State 6—shutdown occurs after the first 100 years but before 200 years
Damage State 7—shutdown occurs in first 100 years

As indicated earlier, the tools for assessing the potential impacts of these damage states on human populations and ecosystems are just beginning to be developed. The work by Higgins and Vellinga[8] indicates that a large weakening (Damage States 4 and higher in our terminology) would likely have global impacts on net primary productivity. The effects would not be uniform, of course, so a disaggregated analysis would be necessary to study impacts in detail. Mastrandrea and Schneider[11] use a linked environmental-economic analysis and aggregate over all sectors and regions to derive an aggregate measure of damage from abrupt climate change. They show that "dangerous anthropogenic interference," defined on the basis of projected ecological impacts, is quite likely under scenarios of global warming at rates considered here. Schneider[9] suggests that abrupt changes in climate can cause much larger impacts than those expected from slow, gradual warming.

Risk assessment involves the development of scenarios that answer the triplet question, what can go wrong. Under what conditions might the Atlantic overturning change in ways described earlier? A scenario starts with an "initiating event," which precipitates a system response. In the case we are considering, the initiating event is the projected increase in GHGs over the next several centuries. These increases in GHGs can be used in an appropriate model to calculate expected changes in the Atlantic overturning, which is the measure of "damage." Different global circulation models (GCMs) give different results for projection of changes in the THC because they have different parameterizations, different grid resolutions, and start from slightly different initial conditions (Figure 6.2).

As the large-scale experiment of adding large amounts of GHGs to the atmosphere proceeds, should we be concerned about the possibility of an abrupt climate change caused by a bifurcation in the THC? The cautious scientific view is given by Wood *et al.*,[12] who say that a rapid change in the THC "due to GHG warming cannot be ruled out but is considered a low-probability, high impact event." Thus, the issue is a good candidate for the methodology advocated in this book. To implement the methodology, a mathematical model for the phenomena must be available.

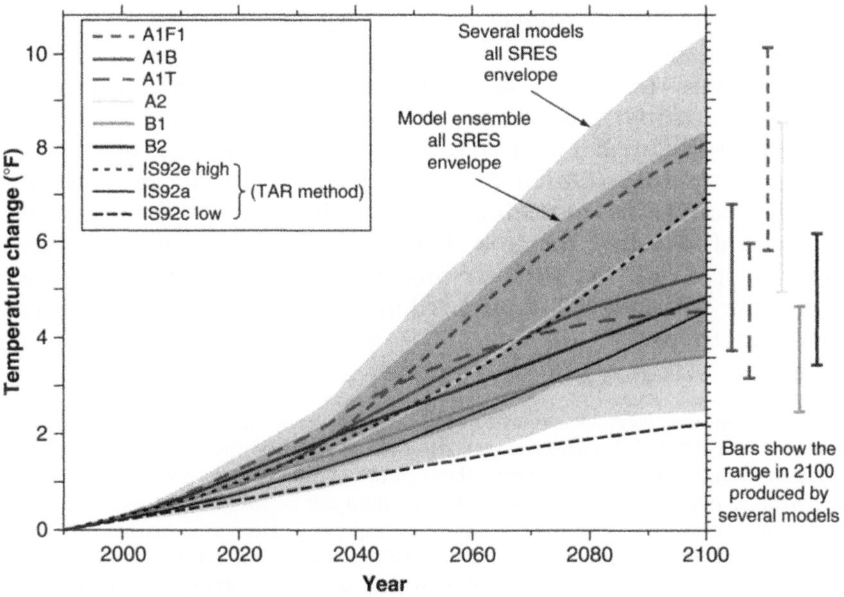

FIGURE 6.2. Projections of temperature changes over this century (IPCC).

A fully adequate model for the coupled atmosphere–ocean system is not available. This is due in no small degree to the complexity of the Earth system and the myriad of feedback mechanisms that can amplify small signals. Bigg *et al.*[13] outline a set of main effects and feedback processes that can affect the THC under global warming. For example, local effects in the north Atlantic include increased heat and freshwater fluxes under global warming scenarios, as both temperature and precipitation are expected to increase at high latitudes. Heating of the surface of the north Atlantic reduces the density of the water thus stabilizing the water column and decreasing the rate of the overturning. If the THC weakens, however, there is a negative feedback caused by reduced transport of salt from lower latitudes—part of the THC itself. The Earth system is replete with such feedbacks.[14]

The central tool for the quantitative assessment of climate variability and change is large, complex computer models known as GCMs. The challenge of modeling the Earth system, even with the largest computers, is daunting. The roots of the current GCMs date back to the 1960s when digital computers became available.[a] Even with the most advanced modern computers, the representation of many of the feedbacks must be done through parameterizations of the physical relationships. Thus, although the models arguably keep being improved, they cannot possibly capture all of the complexities of the natural system. Furthermore, the computational demands of these models make it essentially impossible to exercise them to explore fully all of the uncertainties associated with projections of future changes.

In response to the need to have models that can be run over long time periods—tens of thousands of years—and that can be run many times so sensitivity to various processes can be studied systematically, a class of models referred to as Earth-System Models of Intermediate Complexity (EMICs) has arisen. These models are certainly simpler than GCMs in treating certain feedbacks and in the resolution of the grid covering the globe, but they still include an impressive amount of detail. As with any model, the EMICs have their drawbacks; they mainly can be criticized for oversimplifying some process representations.[16]

[a] Numerical weather prediction can be traced to the early part of the twentieth century. In 1917, L.F. Richardson was attached to a French infantry division on the western front. As World War I raged about him, he embarked on a project to predict the weather over a section of Western Europe for a 6-h period given measurements of winds, pressures, temperatures, etc., as a starting point. His numerical computation approximated the equations describing atmospheric dynamics and was done with the help of only a slide rule. The computation took 6 months to complete and was a spectacular failure because of technical numerical problems. (See Hayes[15] for a delightful exposition of the story.) But the "failure" was short lived, as demonstrated by the GCMs in existence today.

Simple "box" models have been used in climate studies for a long time. Arrhenius[17] made what is probably the first quantitatively based forecast of global warming under increased GHGs. Budyko[18] and Sellers[19] presented box models for the Earth's energy balance. These box models are even easier to criticize for lack of representation of enough detail than are the EMICs, but they do have the virtue of being computationally tractable for extensive sensitivity and uncertainty studies.

As discussed earlier, the computational burden carried by GCMs makes it difficult to use them as tools for risk assessment. It has been suggested that GCMs may not be the tool of choice for assessments in any event.[20] We accept Harte's[21] suggestion that an approach "based on models that capture the essence of the problem, but not all the details, might get us further." We therefore elect to use the box model of Zickfeld *et al.*[22] as the computational engine for this scoping analysis (see Section 6.2.4 for details). The initiating event for the scenarios using this model is the increase in global mean temperature expected over the next several hundred years.

The rate at which the global mean temperature is expected to rise over the next century cannot be estimated with precision, but there is essentially universal agreement that temperatures will in fact rise.[2] We will consider scenarios defined by a distribution of warming rates that incorporates the large uncertainties in projections. Of course, the projections of temperature increase depend on models as well. The IPCC[2] reports that the warming in the twenty-first century is likely to be between 1.5 and 4.5 °C. Wigley and Raper[23] suggested that the likelihood of warming at both the low end and the high end of the IPCC figures is very low and that a 90% probability interval for warming in the year 2100 in the absence of mitigation is about 1.7–4.9 °C. Murphy *et al.*[24] report a 5–95% confidence range as 2.4–5.4 °C from their ensemble model studies assuming a doubling of atmospheric CO_2. Knutti *et al.*[25] suggest that this interval should be set at about 2–7.5 °C, Forest *et al.*[26] suggest 1.4–7.7 °C, and Androvona and Schlessinger[27] suggest 1–9.3 °C. The uncertainties are obviously fairly large, but most of the evidence supports a distribution with a median in the 2–3.5 °C range and with the 95th percentile in the vicinity of 5 °C.

6.2.4 Quantification of the Likelihood of the Scenarios (Step 4)

As indicated earlier, we use the model presented by Zickfeld *et al.*[22] to compute the rate of the Atlantic overturning over time. We adapt the description below from Zickfeld *et al.*,[22] where details of the model and the rationale for it can be found.

The four boxes in the model represent the south and north Atlantic Oceans (Boxes 1 and 2, respectively); the surface ocean in

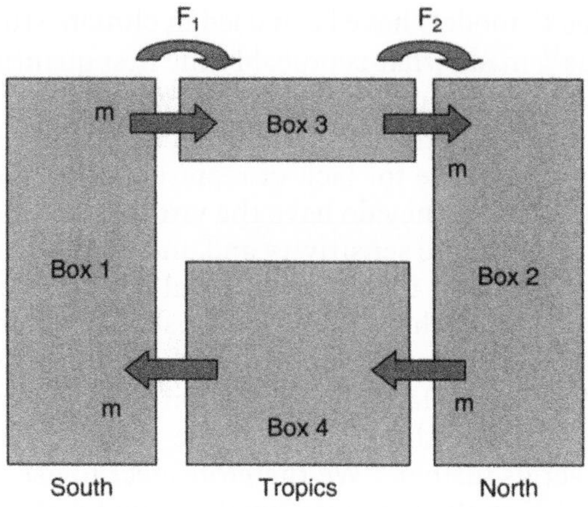

FIGURE 6.3. The box model for the Atlantic THC; from Zickfeld *et al.*[22]

the tropics (Box 3); and the deep ocean in the tropics (Box 4). The overturning rate, *m*, is the surface flow northward and the return flow at depth (Figure 6.3) and is driven by density differences between Boxes 1 and 2.

$$m = \frac{k(\rho_2 - \rho_1)}{\rho_0} = k[\beta(S_2 - S_1) - \alpha(T_2 - T_1)] \qquad (6.1)$$

where ρ is density, k is a constant relating volumetric flow to the density difference, S is salinity, T is temperature, β is a coefficient relating changes in salinity to changes in density, α is a coefficient relating changes in temperature to changes in density, the subscripts 1 and 2 refer to the respective boxes, and ρ_0 is a reference density. Total flows in ocean circulation are large so it is customary to use the unit *Sverdrup* (Sv) to describe them: 1 Sv is 10^6 m^3/s.

Temperatures and salinities of the boxes are conditioned by the transport due to overturning and by two other forcing terms. First, there is a net flux of freshwater between the surface boxes due to atmospheric vapor transport (and subsequent rainfall), wind-driven ocean currents, and evaporation from the surface (a negative flux of freshwater). The net fluxes from Box 1 to Box 3 and from Box 3 to Box 2 are represented by F_1 and F_2 respectively (Figure 6.3). Second, the temperature of the surface boxes is conditioned by interaction with the overlying atmosphere. This surface forcing is represented in the model by specifying the temperatures of the surface boxes in the absence of ocean transport, T_1^*, T_2^*, andT_3^*, and having the temperatures of the respective boxes move toward these restoring temperatures.

The model for the overturning consists of a set of ordinary differential equations derived from heat and mass-balance considerations, treating the boxes as completely and continuously mixed.

$$\frac{dT_1}{dt} = \frac{m}{V_1}(T_4 - T_1) + \lambda_1(T_1^* - T_1) \tag{6.2}$$

$$\frac{dT_2}{dt} = \frac{m}{V_2}(T_3 - T_2) + \lambda_2(T_2^* - T_2) \tag{6.3}$$

$$\frac{dT_3}{dt} = \frac{m}{V_3}(T_1 - T_3) + \lambda_3(T_3^* - T_3) \tag{6.4}$$

$$\frac{dT_4}{dt} = \frac{m}{V_4}(T_2 - T_4) \tag{6.5}$$

$$\frac{dS_1}{dt} = \frac{m}{V_1}(S_4 - S_1) + \frac{S_0 F_1}{V_1} \tag{6.6}$$

$$\frac{dS_2}{dt} = \frac{m}{V_2}(S_3 - S_2) + \frac{S_0 F_2}{V_2} \tag{6.7}$$

$$\frac{dS_3}{dt} = \frac{m}{V_3}(S_1 - S_3) + \frac{S_0(F_1 - F_2)}{V_2} \tag{6.8}$$

$$\frac{dS_4}{dt} = \frac{m}{V_4}(S_2 - S_4) \tag{6.9}$$

The V_i are the box volumes, S_0 is a reference salinity for the freshwater fluxes, and the λ_i are thermal coupling coefficients to describe the surface forcing. The latter coefficients are given by $\lambda_i = \Gamma / c\rho_0 z_i$, where Γ is a constant that reflects radiative and diffusive heat transport between atmosphere and ocean, c is the specific heat capacity of seawater, and the z_i are the depths of the surface boxes.

We use the values for parameters derived by Zickfeld *et al.*[22] Several of the values are well-defined constants (Table 6.1).

TABLE 6.1
Values for physical constants used in the model

Parameter	*Value*	*Description*
c	4000 J kg^{-1} °C^{-1}	Specific heat capacity
ρ_0	1025 kg m^{-3}	Density of seawater
α	1.7×10^{-4} °C^{-1}	Thermal expansion coefficient
β	8×10^{-4} psu^{-1}	Haline expansion coefficient
S_0	35 psu	Reference salinity

TABLE 6.2
Values for model parameters[22]

Parameter	*Value*	*Description*
V_1	1.1×10^{17} m^3	Volume of Box 1
V_2	0.4×10^{17} m^3	Volume of Box 2
V_3	0.68×10^{17} m^3	Volume of Box 3
V_4	0.05×10^{17} m^3	Volume of Box 4
z_1	3000 m	Depth of Box 1
z_2	3000 m	Depth of Box 2
z_3	1000 m	Depth of Box 3
Γ	7.3×10^8 J y^{-1} m^{-1} °C^{-1}	Thermal coupling coefficient
k	25.4×10^{17} m^3 y^{-1}	Flow constant
F_{10}	0.014 Sv	Current ("time zero") value for F_1
F_{20}	0.065 Sv	Current ("time zero") value for F_2
T^*_{10}	6.6 °C	Current ("time zero") value for T^*_1
T^*_{20}	2.7 °C	Current ("time zero") value for T^*_2
T^*_{30}	11.7 °C	Current ("time zero") value for T^*_3

Other parameters, which we take to have fixed values, either are estimated on the basis of physical knowledge or are determined by "tuning" the model (Table 6.2).

Equations (6.1)–(6.9) can be integrated to give the rate of the Atlantic overturning as a function of time in response to increasing global mean temperature. For our scenarios, we adopt the approach used by Zickfeld *et al.*[22] and use a linear increase in global mean temperature, T^{GL}, for 150 years with steady temperatures thereafter. The transient behavior of the Atlantic overturning is driven by time evolution of the restoring temperatures, T^*_1, T^*_2, and T^*_3, and of the freshwater fluxes, F_1 and F_2. These evolve in time according to the following equations[22]:

$$T^*_1(t) = T^*_{10} + p_1 t \frac{dT^{GL}}{dt} \tag{6.10}$$

$$T^*_2(t) = T^*_{20} + p_2 t \frac{dT^{GL}}{dt} \tag{6.11}$$

$$T^*_3(t) = T^*_{30} + p_3 t \frac{dT^{GL}}{dt} \tag{6.12}$$

$$F_1(t) = F_{10} + h_1 p_{SH} t \frac{dT^{GL}}{dt} \tag{6.13}$$

$$F_2(t) = F_{20} + h_2 p_{NH} t \frac{dT^{GL}}{dt} \tag{6.14}$$

Time, t, is measured in years with present being zero. The parameters p_i are constants that disaggregate global mean temperature changes to regional levels. The parameters p_{SH} and p_{NH} are coefficients to relate mean hemispheric temperature changes to global changes. The parameters h_1 and h_2 are hydrological sensitivity coefficients. In the simulations for the scenarios, Equations (6.10)–(6.14) apply for the 150 years of temperature increase; after this time values are held constant at the 150-year levels. We start with the nominal values for these parameters as presented by Zickfeld *et al.*[22] (Table 6.3).

Treatment of Uncertainty in the Risk Model

For purposes of this case study, we take the most important uncertainties to be related to the temporal evolution of the Atlantic overturning. The first aspect of the temporal evolution is the initiating event—the rate of increase in global mean temperature,

$$dT^{GL}/dt^*$$

As indicated in Section 6.2.3, a fairly broad range of possible warming rates is reported in the literature. We consider warming rates of 0.01–0.07 °C/year in steps of 0.01 °C/year. That is, we use a discrete probability density function for rate of warming that is broadly consistent with estimates of the distribution for temperature change in the year 2100 made by a number of experts (see Section 6.2.3). We opted to use a coarse division for warming rate with the largest probability densities in the 2–3 °C/century range but with a 7% chance that the warming would be 6 or 7 °C/century. Our discrete probability density function retains the bulk of warming rates within the IPCC 1.4–4.5 °C/century range but allows for a small probability that the rate might be as high as 7 °C/century (Figure 6.4).

In a more comprehensive risk assessment, we would use the Bayesian methods discussed in Chapter 2 to derive a more rigorous

TABLE 6.3
Nominal values for time evolution parameters

Parameter	*Value (units)*
p_1	0.86
p_2	1.07
p_3	0.79
p_{SH}	0.93
p_{NH}	1.07
h_1	–0.005 Sv °C^{-1}
h_2	0.023 Sv °C^{-1}

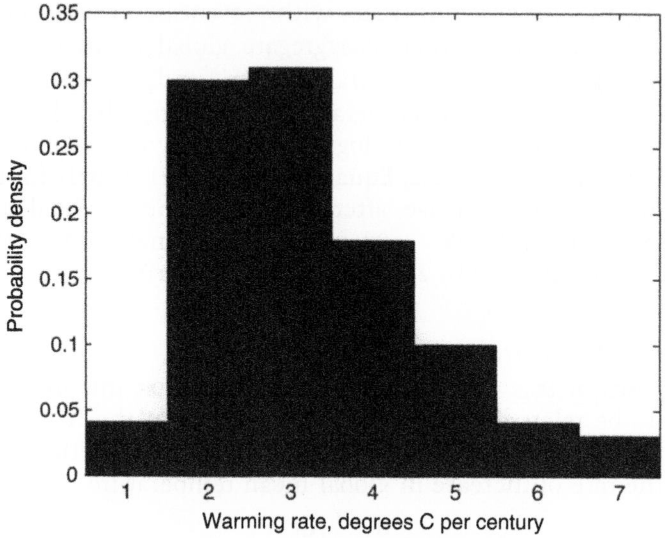

FIGURE 6.4. Probability density used in risk calculations.

representation of the warming rate uncertainty. For example, a systematic Bayesian treatment of the available projections would explicitly account for both the uncertainties in the estimates from each expert, and variability among the individual experts. Sampled historical data for the observed warming rate would then provide additional evidence to further refine the uncertainties from these predictive estimations. Section 6.3 contains examples that illustrate how the uncertainty about the warming rate would change as additional evidence is compiled and integrated through a formal Bayesian process.

The other parameters that affect the temporal evolution of overturning are those in Table 6.3. We adopt a discrete distribution to represent the p_i values; we use the nominal values in Table 6.3 as the mean and a 5% coefficient of variation for a Gaussian distribution and apply a coarse discretization (Figure 6.5). Zickfeld *et al.*[22] report modest sensitivities of calculated overturning to these regional coefficients by considering cases where differences among regions are enhanced or reduced. The distribution that we use allows such variation within reason.

The major source of uncertainty in projections of the rate of the Atlantic overturning made using this model is the hydrological sensitivities. (Note that because p_{SH} and p_{NH} appear as multipliers of the hydrological sensitivities, we can fix the values at those listed in Table 6.3, in effect subsuming uncertainty in these into the uncertainties in h_1 and h_2.) Zickfeld *et al.*[22] argue that values for h_2 of 0.05 Sv/°C are

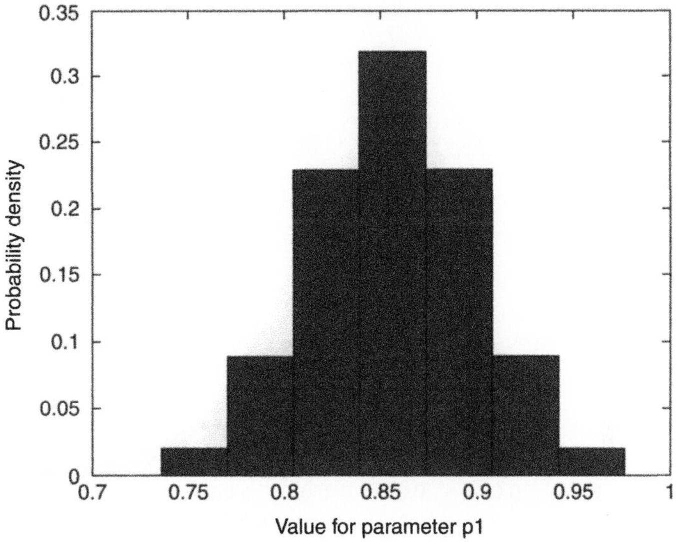

FIGURE 6.5. Probability density for parameter p_1 used in simulations. Distributions for parameters p_2 and p_3 are similar.

not unrealistic and they use values up to 0.06 Sv/°C in their sensitivity analyses. We adopt a lognormal distribution for h_2 as a reasonable approximation of uncertainty and truncate and discretize it into 94 "bins" for our simulations (Figure 6.6). The mean value of the distribution is the nominal value in Table 6.3 and the probability that h_2 exceeds 0.05 Sv/°C is about 0.1%. We use a similar distribution for h_1.

Monte Carlo Simulations

To examine the uncertainties within the context of the scoping analysis, we ran the model choosing values for uncertain parameters and temperature increases at random. We performed 50,000 such Monte-Carlo realizations for the full case, with warming rates and all parameters selected randomly within the distributions specified above. To decompose the uncertainties to show the influence of rates of increase in global temperature, we also ran 50,000 realizations independently for each warming rate from 1 to 7 °C/century.

6.2.5 Assembly of the Scenarios into Measures of Risk (Step 5)

The Monte-Carlo runs were made for a total time of 500 years into the future. The probability weighted mean curves for Damage States 1–4 show declines in the rate of the Atlantic overturning over the first 200 years of the simulations and recovery to ~90% of the current rate

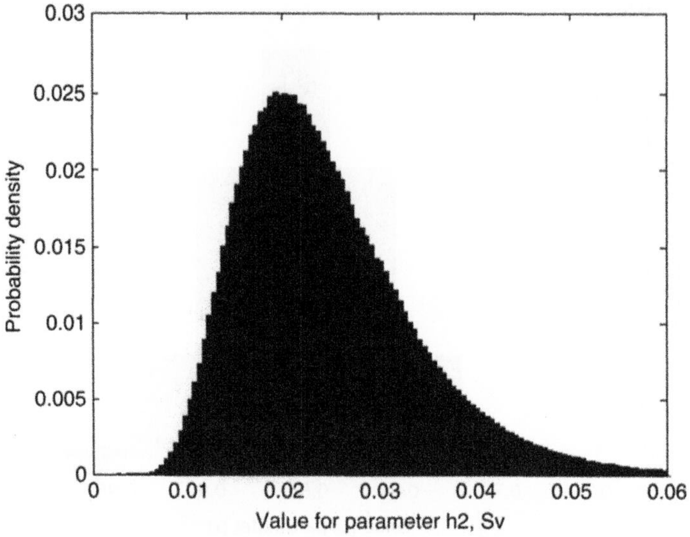

FIGURE 6.6. Probability density for parameter h_2. The distribution for parameter h_1 is similar.

by year 500, whereas Damage States 5–7 indicate shutdown of the overturning between about the 100–400 year mark (Figure 6.7).

The results in Figure 6.7 can also be presented in the format of cumulative risk curves, as described in Chapter 2. Our seven damage states correspond to increasingly more severe potential consequences with respect to THC reduction, from rather benign conditions (Damage State 1) to complete shutdown within the next century (Damage State 7). If we reformulate our results to show the cumulative probability of achieving a specified severity of THC reduction as a function of time, the results take the form of risk curves that display the full uncertainties from our analyses (Figure 6.8). In effect, these curves are derived by examining successive "slices" through the results in Figure 6.7 to determine the time at which each increment of THC reduction is achieved and the associated probability that the reduction is achieved at that time, or earlier.

6.2.6 Interpretation of the Results (Step 6)

As described in the early chapters of this book, these risk curves present a wealth of information in compact form. They can be read in a variety of ways. For example, reading across the curves horizontally at 200 years, we see that for the Zickfeld model with uncertainties as we have characterized them, the expected (mean) reduction in overturning rate is about 40%, the median of the distribution is about 32%

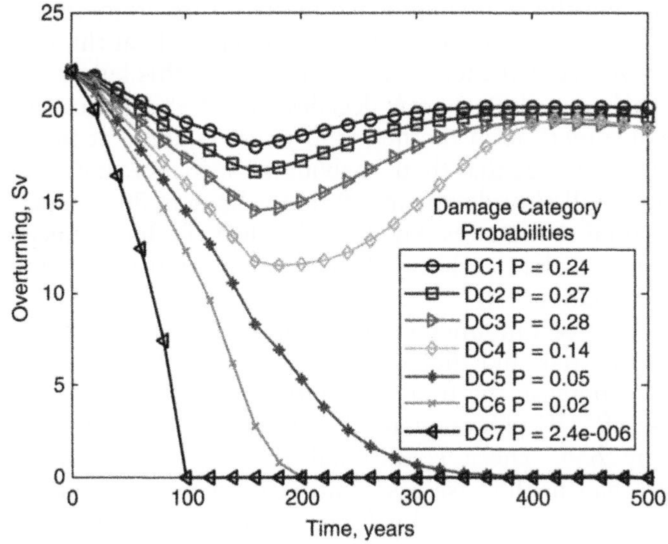

FIGURE 6.7. Probability weighted mean time course for the seven damage states for 50,000 Monte-Carlo realizations.

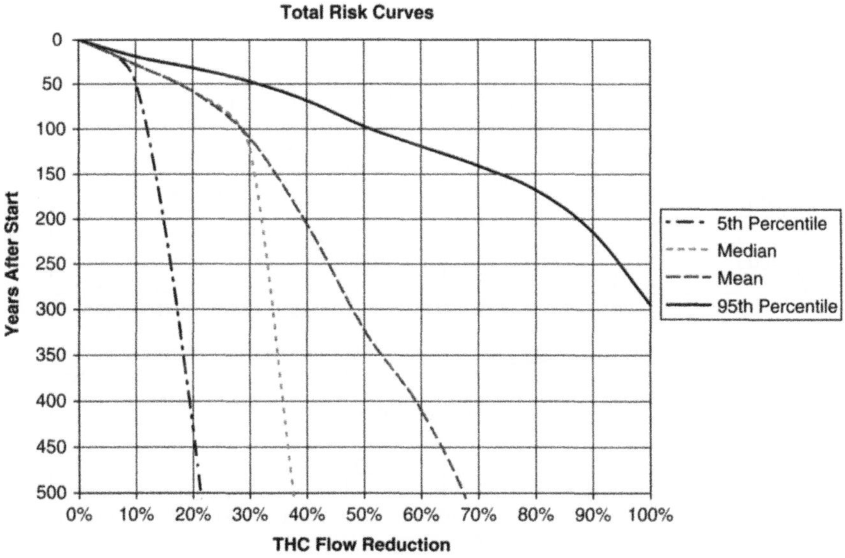

FIGURE 6.8. Risk curves for the base case, all parameters and the rate of warming treated as uncertain.

reduction, and the 90% confidence interval runs from about 15% to about 86% reduction. Reading the curves vertically at the 20% reduction level, we see that there is a 5% chance that this level of reduction will be reached in the next few decades and a 95% chance that it will be reached sometime in the next 450 years. Looking at the 95th percentile curve, we see that there is about a 5% chance that the Atlantic overturning will shut down sometime in the next 300 years.

The total risk curves (Figure 6.8) reflect the broad uncertainties in the approach that we have adopted to represent the Atlantic overturning circulation.

To enhance our understanding of the most important contributors to this risk and their associated uncertainties, it is useful to partially decompose the total risk curves and to examine the conditional risk of THC reduction as a function of the rate of global warming (Figure 6.9). These conditional risk curves give information similar to that for the base case, but with the presumption of a fixed rate of warming. Conditional risk results are not shown for a warming rate of 1 °C/century because our analyses conclude that the maximum reduction in the Atlantic overturning will be less than 10% for that case. The results indicate that for the Zickfeld model, we are 95% confident that reductions in the Atlantic overturning will be less than

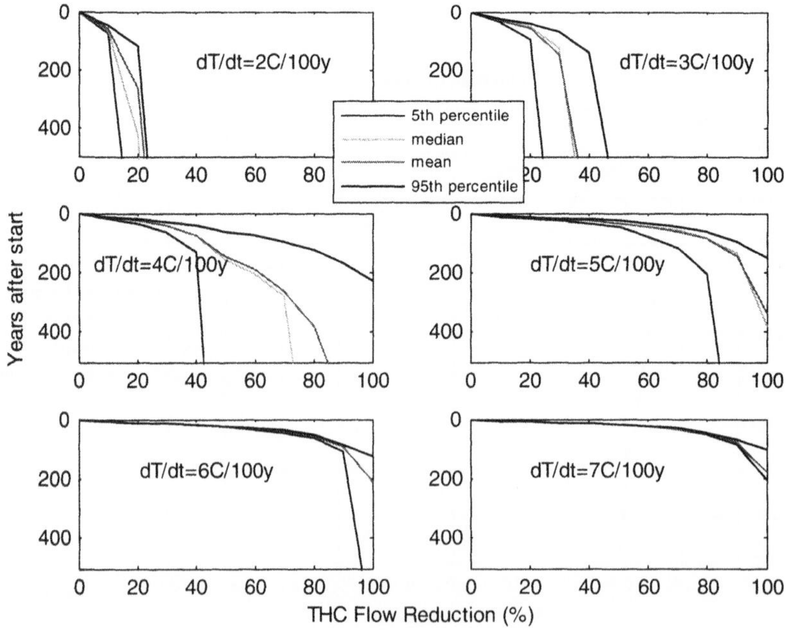

FIGURE 6.9. Risk curves conditional for different rates of warming over 150 years.

about 20% over the next 500 years if the warming rate over the next 150 years is 2 °C/century (or less). Conversely, we are essentially certain that the overturning will shut down sometime in the next several hundred years if the rate of warming is 7 °C/century (or larger).

A comparison of the total and conditional risk results clearly shows that the overall risk of a severe reduction in the Atlantic overturning is most strongly influenced by our uncertainty in the rate of global warming over the next 150 years. Uncertainties in other parameters of the Zickfeld model have a measurable, but clearly second order effect on our characterization and understanding of the risk. Thus, it is evident that we should focus our attention and resources on efforts to better understand the analytical bases for the estimated warming rates and the uncertainties that are associated with those estimates. As indicated previously, Bayesian methods may be used to systematically examine the available evidence, and to develop a more refined assessment of the contributors and associated uncertainties. (See Chapter 2 for an illustration of how such an analysis could be useful.)

Even though our modeling approach is fairly simplistic, our results are not out of bounds with projections that have been made by others. At 100 years in the future (*ca.* 2100) we estimate that the expected reduction in the Atlantic overturning will be in the vicinity of 30% with a 90% confidence range of about 12–50%. The IPCC[2] estimates (Figure 6.2) can be interpreted to indicate a range of about 0–50% reduction with the central tendency at about a 25% or so reduction. Knutti *et al.*[25] estimate a probability density of reduction in Atlantic THC in the year 2100 using an ensemble of different models; they conclude that the 90% confidence interval is 15–85% with a "best guess" value of 65% reduction.

Our main contention in this book is that the risk assessment process is valuable for problems with low probability of occurrence and high potential impact. As indicated previously, we do not have good estimates of ecosystem consequences, so our risk assessment is terminated artificially at a damage state denoted by THC reduction and not carried through to ultimate impacts of most direct concern to people. But despite the lack of precise estimates of ultimate consequences to humans, there is good evidence that climate changes have had very significant impacts on past societies.[28] Weiss and Bradley argue that in the latter half of the present century, world population is likely to be near ten billion people and that a significant fraction of the populace will be engaged in small-scale agriculture. They will be quite susceptible to consequences of climate change, particularly if it occurs abruptly. There are plausibility arguments to indicate that economic and ecological impacts of abrupt climate change are likely to be substantial.[8,9,11]

With respect to the low probability part of the definition of case studies for this book, our results confirm that a complete shutdown of the Atlantic THC is indeed a low probability event, given current estimates of warming and of uncertainties in key parameters. The probability that a shutdown will occur in this century is on the order of one in a million (Figure 6.7). Even going out 300 years, the probability of a complete shutdown is only about 5% (Figure 6.8). Furthermore, the conditional probabilities of a shutdown if the rate of warming is less than about 4 °C/century are very small indeed (Figure 6.9). On the other hand, if the warming rate turns out to be on the high end of those considered (e.g., 6 or 7 °C/century), the conditional probability of a complete shutdown in the next 500 years is close to certainty.

Although a complete shutdown of the Atlantic THC in the next century appears to be an unlikely consequence of increasing GHGs in our assessment, the model that we used did not account for perturbations that affect the Earth system, for example perturbations in solar input or extreme weather conditions. Some concern has been expressed that as the rate of the overturning is lessened, the size of a perturbation needed to cause the system to reach the bifurcation point where the THC shuts down becomes less and less; e.g., see discussion in NRC.[6] Tziperman[10] used model simulations to suggest that a reduction in rate in the THC of about 25% may lead to instability. Thus, it is also instructive to consider the results for weakening of the Atlantic THC in the 25% range. The results indicate that there is a greater than 50% chance that a 25% reduction will occur in this century and a 5% chance that it will occur in the next 40 years or so. If it turns out that a 25% weakening does lead to instability, the risk is quite high. In any event, these results provide a reasonable basis to move forward with a more comprehensive risk assessment.

6.3 Illustration of Bayesian Analyses for Warming Rates

This scoping analysis shows that the risk of a significant reduction in the Atlantic overturning circulation is influenced very strongly by the uncertainty about the rate of increase in global mean temperature. Therefore, to fully understand the risk, we must clearly identify the sources of the uncertainty and develop the best possible expression of our confidence in the predicted warming rate. Bayesian methods provide the analytical framework to consistently account for the strength of the available evidence and to evaluate its influence on our current state of knowledge.

Figure 6.4 shows our assigned probability distribution for the rate of global warming. The uncertainty is represented by a discrete probability density function over the range of predicted warming rates, from

1 to 7 °C/century. This composite estimate is derived from a large number of predictive analyses, with considerable diversity of expert opinion. Our current state of knowledge may be improved by a Bayesian combination of these expert estimates, rigorously accounting for both the uncertainty in each estimate and the variability among the respective experts. This two-stage Bayesian application would provide a more comprehensive treatment of the supporting evidence, and it would clearly show how each expert estimate contributes to our current overall uncertainty.

Derivation of the input data from each expert and demonstration of a full two-stage Bayesian analysis is beyond the scope of this case study. However, this section presents examples of simpler one-stage Bayesian analyses that show how accumulated evidence about the measured rate of global warming can influence our uncertainty and affect our estimates of the overall risk. The analyses accept Figure 6.4 as the starting point for our current state of knowledge about the predicted warming rate.

The record of global mean temperature for approximately the past five decades indicates a continual warming (Figure 6.10). If we do a simple statistical analysis, using linear regression to extrapolate these data to 2025, the expected temperature increase in 2025 relative to 2005 would be about 0.4 °C. We first consider the hypothetical situation that the *observed* temperature increase in 2025 over 2005 is

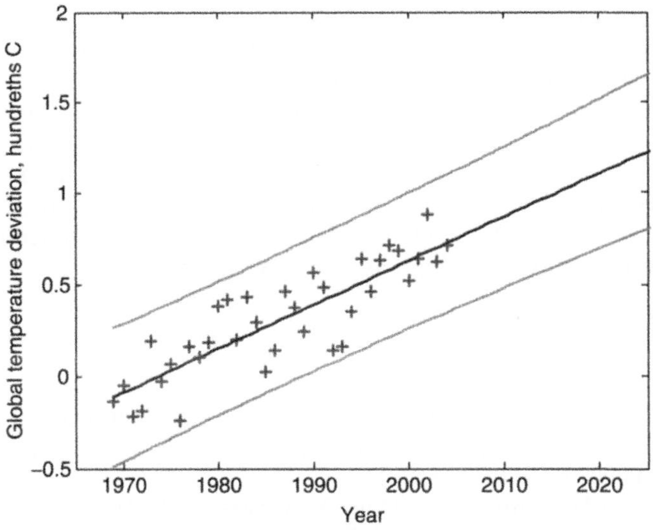

FIGURE 6.10. Regression of observed global mean temperature deviations from 1969 through 2004 with extrapolation in 2025 [data from Angell[29]].

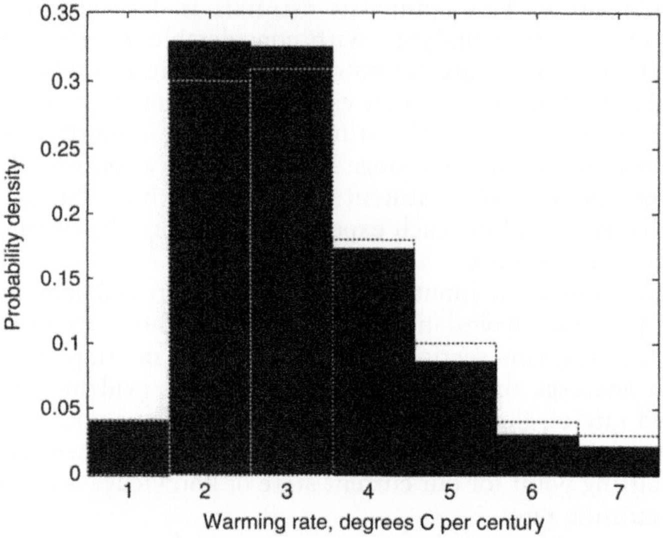

FIGURE 6.11. Bayesian update of the probability density for warming rate given a hypothetical observation in 2025.

actually 0.4 °C. This experience then becomes our evidentiary input for a Bayesian assessment of the predicted warming rate, starting with our current expert state of knowledge.

Figure 6.11 shows the results from that analysis. The solid bars are the updated discrete probability distribution for the warming rate, after we account for the 20 years of experience. The dashed bars show our current state of knowledge that is used in the case study risk models.

The observed evidence of 0.4 °C warming in 20 years reinforces our assessed probability that the actual warming rate is in the range of 2–3 °C/century. The evidence also reduces our confidence that the warming rate may be as high as 5–7 °C/century.

References

[1] Ruddiman, W., "The Anthropogenic Greenhouse Era Began Thousands of Years Ago," *Climatic Change* 2003, *61*, 261–293, doi:10.1023/B:CLIM.0000004577.17928.fa.

[2] IPCC, In Houghton, J. T., Ding, Y., Griggs, D. J., Noguer, M., van der Linden, P. J., Dai, X., Maskell, K., Johnson, C. A. (Eds.), *Contribution of Working Group I to the Third Assessment Report of the Intergovernmental Panel on Climate Change*, Cambridge University Press, Cambridge, UK; New York, NY, 2001, 881 pp.

[3] Stommel, H., "Thermohaline Convection with Two Stable Regimes of Flow," *Tellus* 1961, *13*, 224–241.
[4] Broecker, W. S., "Unpleasant Surprises in the Greenhouse?" *Nature* 1987, *328*, 123–126.
[5] Broecker, W. S., "Thermohaline Circulation, the Achilles Heel of Our Climate System: Will Man-Made CO_2 Upset the Current Balance?" *Science* 1997, *278*, 1582–1588.
[6] NRC (National Research Council), *Abrupt Climate Change: Inevitable Surprises*, National Academy Press, 2002.
[7] NAST (National Assessment Synthesis Team), Climate Change Impacts on the United States—Overview Report: The Potential Consequences of Climate Variability and Change, U.S. Global Change Research Program, http://www.usgcrp.gov/usgcrp/Library/nationalassessment/overview.htm, 2000.
[8] Higgins, P. A. T. and Vellinga, M., "Ecosystem Responses to Abrupt Climate Change: Teleconnections, Scale and the Hydrological Cycle," *Clim. Change* 2004, *64*, 127–142.
[9] Schneider, S. H., "Abrupt Non-Linear Climate Change, Irreversibility and Surprise," *Global Environ. Change Hum. Policy Dimen.* 2004, *14*, 245–258.
[10] Tziperman, E., "Proximity of the Present-Day Thermohaline Circulation to An Instability Threshold," *J. Phys. Oceanogr.* 2000, *30*, 90–104.
[11] Mastrandrea, M. D. and Schneider, S. H., "Probabilistic Integrated Assessment of 'Dangerous' Climate Change," *Science* 2004, *304*, 571–575.
[12] Wood, R. A., Vellinga, M., and Thorpe, R., "Global Warning and Thermohaline Circulation Stability," *Philos. Trans. R. Soc. Lond. A* 2003, *361*, 1961–1975.
[13] Bigg, G. R., Jickells, T. D., Liss, P. S., and Osborn, T. J., "The Role of the Oceans in Climate," *Int. J. Climatol.* 2003, *23*, 1127–1159, DOI: 10.1002/joc.926.
[14] NRC (National Research Council), *Understanding Climate Change Feedbacks* National Academy Press, 2003.
[15] Hayes, B., "The Weatherman," *Am. Sci.* 2001, *89*, 10–14.
[16] McGuffie, K. and Henderson-Sellers, A., "Forty Years of Numerical Climate Modelling," *Int. J. Climatol.* 2001, *21*, 1067–1109. DOI: 10.1002/joc.632.
[17] Arrhenius, S., "On the Influence of Carbonic Acid in the Air Upon the Temperature of the Ground," *Philos. Mag.* 1896, *41*, 237–276.
[18] Budyko, M. I., "The Effect of Solar Radiation Variations on the Climate of the Earth," *Tellus* 1969, *21*, 611–619.
[19] Sellers, W. D., "A Global Climatic Model Based on the Energy Balance of the Earth-Atmosphere System," *J. Appl. Meteorol.* 1969, *8*, 392–400.
[20] Shackley, S., Young, P., Parkinson, S., and Wynne, B., "Uncertainty, Complexity and Concepts of Good Science in Climate Change Modelling: Are GCMs the Best Tools?" *Clim. Change* 1998, *38*, 159–205.
[21] Harte, J., "Toward a Synthesis of the Newtonian and Darwinian Worldview," *Phys. Today* 2002, *55*(10), 29–34.

[22] Zickfeld, K., Slawig, T., and Rahmstorf, S., "A Low-Order Model for the Response of the Atlantic Thermohaline Circulation to Climate Change," *Ocean Dyn.* 2004, *54*, 8–26, DOT 10.1007/s10236–003–0054–7.

[23] Wigley, T. M. L. and Raper, S. C. B., "Interpretation of High Projections for Global-Mean Warming," *Science* 2001, *293*, 451–454.

[24] Murphy, J. M., Sexton, D. M. H., Barnett, D. N., Jones, G. S., Webb, M. J., Collins, M., "Qunantification of Modelling Uncertainties in a Large Ensemble of Climate Change Simulations," *Nature* 2004, *430*(7001), 768–772.

[25] Knutti, R., Stocker, T. F., Joos, F., and Plattner, G. K., "Probabilistic Climate Change Projections Using Neural Networks," *Climate Dyn.* 2003, *21*, 257–272. http://ejournals.ebsco.com/direct.asp?ArticleID=A8HQM9X5NL6YNFHDX36R.

[26] Forest, C. E., Stone, P. H., Sokolov, A. P., Allen, M. R., and Webster, M. D., "Quantifying Uncertainties in Climate System Properties with the Use of Recent Climate Observations," *Science* 2002, *295*, 113–117.

[27] Androvona, N. G. and Schlessinger, M. E., "Objective Estimation of the Probability Density Function for Climate Sensitivity," *JGR* 2001, *106*, 22605–22611.

[28] Weiss, H. and Bradley, R. S., "What Drives Societal Collapse?" *Science* 2001, *291*, 609–610.

[29] Angell, J. K., *Trends Online, A Compendium of Data on Global Change*, Carbon Dioxide Information Analysis Center, Oak Ridge National Laboratory, U.S. Department of Energy, Oak Ridge, Tennessee, 2005.

CHAPTER 7

Examples of Risks Having the Potential for Catastrophic Consequences

In Chapter 1 it was pointed out that this book is targeting risks that are a combination of existential risks and risks that are believed to be rare but of catastrophic consequences (defined as resulting in 10,000 fatalities or greater). As demonstrated in the case studies, the risk sciences can be applied to screen and importance rank risks that meet this criterion. The focus on this class of risks is believed important because of the perceived lack of urgency about them in the minds of the public and our leaders, and the ability of such threats to dramatically change or end life as we know it. This lack of interest is partly because the risks considered here most often do not happen in our lifetimes and in some cases they don't happen in many millennia, or for the case of existential risk they may not happen for tens of millions of years. It is the purpose of this chapter to raise our consciousness about potential catastrophic risks by presenting a cross section of examples.

Four such risks having to do with hurricanes, asteroids, terrorists' attacks, and abrupt climate change were somewhat arbitrarily selected for case studies in the previous chapters. A systematic application of the risk assessment thought process could provide the necessary guidance on which risks should have priority for future case studies and the taking of action. There are many factors to consider besides just the human fatalities involved. These include information on the resources and time needed to mitigate or reduce the consequences of threatening events and conditions. For example, from the asteroid case study it was observed that if an asteroid that is a threat to our planet can be identified far enough in advance, actions might be possible to alter its course and completely mitigate the event.

Knowledge about the risks we face is the key. A structured and systematic approach for developing a roadmap of risks requiring action should include certain fundamental considerations. Of course in the final analysis it should be risk-informed. That is, the risks should be understood in terms of their likelihoods and consequences. Actions to take should then be based on costs and risk benefits, and value judgments, that is, the principles of decision analysis.

It is not reasonable to quantify the likelihoods and consequences of all of the risks facing humanity, even if the target risks are narrowed to catastrophic and existential. It has to be an iterative and phased process. A logical first step would be to sort the risks qualitatively by area of impact and consequence. By area is meant global, regional, and local and by consequence is meant the population at risk. The first cut at the lists of risks to consider would be principally based on expert knowledge and readily available evidence of the level of the threat involved. The lists would then be refined by the application of risk assessment techniques in phases, peer reviewed with public participation, and challenged using the methods of contemporary decision analysis. At some point of the iterative and phasing process, the risks will need to be quantified and decisions made on what to do about them.

Possible criteria for developing a priority list of risks that matches the focus of this book are (1) risks having the potential of catastrophic consequences, (2) the existence of limited knowledge about the risk especially in terms of how likely it is to occur, and (3) a lack of attention and priority by the public and the government. Clearly, there are many risks that could meet these criteria. For example, we frequently hear about the disastrous consequences that could follow a collapse of the world's economies. Some economists fear such an event, which could be followed by global war, famine, and disease. Economic issues of concern are the heavy dependence of countries on foreign oil and its impact on economic stability, possible changes of the role of the U.S. dollar in world currency markets, and changing patterns of economic bases from factories, machinery, etc., to services and paper commodities. Another example of concern to many is the growing dependence of economies and the general conduct of businesses and society infrastructures on electronic communications. We live in an increasingly interconnected world of technology-based systems and networks. An example of the risk of such electronic dependence is presented in Chapter 5 as a case study of a terrorist attack on the national electrical grid.

Other examples that meet our criteria are listed in Table 7.1 and discussed below.

Each of the risks of Table 7.1 is briefly discussed to provide some insight as to why it should be included on any list of future quantitative risk assessments. Nuclear war is discussed in more detail than

TABLE 7.1
Potential catastrophic events

Nuclear war	Super volcanoes	Destruction of the ozone layer
Global water management	Pollution from fossil fuels	Infectious disease pandemic
Species destruction: tropical rain forests	Species destruction: coral reefs	Giant tsunamis
Genetic engineering and synthetic biology	Global warming	Super earthquakes
Industrial accidents	Nanotechnology	Population management

the others because of current concerns about nuclear materials proliferation and its connection with terrorism.

7.1 Nuclear War

It has been said that no one can estimate with any confidence the likelihood of a nuclear war. To be sure it is not possible to calculate the likelihood of a nuclear war with complete confidence, that is, with no uncertainty. But one of the purposes of this book is to demonstrate the value of calculating risks in the presence of uncertainty. The concept of the "set of triplets" definition of risk is scenarios, likelihoods, and consequences and was conceived to be a framework for addressing any kind of risk. The premise is that any risk that can be identified can be characterized in terms of these three factors, recognizing that a part of the process is to embrace the uncertainties involved and make them an intrinsic part of the likelihood calculation. The result is not always precise in the sense of absolute numbers, but can clearly be quantitative in terms of the state of knowledge about the numbers. But before consideration is given to how one would structure a risk model for nuclear war, it is appropriate to discuss some of the characteristics of nuclear wars. In particular, before a risk model can be developed of something, it is important to understand the system that is to be modeled. Much of this discussion is based on a paper by Garrick.[1]

A key consideration in the building of a risk model is an understanding of how the "what can go wrong" scenarios are initiated. In this case, we mean "how does a nuclear war get started." Obviously there are many different types of nuclear wars. The type we worried about during the Cold War was one that involved nation states, for example the U.S. and the U.S.S.R., and occurred deliberately. This might be classified as the conventional nuclear war. Now we know of many other types of nuclear wars or smaller scale nuclear attacks that could escalate to the level of a conventional nuclear war. These include not only nation states, but also amorphous groups and even individuals. Examples are dictators, terrorists, religious fanatics, deranged military commanders, mistakes at nuclear missile launch sites, and unforeseen responses or consequences of pre-emptive actions. The actions of nations or groups can be a major factor in increasing the likelihood of a nuclear war. Many believe we were very close to nuclear war during the Cuban missile crises of 1962. One of the great fears of societies is being drawn into a nuclear war through miscalculation and poor decision-making. These are the types of factors that must be considered in defining a set of initial conditions for different nuclear war scenarios.

What kinds of scenarios describe non-conventional initiators of a nuclear attack or war? Two types of failures could lead to a nuclear attack and possibly an all out nuclear war. One is a *command failure* such as a rogue commander initiating a nuclear attack without proper authorization. An example of a nuclear attack being started by the actions of a deranged military commander could be a nuclear submarine commander with the support of a small part of the submarine crew launching several nuclear missiles. It wasn't so long ago that the U.S. and the U.S.S.R. had a substantial fleet of nuclear submarines roaming the oceans of the world, each with sufficient nuclear fire power to destroy a large number of metropolitan areas. Such an action while not likely because of the selection process and training of crews must be considered in any risk model of a nuclear war. It is a simple fact that no human is completely predictable. A second type of failure that would have to be taken into account in assessing the likelihood of a nuclear war is a *control failure.* An example of a control failure would be a terrorist group gaining access to a nuclear weapon. Another control failure would be the initiation of a nuclear war by accident such as the inadvertent launching of a nuclear missile. Some experts consider this to be one of the more likely risk scenarios. The point is that the initiation of a nuclear war or attack is dependent on much more than the sanity of a few politicians.

Miscalculation is the basis for another set of scenarios that could end in nuclear war. An example would be to underestimate or overestimate the nuclear strength of an adversary. One of the factors of strength the U.S. had during the Cuban crisis was a much larger nuclear arsenal than the U.S.S.R., an advantage that was soon erased in the coming years. A most important issue relating to "miscalculation" is the system of intelligence available to the decision-makers. Intelligence has to be a major consideration in any risk model as was so dramatically illustrated not with respect to a nuclear confrontation, but with respect to the actions taken by the U.S. against Iraq. And of course a nuclear war could be started by deception. One such scenario would be for a nation to fire nuclear missiles at Russian cities from American territorial waters, triggering the Russians to launch a nuclear attack on the U.S. A scenario of increasing concern as nuclear weapons become smaller, lighter, and more transportable is what might be called the "suitcase scenario." A terrorist could either engage in a suicidal nuclear event or strategically place the suitcase at a location of high vulnerability.

Other factors that must be considered in assessing the likelihood of a nuclear war or attack include (1) the progress being made to control nuclear proliferation and the reduction of nuclear arsenals, (2) the number of nations who possess nuclear weapons, (3) the number of warheads and delivery capability in the possession of each nation, (4) the political stability of the nuclear nations, (5) the priority given to reducing the threat of nuclear war, and (6) the safeguards and security in place to control their nuclear stockpiles.

It is worth noting that the treaties in place between such nuclear powers as the U.S. and the Russian Federation to reduce the nuclear stockpiles may have little or no impact on risk, because of the excess number of weapons that each nation possesses. It is estimated that the United States and Russian arsenals still total more than 30,000 nuclear weapons. On the matter of the number of nations that possess nuclear weapons, it is believed to be nine: Pakistan, India, Israel, China, France, Britain, Russia, North Korea, and the United States. A concern that would have to be factored into any nuclear war risk assessment would be the priority nations are giving to reducing the nuclear threat. In many nations, including the U.S., the threat of nuclear war does not seem to be a high priority issue because the public is not demanding it.

On the matter of political stability of nations having nuclear weapons, there are many issues that would have to be taken into account in structuring the risk scenarios. The feuding of India and Pakistan and the efforts going on in the Middle East to obtain such weapons by Iran and other nations where only Israel currently has a nuclear arsenal are good examples of political strife and unrest. The situation in the Middle East and Russia is a good example of concerns about adequate safeguards and security. There is a need to understand the scenarios and rationale where each of these countries may feel "justified" in using nuclear weapons.

An issue that cannot be overlooked in assessing the risk of a nuclear war is something that might be referred to as the deterrence factor of nuclear weapons. There are those who would argue that the existence of nuclear weapons and the realization of their potential to end the human race is the principal reason that societies have gone through the longest period in the history of man without major powers waging war against each other. As pointed out by Roland[2] "consider how many deaths have been prevented by nuclear weapons. An ever-growing body of evidence suggests that the number of people whose lives have been saved by nuclear weapons reaches into the hundreds of millions." Roland went on to observe that "Had there been conventional war in the second half of the twentieth century on the scale seen in the first half, we could have expected more war deaths than occurred throughout recorded history up to the twentieth century—and many more than piled up in the two world wars combined." Not that it is being advocated that peace should be sought through the fear of nuclear war. It is only to acknowledge that the deterrence factor of nuclear weapons must be part of the risk equation.

One further issue important to structuring the risk scenarios of a nuclear war has to do with the actions taken by the nuclear nations to control and safeguard nuclear materials, including nuclear weapons. The actions referred to have to do with agreements and treaties between existing and future nuclear nations. Obviously, such actions are critical to any assessment of the risk of a nuclear war. For example, what is the impact on the risk of a nuclear war of the Nuclear Non-Proliferation Treaty? The 187 countries that are party to the Nuclear Non-Proliferation Treaty, including the United States and Russia, are striving to make elimination of nuclear weapons a part of their agenda. While little real progress to such a lofty goal has been made, the recognition of the need to move in that direction is encouraging. There is evidence that it has raised the consciousness of the need for more binding international agreements—perhaps the most important avenue for reducing the risk of a nuclear war. One organized effort directed at the United Nations by a group consisting of physicians, lawyers, scientists, and engineers involves a Model Nuclear Weapons Convention that "provides a vision of what complete nuclear disarmament might look like in concrete detail." The likelihood of success of such efforts must be a consideration of any comprehensive nuclear war risk assessment. The fact that binding international agreements are not always complied with and there are rogue states suggests the need to take this into account in any attempt to assess the risk of a nuclear attack. For example, the Nuclear Non-Proliferation Treaty did not prevent North Korea or Iran (or earlier Libya) from seeking nuclear weapons.

7.2 Super Volcanoes

Super volcanoes are described as "volcanoes that occur over 'hot spots' in the Earth and erupt every few hundred thousand years in catastrophic explosions sending hundreds to thousands of cubic kilometers of ash into the atmosphere and wreaking climatic havoc on a global scale."

The term "super volcano" is not defined in terms of a specific threshold. Rather, it has come into use as the descriptor for the world's largest and most destructive volcanoes—volcanoes that produce exceedingly large, catastrophic explosive eruptions and giant calderas. For example, Yellowstone has produced three such very large caldera-forming explosive eruptions in the past 2.1 million years and is considered a super volcano. Other super volcanoes are Long Valley in eastern California, Toba in Indonesia, Taupo in New Zealand, and the large caldera volcanoes of Japan and Alaska. The three caldera-forming Yellowstone eruptions, respectively, were about 2500, 280, and 1000 times larger than the May 18, 1980, eruption of Mt. St. Helens in the state of Washington. Together, the three catastrophic eruptions expelled enough ash and lava to fill the Grand Canyon—emitting much more material than the combined eruptions of Mount St. Helens (1980), Mount Pinatubo (1991), Krakatau (1883), Mount Mazama (7600 years ago), and Tambora (1815). During the three giant caldera-forming eruptions, tiny particles of volcanic debris (volcanic ash) covered much of the western half of North America, likely a third of a meter deep several hundred kilometers from Yellowstone and several centimeters thick farther away. Wind carried sulfur aerosol and the lightest ash particles around the planet and likely caused a notable decrease in temperatures around the globe.

These three super volcano eruptions occurred 2.1 million, 1.3 million, and 640,000 years ago, suggesting a recurrence interval of between 600,000 and 800,000 years. The eruption that occurred 2.1 million years ago removed so much magma from its subsurface storage reservoir that the ground above it collapsed into the magma chamber and left a gigantic depression in the ground—a hole larger than the state of Rhode Island. The resulting crater (known as a caldera) measured as much as 80 km long, 65 km wide, and hundreds of meters deep, extending from outside of Yellowstone National Park into the central area of the park. Since it has been 640,000 years since the last eruption, some may argue that the Yellowstone super volcano is about due for another eruption, plus or minus 100,000 years or so.

If another catastrophic caldera-forming Yellowstone eruption were to occur, it quite likely would alter global weather patterns and have enormous effects on human activity, especially agricultural production, for many years. In fact, the relatively small 1991 eruption of Mt. Pinatubo in the Philippines was shown to have temporarily, yet measurably, changed global temperatures. Scientists, however, at this time do not have the predictive ability to determine specific consequences or durations of possible global impacts from such large eruptions.

While not a catastrophic eruption as described above, the most recent volcanic activity consisted of rhyolitic lava flows that erupted approximately 70,000 years ago. The largest of these flows formed the Pitchstone Plateau in southwestern Yellowstone National Park.

The current geologic activity at Yellowstone has remained relatively constant since earth scientists first started monitoring some 30 years ago. Although another caldera-forming catastrophic eruption is theoretically possible, it is very unlikely to occur in the next several thousand years. Of course, smaller eruptions may occur at shorter intervals.

Assessing the risk of a super volcano such as the Yellowstone super volcano is an excellent application of quantitative risk assessment. While there would be uncertainties as there always is with catastrophic events that rarely happen, there are distinct scenarios that could be postulated with considerable supporting evidence. The results of a comprehensive QRA of the Yellowstone super volcano would be very useful in planning consequence-mitigating activities such as an improved technical basis for the monitoring program, exposure of the most likely eruption scenarios, and emergency planning to minimize the devastation.

7.3 Destruction of the Ozone Layer

In 1974 two scientists at the University of California, Irvine, F. Sherwood Rowland and Mario Molina, published an article in the British science journal *Nature* that started the ozone layer discussion. Rowland and Molina found that chlorofluorocarbons, or CFCs, could eventually destroy the earth's stratospheric ozone layer, leading to dire consequences for most plants and animals. CFCs are manufactured substances used in air conditioning, insulation, and elsewhere. CFCs are not the only ozone-depleting substances, but they are the most important and they were the first to come to our attention.

Ozone shields the Earth from the sun's dangerous ultraviolet (UV) radiation. As the ozone layer thins, the sun's unblocked UV rays lead to deoxyribonucleic acid (DNA) damage. That can cause increased levels of skin cancer and cataracts. Stronger ultraviolet rays can also lead to reduced crop yields and damage to phytoplankton and zooplankton, marine organisms which play a crucial role in the marine food chain.

This is not a small problem. If there were no ozone at all, the amount of ultraviolet radiation reaching the Earth's surface would be catastrophically high. For many living things, including humans, life would be impossible without special protection. This is not a problem that can be turned around quickly, either. The ozone layer will eventually reconstitute itself through natural processes, but that will take decades. The U.S. Environmental Protection Agency

(EPA) says that the ozone layer might return to normal in as little as 50 years—if we are able to halt the release of CFCs and other ozone-depleting substances.

There is no longer much debate about the seriousness of the problem. Rowland and Molina won the 1995 Nobel Prize in chemistry for their pioneering work, and the Montreal Protocol (a major international agreement) limits or bans the production of a long list of ozone-depleting substances. In the developed world, many of these substances have been banned for years now. Progress will be slower in the third world.

Eventually these moves should make a difference, but it is not clear that the corner has been turned. Measurements from the winter of 2005 showed that the ozone layer over the Arctic had thinned to the lowest levels since records began. Cambridge University scientists[3] believe that in late March 2005, Arctic air masses drifted over the United Kingdom and the rest of Europe as far south as northern Italy, allowing for much higher than normal doses of ultraviolet radiation and greatly increased sunburn risk. These scientists also believe that global warming and climate change are slowing the ozone layer's recovery.

A comprehensive quantitative risk assessment of this threat would provide much needed insight to the magnitude of the problem.

7.4 Global Water Management

Half of the people in the world will suffer directly from a worldwide water crisis within the next fifty years, according to a 2003 United Nations report. *The Financial Times of London* began a story a few years back by saying that water, "like energy in the late 1970s, will probably become the most critical natural resource issue facing most parts of the world by the start of the next century."

The world is not running out of water and this is not an existential risk. The risk is from inadequate conservation, distribution and quality of potable water, and pollution of freshwater and marine environments that impact ecosystems, food supplies, and human health. This is a problem that will worsen gradually over a period of years. It will not appear all of a sudden like a tsunami or a meteor strike.

Nevertheless, the problem is real and it is serious. The United Nations report estimates that 400 million people are currently threatened by severe water shortages, and that this number will increase to 4 billion by the year 2050. Major third world water sources (like Asia's Aral Sea and Africa's Lake Chad) are shrinking rapidly from overuse and are becoming polluted. Mexico City is sinking because the city is using up its underground water faster than it can be replaced.

Water is also a political problem. Many nations, including Botswana, Bulgaria, Cambodia, the Congo, Gambia, the Sudan, and Syria, receive most of their freshwater from the river flow of hostile upstream neighbors. Syria and Iraq get much of their water from the Tigris and Euphrates Rivers, which flow down from Turkey. Now Turkey has proposed a series of dams that would reduce river flow. There have been threats back and forth, and it remains to be seen how the problem will be resolved.

The impending "water crisis" is very much like the "energy crisis" that has been with us since the 1970s. Water (like oil) has traditionally been cheap, or free, and it has been readily available. Now the least expensive sources of water are being depleted, or polluted, and world population and water demand is increasing. Water is becoming scarce, and people will have to pay more for less.

The developed world will adjust, as it has adjusted to more expensive oil. The market will encourage conservation, and higher prices will lead to new sources of water. If prices get high enough, desalination will become an option. There are already more than 12,000 desalination plants in 155 nations in the world, 60% of them in the Middle East. (More data can be obtained at the Global Water Intelligence website, globalwaterintel.com.) Political issues (like the vast water subsidies that are now used to promote agriculture in places like California) will be resolved peacefully.

Things will not be so easy for the third world, where there is less water to begin with and where political and market mechanisms are much less sophisticated. Right now, 1 billion people lack access to potable water. The World Health Organization says that half of the people in the world suffer from one of the six main diseases (diarrhea, schistosomiasis, or trachoma, or infestation with ascaris, guinea worm, or hookworm) associated with poor drinking water and inadequate sanitation. About 5 million people die each year frompoor drinking water, poor sanitation, or a dirty home environment.

The expectation is that all of this will become worse in the decades to come.

How serious are these risks? How can these problems best be remedied, or at least mitigated? What approach makes the most sense, and what should the timing be? A quantitative risk assessment of this issue would provide some of the answers.

7.5 Pollution from Fossil Fuels

The most difficult kinds of risk to get action on are those that do not involve sudden and singular events such as a catastrophic accident or event, but rather activities that kill people by the hundreds of thousands every year throughout all human society over sustained periods of time. Pollution from fossil fuels may be the best example of the latter. Fossil fuels (coal, oil, and natural gas) account for 85% of the fuel use in the United States. The burning of fossil fuels is the principal cause of such environmental impacts as global warming, acid rain, and water pollution. Burning of fossil fuels also emits substantial quantities of fine particulates and metals naturally present in the fuels (e.g., mercury, arsenic, selenium) into the atmosphere, adversely impacting public health. There are many undesirable side effects as well, such as the national security costs to protect fossil fuel resources.

Besides a major cause of adverse environmental impacts and consequent health effects, there is the problem of dependency on foreign sources and providing the necessary security of those sources. By 2030, if we do not change our energy policy, we may be relying on Middle East oil for two-thirds of our supply.

U.S. consumption rates of fossil fuels are enormous. The United States uses about 17 million barrels of oil every day. Coal is used to produce almost 60% of our nation's electrical power and accounts for 22% of our overall energy consumption. Natural gas, a third form of fossil fuel, accounts for roughly 23% of the United States' energy usage. The United States is home to 4% of the world's population, yet consumes 26% of the world's energy. Coal is our most abundant fossil fuel. The United States has more coal than the rest of the world has oil. In particular, the demonstrated reserve base of coal in the United States is approximately 500 billion short tons or in energy units 1×10^{19} British Thermal Units.

The full cycle of fossil fuel mining, production, distribution, and use must be considered to understand the full magnitude of impacts and consequences. The health effects of fossil fuels derive from air pollution and mining and drilling accidents including black lung disease. Environmental impacts include global warming, acid rain, and water pollution. Pollutants produced by fossil fuel combustion include carbon monoxide, nitrogen oxides, sulfur oxides, and hydrocarbons. In addition, total suspended particulates contribute to air pollution, and nitrogen oxides and hydrocarbons can combine in the atmosphere to form tropospheric ozone, the major constituent of smog. Carbon dioxide, because it traps heat in the Earth's atmosphere (global warming), may be the most significant gas emitted when fossil fuels are burned. For example, scientists estimate that even relatively small reductions in emissions of carbon dioxide worldwide could prevent 700,000 premature deaths a year by 2020. The Earth appears to have the capacity to absorb carbon dioxide emissions at a level of 3 gigatons per year, although the exact level of tolerance and absorption is uncertain. Today's emissions total about 9 gigatons, about two-thirds of which is due to fossil fuels. The remainder is the result of biomass burning. It is estimated that over the last 150 years, burning fossil fuels has resulted in more than a 25% increase in the amount of carbon dioxide in our atmosphere. Global warming caused by increased carbon dioxide levels in the atmosphere could result in rising sea levels as a result of warming of the oceans and the melting of glaciers. The consequences could be the inundation of wetlands, river deltas, and even populated areas. Changed weather patterns may result in more extreme weather events including an increase in the frequency of droughts. Further, fossil fuel use in the current mode presents risks of climate change that are not yet well understood, but may be catastrophic and irreversible.

Coal contains pyrite, a sulfur compound; as water washes through mines, this compound forms a dilute acid (sometimes concentrated acid causing waterbody pHs less than 3), which is then washed into nearby rivers and streams contributing to water pollution. Coal is not the only fossil fuel that presents a health risk. Production, transportation, and use of oil can cause water pollution. Oil spills, for example, leave waterways and their surrounding shores uninhabitable for some time. Such spills often result in the loss of plant and animal life.

Not only must we improve energy conservation and find alternative fuels to protect our health and the environment, we must do so to avoid the disasters that could result from running out of fossil fuels. In particular, proved reserves at current consumption rates are 210 years for coal, 42 years for oil, and 60 years for gas. In this regard, risk assessment techniques could play an important role in making the right decisions about alternative fuels. Energy systems cannot be changed quickly. It is an issue that has to be considered over long time horizons, 50–100 years, which explains why it may not be a high-priority issue for politicians whose careers are linked to short-term election cycles. Not only would risk assessment quantify just what the risks are of continuing to have a high dependence on fossil fuels, but also risk assessments could be made of alternative fuels and fuel cycles to provide a stronger basis for making the right energy decisions. Such fuel cycles as fossil, renewables, uranium, and hydrogen and combinations thereof should be part of the assessment.

7.6 Infectious Disease Pandemic

Until fairly recently, infectious diseases were by far the biggest killers on Earth, including in the developed world. Polio and smallpox—unknown diseases today—were feared in every household. The great flu pandemic of 1918 killed 40–50 million people worldwide in the years following World War I, according to the World Health Organization.

Many of the most deadly diseases have been controlled, or eliminated, with antibiotics or vaccines. Smallpox and polio have largely been eliminated. But many pathogens have evolved into antibiotic resistant strains, antibiotics don't kill viruses, and it takes time to develop effective vaccines for new viruses. There is still no vaccine for AIDS, and the flu virus changes constantly. Could something like the 1918 global pandemic happen again?

Some scientists say that it absolutely will happen again, and that when it does happen, it will be worse. (Pandemic, by the way, means "global epidemic." Pandemics do not have to be particularly serious, but they must be widespread.)

Bird flu pandemics hit at irregular intervals. There was the great one in 1918, and there were lesser ones in 1957 and 1968. Anthony Fauci, director of the National Institute of Allergy and Infectious Diseases, says that we are "overdue" for the next one. And as a matter of fact, there have been concerns for several years about a bird flu strain known as H5N1, which originated in southeast Asia. People have no immunity to the virus. There is not yet a vaccine. Currently it's a bird disease, though some humans have been infected through birds. The fear is that the virus will mutate and become transmittable between humans. "You can get rid of the 'if' because it's going to occur," said Fauci. It may not occur this year, or next, he said, "but [the threat] is not going to go away." Officials quoted by *U.S. News & World Report* in June 2005 said that the death toll from a pandemic of this virus could easily be as high as 360 million worldwide.

What happens if there's a pandemic? Professor Michael Osterholm, from the Center for Infectious Disease Research and Policy at the University of Minnesota, has said that "the arrival of pandemic flu will trigger a reaction that will change the world overnight... Individual communities will want to bar 'outsiders'. Global, national, and regional economies will come to an abrupt halt." Osterholm believes that such efforts will fail "given the infectiousness of the virus and the volume of illegal crossing that occurs at most borders. But government officials will feel compelled to do something to demonstrate leadership."

Not everyone thinks H5N1 will necessarily lead to a pandemic, let alone one that kills millions of people. Michael Fumento, in the November 21, 2005, edition of *The Weekly Standard*, criticizes Fauci and Osterholm. Fumento says no one has any idea how serious H5N1 will turn out to be. He reminds us that back in 1976, America's top health official predicted one million American deaths from the swine flu. The actual death toll was one. (Fumento believes that the next pandemic, when it occurs, will probably be about as severe as the ones in 1957 and 1968, which together killed about 100,000 Americans and 2–5 million worldwide.)

The flu is not the only infectious disease risk. Germs for diseases that have traditionally been controlled by antibiotics, such as cholera, measles, and tuberculosis, are constantly mutating, and antibiotics have been overprescribed. Supergerms are becoming more common.

So what should we do? Should we spend more money, or less? How should our expenditures be allocated? Should it be spent on vaccines or on other forms of medical preparedness and prevention? Obviously, we need to better understand these risks to make the right decisions.

7.7 Species Destruction: Tropical Rainforests

Where are half of the Earth's plant and animal species and vast amounts of its natural resources? They're in rainforests, which covered 15% of the Earth's land surface as recently as 1950. By 2005 that number was down to 6% and in a quarter century, if trends continue, the rainforests will be mostly gone.

The rainforests are disappearing because human beings are destroying them for the most prosaic of reasons: logging, simple agriculture, and to graze cattle. Rainforests are found almost exclusively in poor countries, and to the resident of a poor country it's much more important to make a little money today than to save a resource for perpetuity.

How important are the rainforests? Half of all the plant and animal species in the world are in them, and if the rainforests are destroyed many of these species will disappear. They won't be saved in zoos or laboratories, either—they will simply be gone. In most cases, we won't know what we've lost, because most of the life in the rainforests remains uncatalogued and unstudied. Scientists have only a rough idea of the number of species on Earth. National Geographic News[4] reports that taxonomists have identified less than 2 million distinct species, mostly mammals and birds. It is estimated that from 10 million to more than 100 million species are still undiscovered, mainly fish, fungi, microbes, and insects.

Rainforests have already been extraordinarily beneficial to human life. More than three-fourths of the developed world's diet originates in the tropical rainforest. A quarter of Western pharmaceuticals are derived from rainforest ingredients. Childhood leukemia, which used to be almost always fatal, is now a curable disease thanks to drugs extracted from the rosy periwinkle. Taxol, a promising new cancer drug, comes from the bark of the Pacific yew tree.

What might the future benefits be? In an article published in *Economic Botany*, Dr. Robert Mendelsohn, an economist at Yale University, and Dr. Michael J. Balick, director of the Institute of Economic Botany at the New York Botanical Gardens, conservatively estimate that at least 300 new drugs still await discovery in the rainforest, with a potential value of as much as $147 billion to society as a whole.[5]

Pharmaceuticals are only part of the picture. There are an almost incalculable number of other products and benefits that could come from the rainforests, and there are tremendous climatic implications to rainforest destruction as well.

Species destruction (in the rainforests and elsewhere) won't be stopped unless something is done immediately. Edward O. Wilson, Pellegrino University Research Professor, Emeritus, at Harvard University and two-time winner of the Pulitzer Prize, estimates that we are losing 137 plant and animal species every single day, or 50,000 species a year. In his book *The Diversity of Life* (Harvard University Press, 1992), Wilson compares today's species destruction to five past "natural blows to the planet," such as meteorite strikes and climactic changes. These changes required 10–100 million years of evolutionary repair. Wilson says that what's happening now—the "sixth great spasm of extinction on earth"—could "break the crucible of life." Wilson says that much of the damage will have already been done by 2020.

What are the real costs of rainforest destruction? What would it take to prevent, or slow down, the deforestation? Would it make sense for wealthy nations, or drug companies, to pay poorer nations to preserve their rainforests?

We need to grapple with these questions in a systematic way.

7.8 Species Destruction: Coral Reefs

Coral reefs occupy less than half of 1% of the ocean's area, but they play a critical, outsize role in the ocean environment. They are being destroyed, mostly by humans, at a stunning rate.

Coral reefs have been called the "rainforests of the sea," and like rainforests they are a laboratory of concentrated species creation and evolution. Coral reefs are home to about one-third of all marine fish, and about 1 million species (fish and non-fish) altogether. Only about 10% of these species have been studied and described,[6] according to University of Maryland zoologist Marjorie L. Reaka-Kudla (U.S. State Department). The EPA says that only rainforests support more biodiversity. Like the rainforests, coral reefs have led to pharmaceutical breakthroughs and to substances that have helped to cure certain types of cancers. The most important benefits of coral reefs are probably still undiscovered.

The world's coral reefs are disappearing rapidly, and this is a relatively new phenomenon. In the Florida Keys alone, monitoring by the EPA shows that the reefs lost more than 38% of their living coral cover from 1996 to 1999. The Global Coral Reef Monitoring Network, the single largest coral reef monitoring effort in the world, reported in October 2000 at the 9th International Coral Reef Symposium in Bali, Indonesia, that of all the reefs they monitor worldwide, 27% have been lost and another 32% could be lost in the next 20–30 years.[7]

Why is this happening? It's happening mostly because of human activity. Sewage (off the coast of Florida, for example) disrupts the ecological balance between the coral polyps that form reefs and the special, single-celled algae that live in the polyp's own tissue. Destructive fishing practices, using cyanide and explosives, can destroy reefs overnight. In the Philippines, blast fishing has destroyed 2000 square miles of reefs. The same type of thing is happening to reefs in Malaysia, Thailand, Papua New Guinea, Indonesia, and elsewhere.

From a long-term perspective, this type of human activity is senseless. Coral reefs are extraordinarily valuable: The U.S. Fish and Wildlife Service reports that reef habitats provide humans with products and services worth about $375 billion each year. The reefs are being destroyed for pocket change, but pocket change is important to the third-world fishermen who support their families through blast fishing.

Would it make sense for richer nations to pay these fishermen to stop fishing? Are there other strategies that might make sense if we were able to consider all the variables? QRA can help to answer these questions.

7.9 Giant Tsunamis

When it comes to modern day natural disasters, the kingpin of them all has to be the Sumatra Tsunami that occurred on December 26, 2004. The numbers tell the story, almost 300,000 killed, thousands still missing, and an estimated 800,000 people made homeless in Aceh and North Sumatra as a result of the disaster. The tsunami was triggered by an earthquake estimated as approximately 9.0 on the Richter scale in the Indian Ocean off the west coast of northern Sumatra. This is the fourth largest earthquake in the world since 1900 and is the largest since the 1964 Prince William Sound, Alaska, earthquake. Alaska is the site of the largest recorded tsunami. It happened in 1958 and was caused by the collapse of a high mountain cliff in Letuya Bay. The height of the wave was just over 1600 feet as evidenced by pushed up soil and trees from much lower elevations. The consequences of this tsunami are pale compared to the Sumatra event, primarily because of the differences in human population.

Would a quantitative assessment of the risk of a giant tsunami in this region have made anything different? It certainly wouldn't have stopped the tsunami from happening, but it could have made the area much more able to cope with such a disaster and possibly saved tens of thousands of lives, if not more.

Giant tsunamis are sea waves that have the capability to do catastrophic damage to populated shorelines and the environment. Tsunamis that involve heights of 50 feet and greater are considered "giant" in this discussion. Tsunamis seldom exceed 50 feet, but the target of this book is rare and catastrophic events. For a long time scientists believed earthquakes could only cause tsunamis. We now know that giant tsunamis can be caused by many different mechanisms: undersea earthquakes, collapsing undersea volcanoes, asteroid collisions with the oceans, collapsing mountains near or under the sea, and coastal and undersea landslides. For example, asteroids have been identified that if they hit the Atlantic Ocean could create waves as high as 400 feet onto the Atlantic Coast. Given the size and collision point of the asteroid, it is possible to calculate the height of the resulting tsunami if the collision point turns out to be an ocean. Determining the size and collision point is a matter of finding the asteroid in space and calculating its trajectory, which generally can be done. Programs exist to monitor space for the purpose of tracking asteroids that might collide with the Earth. It has been estimated that by 2030 approximately 90% of the impacts that could trigger a global catastrophic event will be quantified.

There is evidence in some locations, including the west coast of the United States, that tsunamis as high as 1000 to over 2000 feet have hit land in prehistoric times. The mechanisms most likely responsible for such tsunamis in the 1000–2000-foot range are collapsing volcanoes and landslides. For example, there is a dangerous situation in Hawaii where a large section of an island is cracking apart. Should it ever let go, the entire west coast of the United States, Canada, and Mexico could be affected. As to other examples, the following excerpt is taken from the Internet: "Dr Simon Day, of the Benfield Greig Hazard Research Centre at University College London, United Kingdom, believes one flank of the Cumbre Vieja volcano on the island of La Palma, in the Canaries archipelago, is unstable and could plunge into the ocean. Swiss researchers who have modeled the landslide say half a trillion tonnes of rock falling into the water all at once would create a wave 650 m high (2130 feet) that would spread out and travel across the Atlantic at high speed. The wall of water would weaken as it crossed the ocean, but would still be 40–50 m (130–160 feet) high by the time it hit land. The surge would create havoc in North America as much as 20 km (12 miles) inland."

An interesting question is, what are the prospects of learning more about how to manage and cope with giant tsunamis as a result of a more systematic assessment of the risks involved? As already discussed, a risk assessment is basically a structured set of scenarios, their likelihoods, and consequences. Usually, the most challenging part of a risk assessment is the calculation of likelihoods. So, what can be observed about the likelihood of giant tsunamis? As it turns out, there is a considerable amount of evidence on the frequency of tsunamis, including those of the giant variety. Like asteroids, tsunamis are real, but fortunately remote. But not so remote that they shouldn't be better understood, including having better knowledge about their risks. The evidence is that the giant variety of a tsunami may occur every few hundred thousand years. On the other hand, the variety that creates waves in the 50–200-foot range occurs more frequently as they can be triggered by less than super earthquakes and can result in catastrophic consequences. For example, an earthquake that measured only 6.2 on the Richter scale triggered the Papua, New Guinea, tsunami noted above that killed over 2000 people. A 6.2 earthquake is not a particularly infrequent earthquake. And, of course, the recent Sumatra tsunami is dramatic evidence that the wave does not have to be in the hundreds of meters to result in catastrophic damage. It is mostly a matter of emergency preparedness and that's where quantifying the risk of such events can provide major benefits.

There are some distinct differences between asteroids and giant tsunamis that enter into assessing their risks. The one common area is where asteroids are the cause of the tsunami. Asteroids having the potential to cause catastrophic damage are easier to detect in advance than most giant tsunamis. This is because many of the causes of tsunamis are out of sight so to speak. The risk scenarios of giant tsunamis can be categorized by the initiating event. The initiating event categories could be (1) earthquakes, (2) volcanoes, (3) landslides, (4) collapsing mountains, etc. In fact, this is exactly the way a quantitative risk assessment is structured. Such a display of the risks could serve management of the risks of giant tsunamis in many ways. Priorities would become much clearer and risk management decision-making would be solidly based.

7.10 Genetic Engineering and Synthetic Biology

Genetic engineering, sometimes referred to as genetic modification or genetic manipulation, has to do with the artificial transfer of genetic material, or DNA, between unrelated species such as plants, animals, bacteria, viruses, and humans. The modern version of the field of genetic engineering started in the late 1960s with the promise of a revolution in how we obtain food through sustainable farming, alter environments, and perfect the human species. Transformation of plants to produce medicines and cosmetics and the transfer of human genes into animals to provide hearts and other critical organs for seriously ill people are other possibilities of this new science. The supporters of genetic engineering are convinced that it can provide medical products that are difficult if not impossible to make in other ways. The transfer of genes between microbes, plants, and animals provides opportunities for altering life forms and even creating new ones. Scientists have cloned animals, and human cloning may not be far behind. Gene therapy research is exploring ways to treat cystic fibrosis, fragile-X syndrome, and other devastating genetic diseases. The use of genetic engineering and gene sequencing in research has already produced important advances in understanding the nature of genes. For example, we now understand a great deal more about the complexity, fluidity, and adaptability of genomes.

Synthetic biology has been described as the process of constructing systems on the molecular scale. Others have described synthetic biology as "the blanket term for a multidisciplinary attempt to identify a class of standard operational components that can be assembled into functioning molecular machines." Synthetic biology was established to design and construct novel organisms to help us solve problems that our natural systems cannot solve. The hope is to create new living systems making possible cancer-destroying organisms, microbial factories, detectors of biological contaminants, and organism health monitors. The promise of synthetic biology is to provide biological solutions to some of society's most pressing and difficult problems. If successful, it should greatly increase our understanding of the "living" world.

Synthetic biology differs from genetic engineering and biotechnology, as the goal is not about changing or tampering with biology, but about remaking it. The field embodies scientists who want to understand biology and engineers who want to rebuild it.

Whether it's synthetic biology or genetic engineering, both have their detractors. There are those who believe that too little is known about the dangers of genetic engineering in humans, about the ecological dangers of genetically engineered agricultural plants, and the possible applications of genetic engineering in industry, agriculture, and the military. For example, there is evidence that the former Soviet Union pursued a major biological weapons program in the 1970s and 1980s that involved the genetic engineering of novel pathogens. Also, the United States has proposed genetic engineering research on novel biological warfare agents. The ecosystems of New Zealand have been ravaged irretrievably by such horizontal transfer of genes in the form of introduced species. Runaway horizontal transfer alone has the potential to reduce biodiversity to a few rampant weedy species. The implications for runaway horizontal transfer of genes carry similar implications in terms of new disease vectors and the destruction of many non-target species by engineered factors designed to provide resistance to major pests. There have already been movements calling for a ban on the construction of genetically altered organisms for any military purpose such as biological warfare.

Embedded disarming mechanisms as part of genetic engineering are often referred to as a way of providing protection against many of these dangers. But many are skeptical about the ability, for example, to create a safe bacterial host that would solve the safety problem. But, what is the evidence of a real health and safety problem? So far, no serious events have resulted from genetic engineering activities. On the contrary, many benefits have come from genetic engineering activities to date. Applications of genetic engineering include healthcare, agriculture, and bioremediation. A well-known agriculture application has been the development of a genetically engineered tomato produced by the biotechnology company Calgene. This product, known as the Flavr Savr tomato, claims an improved shelf life and flavor and is currently sold in food markets in California and Illinois under the brand name "MacGregor's." This genetically engineered tomato was the first such crop plant approved for general, unregulated release into the environment.

Genetic engineering has advanced to the point where scientists can genetically engineer not only microorganisms, but also plants and animals. A major goal of the scientists is to not only engineer crop plants with improved growth characteristics, shelf life, and taste, but increase their resistance to infestation and the need for pesticides—a major advancement in reducing environmental impacts. Genetically engineered species can provide drugs, or even human monoclonal antibodies, which could help treat deficiency diseases or cancer.

There is a risk that such substantially transformed types could accidentally escape and reproduce in the wild, particularly if they are viable and disseminate, e.g., by spores, or genes, or are carried by viruses. Adequate measures have to be established to protect the natural biosphere from genetic pollution by genetically modified organisms. The question is, what safeguards can be provided against a runaway genetic event that could have disastrous human or environmental consequences. For sure, one thing that could be done is a quantification of the risks involved for each application being investigated. An assessment of the risks in a comprehensive way would not only assure the public of the safety risk of specific genetic engineering activities, but also serve to expose the uncertainties and contributors to the risk, thus greatly enhancing the research itself. But the real reason for performing quantitative risk assessments of each new application of genetic engineering is to make better decisions on what actions should be taken in the field of genetic engineering that are in the public interest.

The synthetic biology movement also has its critics. Industry pioneers such as Sun Microsystems co-founder Bill Joy have urged caution as engineers pursue self-replicating machines through biotechnology and nanotechnology. Joy and others worry that inventors could eventually lose control of these processes. For example, there is always the risk that a particular synthetic biology project could lead to the accidental or deliberate creation of pathogenic biological components.

It is possible for engineers to represent the structure of molecular systems by standard engineering logic models, in which case it should be possible to build comprehensive risk models as they employ similar logic diagrams in the structuring of risk scenarios. The idea would be to develop scenarios of the "what can go wrong" variety to assure that particular molecular systems could not mistakenly or easily become pathogenic systems.

7.11 Global Warming

Global warming has to do with whether the emission of certain gases, called greenhouse gases, into the atmosphere could alter the Earth's climate to the point of having disastrous consequences over time. The main greenhouse gases are carbon dioxide (primarily produced by fossil fuel combustion), methane (produced by biological decay, animal waste, biomass burning), chlorofluorocarbons (produced by industrial processes), and nitrous oxide (produced by fertilizer use and the burning of fossil fuels). Ozone is another greenhouse gas, but is not considered a major player in the warming phenomena.

The concern of many scientists is that greenhouse gases upset the sun's radiation cycle with the Earth, resulting in a warming trend of the Earth. Their fear is that the warming could lead to widespread extinction of plant and animal species, cause sea levels to rise, adversely impact agriculture, and increase the severity and frequency of hurricanes, cyclones, and typhoons. The fear of global warming resulted in something called the Kyoto accord whose aim is to curb the air pollution blamed for global warming. The accord is named after the ancient Japanese capital of Kyoto, where the pact was negotiated in 1997. Some 141 countries have ratified the treaty, but the United States is not one of them. The United States' position is that the changes required to comply with the treaty would be too costly and that the agreement is flawed. There continues to be diverse opinions on the causes and state of global warming.

How can the evidence for global warming be assessed to get to the truth? This would seem to be an ideal problem for applying the principles of quantitative risk assessment. It has all the elements of what gave birth to quantitative risk assessment including uncertainty, rare events, limited data, different possible scenarios, and controversy.

7.12 Super Earthquakes

What is a super earthquake? The most common measure of earthquake strength is the Richter scale, which measures the "moment magnitude" and describes horizontal movement. It is a logarithmic scale. A magnitude 6 earthquake has ten times more energy intensity or movement than a magnitude 5 on the Richter scale. The Richter scale has its limitations, as it does not reflect the impact of vertical movement, which can be the wave movement causing the greatest amount of damage. However, for most earthquakes the Richter scale has provided reasonably well correlation with the resulting damage. Other measures of earthquakes could be number of fatalities or injuries, the peak value of the shaking intensity, or the area of intense shaking. As to a formal definition of a super earthquake, there does not appear to be one. So we will loosely take a super earthquake to mean one that has a Richter value of 8 or more or results in thousands of fatalities or billions of dollars of property loss and damage. This is in keeping with the theme of this book to target high consequence, low probability events.[a]

History would suggest that earthquakes have been one of the more serious risks to societies. The U.S. Geological Survey has cataloged the ten most deadly earthquakes since the 9th century. The number of fatalities is staggering and varies from 830,000 in Shansi, China, in 1556 to 100,000 in Messina, Italy, in 1908. The most recent on this top ten list occurred in Tangshan, China, in 1976 and killed 242,000 people. The total number of fatalities for the ten is a staggering 4,810,000. Of course, this is just the top ten over that period, not the total earthquake fatalities, which has to be a much larger number. Four of the top ten were in China, two in Iran, and the rest were in Syria, Japan, Turkmenistan, and Italy. None were in the Americas. While the magnitudes of many of these earthquakes were not known from

[a] It should be noted that the greatest risk from earthquakes is probably not a super earthquake, but the more frequently occurring earthquakes of less magnitude. This characteristic was demonstrated in the previous case studies on hurricanes and asteroids.

direct measurements, as there was no such capability to do so for most of them, earthquake experts have been able to reconstruct estimates by analyzing the descriptions of the damage incurred.

Earthquake damage (fatalities and property damage), like hurricane damage, is strongly dependent on where it occurs, the resistance of buildings and services to damage, and emergency preparedness. For example, the earthquake that took place in Bam, Iran, on December 26, 2003, having a magnitude of 6.6, resulted in killing over 26,000 people, injuring 20,000, leaving 60,000 homeless, and destroying most of the city. Meanwhile, a similar magnitude earthquake, which took place in central California about the same time, did not cause any dramatic damage or loss of life. It has been estimated that half of the 6 million people in the capital cities of the five central Asian republics occupy buildings that are extremely vulnerable to collapse during earthquakes with death tolls up to 135,000 people and at least 500,000 injuries. The difference appears to be in the vulnerability of structures and infrastructure; in California they were ready for such earthquakes and in Bam, Iran, and apparently many other places, they were not. The good news is there is strong evidence that it is possible to greatly reduce the risk of super earthquakes with better information on where they are likely to occur, better building codes, improvements in design and construction of housing and facilities, emergency preparedness, and greater government involvement.

Perhaps the greatest risk of earthquakes, given the increased ability to lessen their impact, is not the earthquake itself, but its ability to trigger other catastrophic events such as volcano eruptions, landslides, and tsunamis. A case in point is the 9.0 earthquake in the Indian Ocean near Sumatra on December 26, 2004, that created a tsunami resulting in almost 300,000 people dead and thousands missing. This same earthquake is believed to have resulted in a flurry of events in the Mount Wrangell volcano in Alaska 7000 miles away. The 7.9 Denali Fault earthquake in 2002 triggered similar volcanic activity at Yellowstone and northern Mexico. Fortunately, neither of the flurry of events resulted in any serious damage to property or life.

Methods for quantifying the risk of earthquakes have greatly advanced primarily because of the need to quantify their occurrence at nuclear facility sites. There is great opportunity for the risk sciences to reduce the risk of earthquakes as triggering events for tsunamis and volcano eruptions.

7.13 Industrial Accidents

The desire to prevent catastrophic industrial accidents has been the principal driver for the development of quantitative approaches to risk assessment. Large dams, water supply systems, large bridges, chemical and petroleum process facilities, and nuclear power plants are all examples of industrial activities seeking better methods of safety analysis. Dam failures are included in this discussion as they are engineered structures, but mining accidents are not as they are a special class of disasters requiring special treatment.

Even though the public has a great fear of industrial accidents, such accidents lag considerably behind other threats to human life.[8] For example, famines have killed about 75 million people in the last century. This is ten times more people than competing sources like earthquakes, severe storms, and floods combined. Severe storms and storm-induced flooding have killed hundreds of thousands of people at a time in the most extreme cases, but probably a million altogether in the last century. Earthquakes have produced deaths also running above 100,000 in the worst case, with perhaps under a million in total for the 20th century.

While industrial disasters don't rival those from famine and acts of nature, they can be catastrophic. If disaster is defined in terms of loss of life, among the worst in recent times was the Bhopal, India, chemical (pesticides) plant accident on December 3, 1984, that released a massive amount of methyl isocyanate killing 2800 people, injuring over 50,000, requiring the evacuation of some 200,000, and rendering extensive damage to livestock and crops. The Vaiont Dam disaster in northern Italy that occurred on October 9, 1963, resulted in 2500 deaths. Accidents of this magnitude are rare. Accidents that have resulted in as many as 500 deaths are not so rare worldwide as several have occurred since about 1975. An examination of the Wikipedia Disaster Data Base reveals that since 1975, not counting the two disasters just noted, there have been four disasters resulting in 400–600 deaths, seven resulting in 200–400 deaths, and some thirteen resulting in 100–200 deaths. The types of facilities involved are mostly chemical and petroleum storage and process plants, but other types of industrial facilities include factories and plants manufacturing explosives. The locations vary all over the world, but most of the big accidents seem to occur in such countries as India, China, Russia, and developing nations where safety regulations and standards have either not yet been developed or are less restrictive than, for example, in such nations as the United Kingdom, France, Germany, and the United States.

Of course, killing a large number of people is not the only criteria for what constitutes a disaster. Other measures of a disaster are environmental impact, dollar loss, and fear, including the resulting consequences of fear, be it real or perceived. An example of an environmental impact disaster is the *Valdez* oil spill in Prince William Sound. The oil spillage from the *Exxon Valdez* tanker in 1989 contaminated 2100 km of beaches in Alaska and caused extensive harm to wildlife. There were no human casualties, but the cost of cleanup has been estimated at 3 billion dollars. Perhaps the industry disaster of recent times that best combines the three factors of fatalities, environmental impact, and cost was the explosion of the Piper Alpha offshore oil production platform located in the North Sea 110 miles northeast of Aberdeen, Scotland. The explosion in the gas processor of the platform, which was initiated by a fire, literally destroyed the entire platform resulting in 167 deaths, considerable oil spillage, and a staggering cost estimated at 4 billion dollars. The 167 deaths was the highest death toll in the history of offshore operations, and the estimated cost of the accident of 4 billion dollars is considered the most costly of any industrial disaster.

The best example of the "fear" of an industrial disaster is a nuclear power plant accident. Contrary to the fact that nuclear power has one of the most impressive safety records of any major industry of similar benefit to society, there remains the perception of it being a high-risk industry. The loss of life associated with the entire history of nuclear power is minuscule when compared to other industries of comparable benefit to society. The root cause of this fear is radiation and nuclear power's unfortunate association with nuclear weapons. While the evidence does not support the level of fear many have of nuclear power, there have been accidents. Two accidents dominate nuclear power plant operating experience, Three Mile Island and Chernobyl. The Three Mile Island accident, which occurred on March 28, 1979, did not result in any known fatalities, but did destroy the plant itself.

The Chernobyl accident, which occurred on April 26, 1986, was a different matter. It also resulted in the destruction of the plant, but in addition involved fatalities and environmental damage. There were 30 fatalities as a result of acute doses of radiation and some 300 others required hospital treatment for radiation and burn injuries. There were no known acute injuries or fatalities off-site, but the accident did have off-site consequences. For example, 45,000 residents of Pripyat were evacuated a day after the accident. There was ground contamination as well, but no injuries or fatalities have been attributed to the contamination.

Following the lead of the nuclear power industry, most industries, including space and defense, have adopted more sophisticated methods of safety analysis to enhance decision-making on the design, construction, and operation of industrial facilities. While few have gone so far as to require the full scope of probabilistic risk assessments, the methods are clearly moving in the direction of more quantitative approaches to assessing the safety of proposed and operating systems. Much progress is expected in the years to come.

7.14 Nanotechnology

Nanotechnology is a branch of engineering that deals with the design and manufacture of extremely small electronic circuits and mechanical devices built at the molecular level of matter. The benefits of nanotechnology should be extraordinary. As Andrew Chen writes, "Imagine a world in which cars can be assembled molecule-by-molecule, garbage can be disassembled and turned into beef steaks, and people can be operated on and healed by cell-sized robots." Nanotechnology could lead to spectacular advances in computing, medicine, and manufacturing.

The risks of nanotechnology are equally awe-inspiring, and they are potentially existential. The best known risk is the "gray goo" problem. "Gray goo" refers to a variety of nanotech-related disaster scenarios, all of which involve the rapid and destructive replication of tiny nanotech molecules. In the worst case, these molecules could turn the oceans, or plant life, or even humans, into "goo." As Bill Joy says: "Gray goo would surely be a depressing ending to our human adventure on Earth, far worse than mere fire or ice, and one that could stem from a simple laboratory accident."

The military (or terrorist) applications of molecular engineering are something to be concerned about as well. Nanomachines could be designed to disassemble buildings, or other structures, or even organic matter (like human beings).

Then there are the "Big Brother" implications. Nanotechnology could be used by the government, or by our enemies, or by misguided individuals, to spy on us and monitor all of our activities with molecular-sized cameras.

Nanotech enthusiasts, like Ray Kurzweil, author of *The Age of Spiritual Machines*, scoff at these risks and focus on the stunning benefits that will almost certainly come from nanotech research. Robotics expert Hans Moravec, of the Robotics Institute at Carnegie Mellon University, says that humans should keep researching and just get out of the way when robots begin to supplant humans as Earth's superior species. Bill Joy takes the other view. He advocates strict controls on research and the abandonment of research into the most dangerous areas.

Should nanotech research be subsidized? Should it be restricted? Should much of it be classified? QRA can't help with all of these decisions, because there's a metaphysical component to this issue. But it can sort out some of the more predictable risks, like the "gray goo" worry. Some tough decisions will be necessary, and if we're smart we'll start making them now.

7.15 Population Management

Any population of organisms becomes excessive when it exceeds the carrying capacity of its environment. For example, if there are a million humans in a country, but only food for 900,000, there is overpopulation.

Is runaway population growth a real threat? Two hundred years ago Thomas Malthus argued that human populations, if left unrestricted, would grow geometrically until they overwhelmed the food supply. Malthus argued that famines and wars would be required to keep population growth in check. He argued for managed population control.

As a matter of fact, scores of millions of humans have died as a result of famines and wars since Malthus wrote those words. It's also true, though, that today there is enough food in the world to feed every living person.

This is a complicated issue. Over the past few decades Paul Ehrlich has been the leading proponent of the notion that the world is seriously overpopulated. Ehrlich believes that the world's developed nations are the most overpopulated, because (he says) they use more resources than they are able to replace. The late Julian Simon sparred with Ehrlich and argued the opposite: that human beings are actually a great resource and population growth is to be desired, not avoided.

The United Nations recognizes that population growth is slowing, but it still estimates that there will be about 9 billion people on Earth in the year 2300, up from 6.3 billion today. This population will be older than the current one, as people will be living longer.

QRA can help to assess whether population growth is a great risk or even a problem. The largest country on earth, China, has for years restricted its population growth by law. But birth rates in the most developed nations are now falling. As their citizens get richer, they tend to have fewer children. These countries would be decreasing in size but for immigration. Is this a good thing, or should these wealthy nations be taking steps to increase the rate of procreation?

The 16 risks described above are just a few of the possible catastrophes facing humanity. A comprehensive list of catastrophes that are a threat to society needs to be developed and prioritized, both in terms of potential global impacts, and by region using the principles of quantitative risk assessment. The purpose of these short descriptions is simply to add clarity to what is being described in this book as rare, but catastrophic events, and to raise our consciousness of such events to facilitate taking actions to either mitigate them or do the best that we can do to better manage their consequences. The idea is not to preach doom and gloom, but to understand the threats that can bring great misery and even extinction to life as we know it, so that we can act in time to avoid them or greatly reduce their consequences.

References

[1] Garrick, B. J. Technological stigmatism, risk perception, and truth, *Reliabil. Eng. Syst. Saf.* 1998, *59*, 41–45.
[2] Roland, A., Keep the Bomb, *Technology Review*, August/September, 1995.
[3] Brown, P., Ozone layer most fragile on record, *The Guardian*, April 27, 2005.
[4] National Geographic News, 2002. http://www.news.nationalgeographic.com/news/2002/03/0305_0305_allspecies.html.
[5] Mendelsohn, R., Balick, M. J. The value of undiscovered pharmaceuticals in tropical forests, *Econ. Bot.* 1995, *49*(2), 223–228.
[6] U.S. State Department. Coral Reefs, Fertile Gardens of the Sea, 1997. http://usinfo.state.gov/products/pubs/biodiv/coral.htm.
[7] NASA, Corals in crises, 2000. http://www.earthobservatory.nasa.gov/Study/Coral/coral2.html.
[8] Lewis, H. W., *Technological Risk*, W.W. Norton & Company, New York, 2000.

CHAPTER 8

The Rational Management of Catastrophic Risks

What we have tried to do in this book is call attention to a class of risks that could threaten the health and safety of significant populations of humans—a class of risks that generally has not had the benefit of consistent, quantitative, and comprehensive analyses for a realistic assessment and prioritization of their threat to society. We have emphasized how essential it is to have a systematic and integrated process for assessing the likelihood and consequences of such risks to maximize the opportunity for their mitigation. We have presented a methodology designed to assure that the risk assessments are based on real and visible evidence, including the lack of evidence, and explicitly account for the uncertainties involved. Four case studies were presented in Chapters 3–6 to demonstrate the methodology. Two of the case studies (Chapters 3 and 4), while limited in scope, were reasonably complete in terms of a location-specific application that addresses the issue of fatality risks to humans. The scope limitations resulted in greater uncertainties in the risk measures than would be the case for a more comprehensive effort. In spite of the considerable uncertainties involved, the results demonstrate the power of the methodology in providing insights on what the risks are and how to manage them. The other two case studies (Chapters 5 and 6) are "scoping analyses" to guide location-specific quantitative risk assessments which are yet to be completed.

The four case studies were (1) the risk of a major hurricane in New Orleans, LA, (2) an asteroid colliding with the contiguous 48 states of the United States and an asteroid impacting New Orleans, LA, (3) a hypothetical risk assessment to scope a future quantitative risk assessment of a terrorist attack of a regional electric grid serving major U.S. metropolitan areas, and (4) a scoping study to assess the risk to the north Atlantic U.S. coastal states of an abrupt climate change due to a shutdown of the thermohaline circulation system.

8.1 Benefits of Quantitative Risk Assessment

The principal benefit of a quantitative risk assessment is that the information such analyses generate can greatly enhance the making of decisions that can save lives and protect our environment. In some cases it may provide a basis for taking action to completely mitigate an existing threat. Among the results of a quantitative risk assessment are the likelihood and consequence of specific threats, including the uncertainties involved, and the identification of the contributors to the risk and their relative importance. The output of the risk assessment can provide a basis for risk management actions to mitigate the risk or at least reduce the consequences.

For industries that develop quantitative risk models as a fundamental part of their operations strategy, the benefits go beyond just calculating the risk. For example, in the commercial nuclear electric power industry, the risk assessment models are kept current and used as a basis for scheduling and prioritizing maintenance activities, the planning and management of outages, decision-making on plant modifications, the monitoring of risk as a function of plant status, and the establishment of accident recovery and emergency response capabilities. See Appendix A for an example.

8.2 The Role of the Case Studies

The principal purpose of the case studies was to illustrate how to apply the methods of quantitative risk assessment as presented in Chapter 2. Having the case studies provides a more visible basis for demonstrating how to interpret risk assessment results. The actual numerical results of the case studies, while providing some insights as discussed in the following sections, are of secondary importance to demonstrating the process of risk quantification.

An important fact about the methodology presented in Chapter 2 and employed in the case studies is that it has been tested. The methodology has been employed for almost four decades in complex systems, both engineered and natural, and has demonstrated its effectiveness in assessing the risk of rare and catastrophic events as well as its central role in risk management (see Appendix A).

8.3 Comparing Quantitative Risks Using the Case Studies

This section will attempt to cast the results of the case studies in a form that manifests the value of quantitative risk assessments. In particular, we now ask the question, based on the case studies alone, what can be said about the importance of the risks to the affected regions that couldn't be said before they were analyzed?

To make location-specific quantitative risk comparisons have meaning, it is necessary to be consistent in terms of populations at risk, methods of analyses, and in the representation of the supporting evidence. To facilitate making a comparison for a particular region, we chose the metropolitan New Orleans area as our population and region at risk, and we used the methodology of Chapter 2 to process the supporting evidence. The risks we chose to compare are hurricanes, asteroids, and a nuclear power plant accident. The hurricane and asteroid risks are based on the completed case studies of Chapters 3 and 4. A representative nuclear power plant (see Appendix A) is used as a surrogate for the risk of a nuclear power plant accident near New Orleans. The surrogate selected is based on the results of a comprehensive risk assessment of a typical U.S. nuclear power plant within 50 miles of a population center involving several million people. The risk results of the surrogate are believed to be typical of plants licensed to operate in the United States.[1]

Figure 8.1 compares the fatality risk of a hurricane landfalling in New Orleans, an asteroid impacting the New Orleans area, and a nuclear power plant accident within 50 miles of the city.[a] Figure 8.1 measures the fatality risk in terms of the frequency per year of the event. Table 8.1 contains the mean value fatality risk in terms of the recurrence interval in years. It is important to recognize that for Figure 8.1, the risk is higher with increasing frequency. In particular, the risk of fatalities from hurricanes in New Orleans is much higher than the risk of fatalities from asteroids, which is much higher than the risk of fatalities from the nuclear power plant when considering catastrophic events resulting in 10,000 early fatalities or greater.[b]

The 90% confidence intervals are shown for the risk of 10,000 fatalities or more to remind the reader of the uncertainties in the

[a] It should be pointed out that there is an operating nuclear power plant within 50 miles of New Orleans.

[b] These are fatalities that are expected to occur essentially during the time of the event and immediately following. No attempt was made in the comparison to evaluate the long-term consequences of the events such as the onset of health effects that might occur much later in time as a result of residual effects, for example, the degradation of the environment.

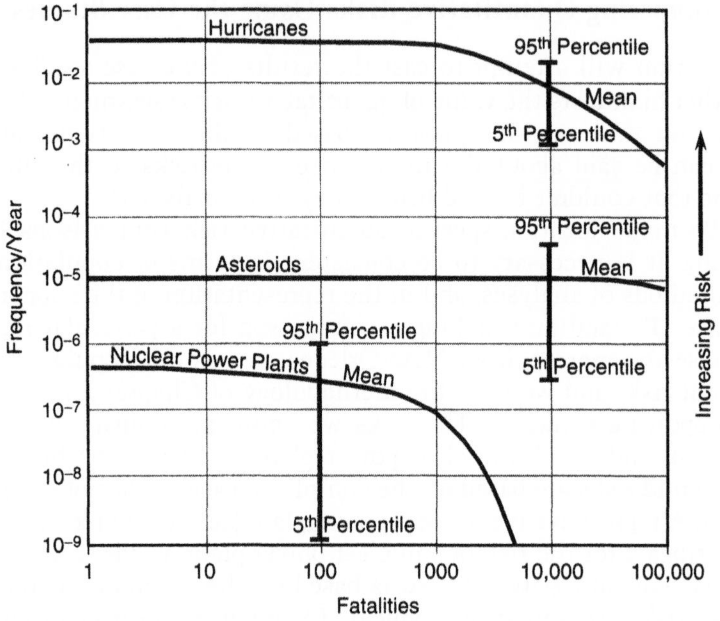

FIGURE 8.1. Comparison of the risk to the greater New Orleans area of hurricanes, asteroids, and nuclear power plants.

hurricane and asteroid analyses. The 90% confidence interval for the nuclear power plant risk is shown for 100 fatalities or more as the risk is essentially zero for 10,000 fatalities. For a review of the uncertainties over the entire range of consequences, the reader is referred to Figures 3.2 and 4.10.

Table 8.1 presents the same key results from Figure 8.1, but this time the results are presented in terms of the recurrence interval in years, which are the inverse of the frequencies presented in Figure 8.1.

The above results certainly suggest that the first priority for New Orleans ought to be to protect the city against hurricanes. The

TABLE 8.1
Mean recurrence interval of catastrophic events for the metropolitan New Orleans area (years)

Threat	*1000 Fatalities or Greater*	*10,000 Fatalities or Greater*	*100,000 Fatalities or Greater*
Hurricane	30	130	2200
Asteroid	110,000	110,000	140,000
Nuclear Power Plant	7,700,000	>1 Billion	>1 Billion

analysis led to this conclusion even without considering hurricane Katrina. The results in Figure 8.1 and Table 8.1 may be surprising to many readers who depend on the news media for information on risk.

The case studies revealed that the asteroid risk is the only risk of the ones analyzed that has the potential for catastrophic global consequences. The analysis in the asteroid case study indicates that we are 90% confident that an asteroid event having global and species extinction consequences has a frequency range of one every 20 million years to one every billion years with a mean value of approximately one such event every 50 million years. There is no such consequence of a threat to any region from hurricanes, power outages, abrupt climate changes, or nuclear power plant accidents. While the risk of such a global event is extremely rare, the extinction of the human species is a very serious consequence. The fact that an asteroid event could result in such profound consequences is something that needs attention, quantification, and investigation. It should be recognized that the asteroid risk of regional catastrophic consequences is greater than the global risk for the regions analyzed as discussed in Chapter 4 and protective measures against these more frequent events may be important.

For the case of the scoping study for a terrorist attack on a region of the national electrical power grid, some insights into what to look for were developed. Of course, the scoping study was only carried out to the risk of power outages and not to the desired result of the risk of human fatalities. Clearly, an outage in a metropolitan region of greater than 24 h could begin to have consequences involving human suffering and possibly significant fatalities. The scoping study indicated that the mean frequency of an outage of 24 h or greater is one in 250 years. This is a low risk with respect to the historical record of electric power in most areas of the world, but is nevertheless a more frequent event than the thermohaline circulation flow reduction event that could trigger an abrupt climate change. Much more work is necessary to pin down the economic and public safety consequences of such an electrical power outage and complete the case study, but it is clear that in an urban area heavily dependent on electric power for all basic services, the consequences could be severe.

What about the abrupt climate change event? Where does that stand in the grand scheme of risks to society? Of course, there is much more work to be done before that question can really be answered, but the scoping study resulted in some important insights. The key issue here is the likelihood and consequences of a complete shutdown of the thermohaline circulation in the Atlantic Ocean. Major assumptions were made in the assessment of Chapter 6 with perhaps the most critical one being warming rates brought about by greenhouse gases. Changes in the warming rates will have a major effect on reductions in flow of the ocean currents. There are major uncertainties associated with the relationship of flow reduction and the onset of an abrupt climate change. The scoping analysis indicated that there is greater than a 5% chance (a one in twenty chance) that the circulation will shut down in the next 300–500 years. The analysis further indicated that 300 years or more into the future, we are 90% confident that the reduction in the thermohaline circulation flow will be between approximately 17% and 100%. The mean value is approximately 48%. There is some evidence that a complete shutdown of the thermohaline circulation is not necessary to result in an abrupt climate change. In particular, some scientists speculate that as little as a 25% reduction in flow might trigger an abrupt climate change. The real conclusion here is that even with only having done a scoping study there is strong evidence that abrupt climate change due to global warming should be a high-priority candidate for a quantitative risk assessment.

8.4 Observations from the Case Studies

Again, the primary purpose of the case studies was to illustrate the methods of quantitative risk assessment. Of particular interest was to expose those issues of risk assessment that are essential for addressing rare and catastrophic events in a manner that can support effective risk management. Knowing the contributors to risk and the respective importance of the contributors and how the contributors are affected by new information are beacons for effective risk management. In spite of the effort to put the emphasis on the methodology and not the numbers of the case studies, it is still interesting to examine the results.

First, it is possible to quantify the risk of specific regions using the methodology described in Chapter 2. Embracing uncertainty in the risk measure allows for greatly increased understanding and action where paralysis and speculation is otherwise the only option.

Second, as seen in the case studies, risk is a dynamic phenomenon. Not only is it a matter of assessing the risk, it is important to know how the risk is changing with time and circumstances. For example, Figure 8.2 (Figure 3.7 from Chapter 3) illustrates how the hurricane risk in New Orleans changes when considering two different time intervals of hurricane observations and changes in population, emergency procedures, and building codes.

Figure 8.2 (Figure 3.7, Chapter 3) illustrates the difference in the hurricane risk based on a half-century of data (from 1900 to 1950) or over a full century of data (from 1900 to 2004). The risk of

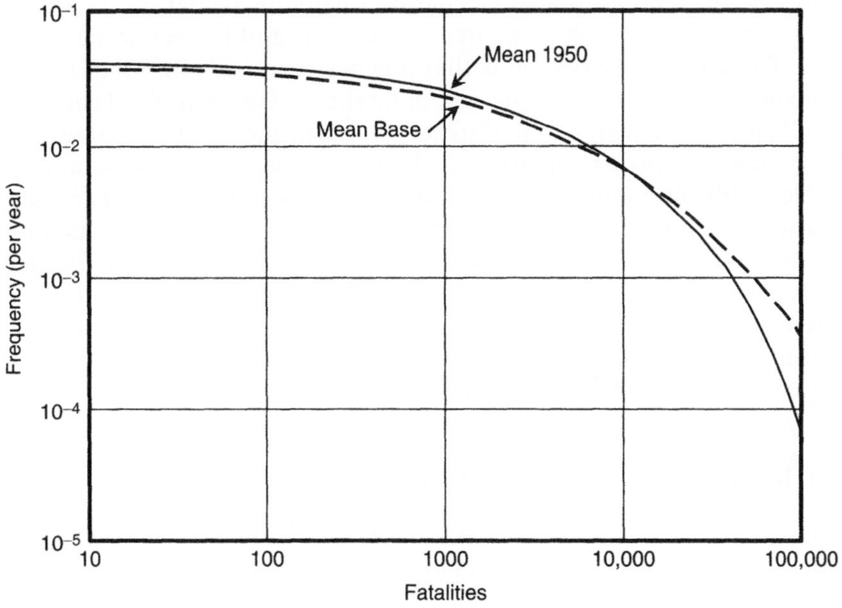

FIGURE 8.2. Hurricane fatality risk in New Orleans: Comparison of results based on different data sets (1900–1950 vs 1900–2004).

having 10,000 fatalities from a major hurricane impacting New Orleans was essentially the same in 1950 as it was in 2004. However, as can also be seen from Figure 8.2, the risk of having 100,000 fatalities from a major hurricane impacting New Orleans was much higher (six times as high) in 2004 than it was in 1950, principally due to a dramatic change in the metropolitan area population. These results are based on the implementation of Bayes theorem, the fundamental theorem for inferring the impact on states of knowledge of additional information. This is an excellent example of how time dependencies can be tracked in quantitative risk assessments.

There are numerous other observations and insights that resulted from the case studies. Two are noted. The first has to do with the observation that equal threats result in different risks depending on where they occur. The risk of fatalities from an asteroid impacting the state of Kansas in the United States or the states surrounding Kansas is much different than the risk of fatalities from a same size asteroid impacting coastal regions that are vulnerable to both a direct impact and a tsunami resulting from an offshore impact. The target area for coastal regions is larger as a result of a greater possibility of collateral damage. Supporting evidence for the higher vulnerability of coastal regions comes from Table 4.2 of Chapter 4. As can be observed from Table 4.2, for asteroids impacting the 48 contiguous states of the United States, scenarios involving tsunamis comprise 47% of the fatality risk compared to 28% for land impact scenarios.

The second observation is that it is possible to identify significant and subtle contributors to the risk of fatalities. For example, Table 3.2 in Chapter 3 indicates that Category 4 hurricanes coming ashore at New Orleans represent approximately 67% of the total risk of fatalities in New Orleans compared to approximately 24% of the Category 5 hurricanes. It is not necessarily true that the most powerful hurricane has the highest risk of fatalities. One must take into account all the parameters of interest. The parameters include the frequency of occurrence (there are more Category 4 than Category 5 hurricanes), the probability of impacting New Orleans, the strength at landfall, the amount of evacuation prior to landfall, and the consequences.

8.5 Insights from Comparing Risks

Even with all of our qualifications about being cautious to not reach conclusions based on the case studies because of their limited scopes, many insights were developed. Of course there are those who would say that most of the insights were already known, which may be true. But now there is some evidence based on the risk sciences to support them. Among the insights that were identified by the case studies were the following:

- Regional risk assessments are important, as many risks are very location dependent even with respect to the same threat. The same event may result in very different consequences depending on whether you are in St. Louis, MO, or New Orleans, LA.
- In some regions of the United States, hurricanes are by far the greatest risk of those investigated. The strongest hurricanes are not necessarily the highest risk as it depends on such other factors as their frequency of occurrence, demographics of the region, topography, building codes, evacuation routes, emergency preparedness, etc.
- The collateral effects of asteroids, for example the creation of tsunamis, make coastal regions generally much more vulnerable than inland locations.
- The risk of a terrorist attack on a large electrical grid can only be catastrophic if it results in a long duration outage in a region of dense population. Coordinated physical attack scenarios of critical components of the grid such as substations and transmission lines are much more likely to result in long duration regional outages then cyber attacks on such critical systems as supervisory control and data acquisition systems.
- A scoping analysis for doing a quantitative risk assessment of an abrupt climate change in the north Atlantic region provides evidence to make it a prime candidate for further study.

- There is sometimes a major discrepancy between perceived risks and risks that are assessed scientifically. An example is the risk from commercial nuclear electric power plants. A nuclear power plant is perceived by some as a major risk, a fact that cannot be supported with scientific evidence.

8.6 Where Do We Go from Here?

Knowing the catastrophic risks that can impact humans is critical for assuring future generations of a life-sustaining planet Earth. The risk sciences have advanced to the point that it is now possible to prioritize catastrophic risks to society in a logical and systematic manner for risk management that may save millions of lives. Society has been reasonably accountable to those risks that occur over a lifetime or a few generations, but not nearly so accountable to the more rare risks that may have the potential for catastrophic consequences, even the possible extinction of the human species. The public is entitled to have better information on the catastrophic risks they face, especially if there is a chance of saving millions of lives. A sufficient amount of analyses has been presented in this book to point out the inconsistencies in the way in which we address catastrophic risk and how easy it is to grossly misrepresent the real risks. Others have made the same point with respect to technological risks versus other kinds of risk.[2] Meanwhile, the absence on a national or global scale of a consistent and systematic process for characterizing the catastrophic risks we face is denying us a clear picture of the options and priorities for protecting our health and safety as well as that of future generations.

The vision we have is an elevation of public consciousness about the risk of events that may have severe human and environmental consequences and to provide information that will enhance making rational decisions to mitigate or reduce their impact. Implementation of the vision depends on a method for identifying, quantifying, and prioritizing of such risks on the basis of sound science. The method advocated is quantitative risk assessment, where *quantification* refers to the credibility of a proposed event, based on the totality of the supporting evidence. Quantification involves exposing the state-of-knowledge about a hypothesis. Risks having the potential for "catastrophic" consequences have been the focus. Two important incentives for addressing catastrophic events are: (1) they tend to be ignored as they fall outside of most human experiences and thus suffer from the syndrome of "out of sight and out of mind;" and (2) because of their low frequency of occurrence, especially compared to "terms of office" of political leaders, there is a lack of motivation for decision-makers to address them. The hope is that a better-informed public will lead to better societal decisions on how to manage catastrophic risk.

How might adopting a systematic and risk-based approach to assessing threats to society aid the decision making process of our leaders and why would we want to do this? With respect to the "why," it is a matter of a return on our investment. All taxpayers would like to believe that their hard-earned money is being wisely spent and that the decisions being made by our leaders are based on effective cost/benefit analyses using the best available information. Decisions made about how to minimize the risk of rare but catastrophic events, natural and manmade, require information of a very special kind—special in the sense that to understand the risk of rare and catastrophic events requires the application of the risk sciences in a systematic and comprehensive manner. An important question in this regard is what form such information should take to facilitate decision-making. Even though the form of the results is not nearly as important as the substance of the analyses, it is important that the results be easily accessed and user friendly.

There are many forms that the results could take. The desire is that the decision-makers have the capability to quickly and easily access the information in its proper context. Suppose there existed simple spreadsheets by geographical region defining the risks threatening that region in a form that easily allowed for comparison and updating of the different risks to that region. This would add enormous confidence to the decision-making process. Such a spreadsheet might take the following form.

Comparative Risk Levels (Region: Los Angeles Basin)

Asset	*Terrorism*	*Severe Storms*	*Earthquakes*	*Asteroids*	*Climate Change*	*Industrial Accidents*	*Nuclear War, Famine, Disease, Other*
People	Risk data (likelihood of fatalities, injuries, etc.)	etc.					
Electricity	Loss of supply (likelihood of different durations)	etc.					
Water Supply	Etc.						
Food Supply							
Transportation							
Other							

The format is envisioned as tables and curves that display risks in the variety of forms that are now associated with quantitative risk assessment, especially information on the levels of uncertainty involved. It is the uncertainty part of the entry that will tell the most about the scope of the assessments and the quality of the supporting evidence.

The idea of the above spreadsheet is to have immediate access to a quantification of the risks by region on a consistent basis and to check on actions being taken to manage them. The entries in the spreadsheets would be consistently developed to permit the comparison of different risks to different assets and to cross compare risks in different regions of the country based on the supporting information. Of course, the entries and priorities would be region-specific. The user would identify a region and then on a specific entry gain access to the details of the analysis, their level of depth, who performed the analysis, peer reviewers, the methods employed, and the supporting evidence. The information would include summaries of actions being taken to deal with specific risks and future directions of the effort.

With the above type of information being the vision, the question is where are we now and how do we move forward. As to where we go from here, we should immediately perform a top-down assessment of global threats using quantitative risk assessment. A list of potential threats for future quantitative risk assessments is listed in Chapter 7. For example, one global threat that is a headline issue at this time is global warming. The goal is to quantify the risks of global warming, including all of its ramifications. While global warming is on a scale of a century or more in terms of causing major changes in our climate, there are ramifications of global warming that could trigger events that could result in a major climate change in much shorter periods of time, perhaps a decade or so in certain regions of the world. These impacts have to be part of the overall global warming risk assessment. A subset of such global warming is the possibility of greenhouse gases (an alleged cause of global warming) triggering changes in some part of the ocean circulation system that could result in an abrupt climate change in such regions as the north Atlantic and the bordering nations. Given the much shorter time for this event to occur, this component of global warming is a candidate for an immediate risk assessment.

References

[1] Garrick, B. J., "Lessons Learned from 21 Nuclear Plant Probabilistic Risk Assessments," *Nucl. Technol.* 1989, *84*, 319–330.

[2] Lewis, H. W., *Technological Risk*, W.W. Norton & Company, New York, 1990.

Appendix A: Roots of Quantitative Risk Assessment with an Example

The general framework of quantitative risk assessment presented in this book is based on the "set of triplets" definition of risk. The *triplet* refers to scenarios, consequences, and likelihoods. In Chapters 1 and 2 we are explicit on the meaning given to "likelihood." The concept is illustrated in the case studies (Chapters 3–6). From a methodology standpoint, calculating likelihoods, or probabilities, is the central issue in "quantifying" risk. From an application perspective, the most important exercise is the structuring of the scenarios. Thus, when we consider the roots of quantitative risk assessment, most of the attention is on calculating likelihoods and structuring scenarios. Calculating likelihoods primarily evolved in the fields of mathematics and mathematical physics and has a history of several hundred years. On the other hand, the formal process of structuring scenarios is a much more recent discipline developed primarily by engineers involved in designing and analyzing complex systems evolving from 20th century technology. For example, the discipline of reliability analysis and engineering has made a significant contribution to the development of integrated models of engineered systems, including the development of a variety of graphical methods for displaying interdependencies of components, subsystems, and systems.

A.1 Calculating Likelihoods

Likelihood is interpreted in a Bayesian sense. Probability, as defined in Chapters 1 and 2, is the credibility of a hypothesis based on the totality of the supporting evidence. The definition of probability has been argued for well over 200 years among the so-called subjectivists and the frequentists, sometimes referred to as the Bayesians and the frequentists, or the Bayesians and the classical statisticians. This debate is not only between Bayesians and classical statisticians, but there are different interpretations of probability within each of these groups. For example, some groups interpret Bayesian probability as a "degree of belief." Such an interpretation has a connotation of "faith" and is not the interpretation of Bayesian probability used in this book, nor is it the interpretation of E. T. Jaynes, a noted contemporary on Bayesian inference. It is not a matter of "belief" or "faith," but rather "the credibility of the hypothesis based on the totality of available evidence." Thus, the probabilities are objective in the sense that they completely depend on the "totality of available evidence." Having established our interpretation of probability, it is important to back up and make a few observations about how the principles behind probability and risk assessment evolved.

Probability theory, the foundation of contemporary risk assessment, was principally developed during the 100-year period between the mid-1600s and 1700s. Of course, there were many contributing events prior to the 1600s. For example, probability involves numbers, ranges of numbers, curves, and families of curves. Thus, the Arabic numbering system known to western civilizations for some 700–800 years has to be a profound event in the history of all science, including probability and risk assessment. But it was during the Renaissance period (14th–17th centuries), a time of the separation of new thoughts from the constraints of medieval cultures, that the contemporary thoughts about probability and risk were essentially formulated.

Cardano and Galileo made important contributions in the 1500s on how to express probabilities and frequencies of past events, Pascal contributed to concepts of decision theory and statistical inference in the mid-1600s, and Fermet and de Méré made major contributions to the theory of numbers about the same time. Other major contributors during the 17th century were Christen Huygens, who published a popular textbook on probability theory; Gottfried Wilhelm von Leibniz, who suggested applying probability methods to legal problems; and members of a Paris monastery named Port Royal. The Port Royal group produced a pioneering work of philosophy and, probably, the first definition of risk: "Fear of harm ought to be proportional not merely to the gravity of the harm, but also to the probability of the event." Jacob Bernoulli produced the Law of Large Numbers and methods of statistical sampling forming the basis of many methods of product testing and quality control. Abraham de Moivre developed the concept of the normal distribution and standard deviation in the early 1700s.

Many modern day probability practitioners, decision analysts, and risk assessors consider Thomas Bayes, an English minister, the real father of contemporary risk assessment. In the mid-1700s, he developed a theorem rooted in fundamental logic for combining old information with new information for the assignment of probabilities. Bayesian inference reduces to the simple product and sum rules of probability theory developed by Bernoulli and Laplace. Bayes theorem provided the foundation for a unified theory of probability not bound to such properties as "randomness" and "large numbers." The doors were opened for employing probability to address problems involving limited information. Bayes theorem, followed by the publication in 1812 of *Théorie analytique des probabilities* by the French mathematician Marquis Pierre Simon de Laplace, provided the primary basis of contemporary probability theory. Diverse problems, such as gambling strategies, military strategies, determining mortality rates, and debating the existence of God, were the subjects of early analytical explorations and precursors to the new science of risk assessment.

Among the 20th and 21st century scholars who contributed to probability and risk assessment as advocated in this book are Harold Jeffreys,[1] R. T. Cox,[2] C. E. Shannon,[3] George Pólya,[4] Howard Raiffa,[5] and E. T. Jaynes.[6] Probability theory in the context of Jaynes' extended logic includes as special cases all the results of the conventional "random variable" theory, and it extends the applications to the useful solution of many problems previously considered to be outside the realm of probability theory. The goal of such investigators as Pólya and Jaynes was to formulate a probability theory that 'could be used for general problems of scientific inference, almost all of which arises out of incomplete information rather than "randomness".' This interpretation of probability is the nugget that makes it possible to perform meaningful risk assessments of complex systems about which there is little information on their threat environment and vulnerabilities.

The widespread, formal application of risk assessment to critical infrastructure began in earnest in the late 1900s. Applications in the insurance and financial fields were more statistical (actuarial) than probabilistic, more experience-based than inferential, more qualitative than quantitative. Only when societies began depending more on technological systems involving large inventories of hazardous materials did investigators begin to look for more scientifically based ways to assess risks. The particular need was for a method of assessing the likelihood of catastrophic events that could do great harm to public health and the environment.

A.2 Structuring the Scenarios

In many respects the heart of a quantitative risk assessment is the scenarios that arises from trying to answer the risk triplet question, what can go wrong. The scenarios are what tie the physical aspects of the risk assessment to the analytical process. A scenario is defined as a sequence of events, starting with an event known as the initiating event (an event that upsets an otherwise normally operating system), or an initial condition, and then proceeds through a series of events until the system either corrects itself or the scenario of events is terminated at a damaged, degradated, or destroyed state. Thus, structuring the scenarios is a representation of the logic of how a system responds to different types of threats, that is, to different initiating events or initial conditions. Of course, the goal of the analyst is to define the initiating event or initial condition set such that it is complete, or at least complete in the practical sense that all of the important initiators have been identified. The practice of developing scenarios has evolved into a general theory of structuring scenarios.

The theory of structuring scenarios has its primary roots in the field of reliability engineering and analysis developed initially by the United Kingdom and Germany in the 1930s and 1940s. The United Kingdom introduced the concept of mean time to failure for aircraft in the 1930s.[7] They used such information to infer reliability criteria for aircraft and the proposing of maximum permissible failure-rates as a basis for establishing levels of safety for aircraft. For example, in the 1940s performance requirements were being given for aircraft in terms of accident rates that should not exceed, on average, one per 100,000 h of flying time.

The Germans applied such principles of reliability analysis as the product rule to the development of the V1 missile during World War II to solve serious reliability problems with that weapon system. The product rule has to do with the reliability of components in series and accounting for differences in reliability of individual components. Until the methods of reliability were employed, the design philosophy of the V1 missile was based on the "weak link" theory—a chain cannot be made stronger than its weakest link. Eventually, it was realized that a large number of fairly strong "links" could be more unreliable than a single "weak link" if reliance is being placed on them all. Going from a design philosophy based on the weak link theory to basing it on reliability theory resulted in a vast improvement in the V1 missile reliability.

In the years following World War II the United States adopted the analysis methods of the United Kingdom and Germany and greatly expanded the use of practical tools to improve the reliability of missiles.[8,9] Activities where the methods of reliability analysis and engineering made a major impact were defense systems, commercial aerospace, electronics, chemical plants, power plants, and electrical distribution systems. Reliability analysis became a cornerstone discipline of the evolving field of systems engineering because of its ability to link interacting systems.

Reliability analysis involved many of the steps of risk assessment: (1) defining of the system and its success state, (2) characterization of the hazards involved, and (3) evaluation of the probability of failure of subsystems and components. Reliability analysis and engineering evolved rapidly in such fields as defense and aerospace, making major contributions to the performance of complex systems. Examples are methods for importance ranking of components and subsystems of total systems, the determination of the life of components subject to such phenomena as fatigue, the quantification of the impact of repair, and the development of solid technical foundations for preventive maintenance programs to optimize system performance. Reliability analysis and risk assessment have much in common. The emphasis is what's different. In risk assessment the emphasis is on "what can go wrong" and in reliability analysis the emphasis is on what to do to make the system run the way it is designed. The output of a reliability analysis is *reliability* and what's driving it, while the output of a risk assessment is the *risk* of something bad happening and what contributes to it. Both employ many of the same analytical tools; the perspectives are just different and thus each has a set of analytical tools unique to that perspective. For example, risk assessment emphasizes the quantification of uncertainties because often the events of interest have little data to support them. On the other hand, reliability analysis focuses much more on factors contributing to good performance than on risk assessment.

One of the most important contributions that the reliability sciences have made to systems engineering in general and risk assessment in particular are transparent methods for graphically representing complex systems that can be transformed into analytical models. For example, reliability analysis used block diagrams to describe how components in a large system were connected. From these block diagrams, Watson at Bell Laboratories developed the fault-tree technique, which he applied to the Minuteman Missile launch control system, and which Boeing later adopted and also computerized. These diagrams in combination with the tools of switching algebra and probability theory have provided a powerful tool for displaying and quantifying the "fault paths" of systems, subsystems, components, human actions, procedures, etc.

The advantage of these techniques in the field of safety analysis, as opposed to just reliability, began to be recognized in the 1960s. F. R. Farmer of the United Kingdom proposed a new approach to nuclear power plant safety based on the reliability of consequence-limiting equipment.[10] Holmes and Narver, Inc., a U.S. engineering firm under contract to the then U.S. Atomic Energy Commission, performed a series of studies on nuclear reactor safety and reliability. The final report in the series advocated, with examples, the need for much greater use of advanced systems-engineering methods of modeling the reliability of safety systems. The authors made explicit reference to the use of logic tools, such as fault-tree methodology, which has its roots in "switching theory" developed by the telecommunications field.[11] At about the same time, a Ph.D. thesis was published that proposed a methodology for probabilistic, integrated systems analysis for analyzing the safety of nuclear power plants.[12]

The breakthrough in quantitative risk assessment (or probabilistic risk assessment as it is generally labeled in the nuclear field) of technological systems came in 1975 with the publication of the *Reactor Safety Study* by the U.S. Atomic Energy Commission under the direction of Professor N. C. Rasmussen of the Massachusetts Institute of Technology.[13] This project, which took 3 years to complete, marked a turning point in the way people think about the safety of complex facilities and systems. The *Reactor Safety Study* provided a basis for a wide range of applications for risk assessment, not only for nuclear power plants and other technological systems (e.g., chemical and petroleum facilities, transportation systems, and defense systems), but also for environmental protection, healthcare, and food safety.

Besides fine-tuning fault-tree analysis for safety applications, the *Reactor Safety Study* introduced another extremely important graphic tool to facilitate the structuring of scenarios, the *event tree.* Fault trees and event trees in combination provided a critically important one-two punch in the theory of structuring scenarios. An event tree starts with an initiating event and proceeds to identify succeeding events, including branches that eventually terminate into possibly undesirable consequences. An event tree, therefore, is a cause-and-effect representation of logic.

A fault tree starts with the end-state or undesired consequence and attempts to determine all of the contributing system states. Therefore, fault trees are effect-and-cause representations of logic. An event tree is developed by inductive reasoning, while a fault tree is based on deductive reasoning. A key difference in the two representations is that a fault tree is only in "failure space," and an event tree includes both "failure and success space." The choice between the two is a matter of circumstances and preference, and they are often used in combination; the event tree provides the basic scenario space of events and branch points, and the fault tree is used to quantify the "split fractions" at the branch points.

The *Reactor Safety Study* inspired many first-of-a-kind risk assessments in the commercial nuclear power industry that led to major advancements in the application of quantitative risk assessment. One important example was the probabilistic risk assessments of the Zion and Indian Point nuclear power plants sponsored by the owners and operators of the plants. New methods were introduced in those assessments that have become standards of many quantitative risk assessment applications.[14,15] The methods included the treatment of uncertainty, a framework of risk assessment embedded in the set of triplets definition of risk,[16] common-cause failure analysis, importance ranking of risk contributors, models for calculating source terms, and improved dispersion models for calculating off-site health effects.

These and other studies have evolved into a contemporary theory of structuring scenarios that is part science and part art. Elements of the theory are a set of principles having to do with issues of completeness and the general structure of scenarios. More details are covered in Chapter 2.

A.3 Steps That Have Evolved for Integrated Quantitative Risk Assessment

Although the scope, depth, and applications of quantitative risk assessments vary widely, they all follow the same steps:

1. Define the system being analyzed in terms of what constitutes normal operation to serve as a baseline reference point.
2. Identify and characterize the sources of danger, that is, the hazards (e.g., stored energy, toxic substances, hazardous materials, acts of nature, sabotage, terrorism, equipment failure, combination of each, etc.).
3. Develop "what can go wrong" scenarios to establish levels of damage and consequences while identifying points of vulnerability.
4. Quantify the likelihoods of the different scenarios and their attendant levels of damage based on the totality of relevant evidence available.
5. Assemble the scenarios according to damage levels, and cast the results into the appropriate risk curves and risk priorities.
6. Interpret the results to guide the risk management process.

These steps provide the answers to the three fundamental questions of risk (the "triplet definition"): what can go wrong, how likely it is to go wrong, and what the consequences will be.

Risk assessments are routinely used in many settings, including the electric nuclear power industry, the chemical and petroleum industries, defense industries, the aerospace industry, food sciences, and health sciences. Industries that are increasingly using formal, quantitative methods of safety analysis include marine transportation and offshore systems, pipelines, motor vehicle, and recreational systems. The space program has stepped up its use of quantitative risk assessment since the *Challenger* accident.[17] Other less publicized applications include the risk management program used by the U.S. Army for the disposal of chemical agents and munitions.[18]

The government agencies most involved in using risk assessments are the U.S. Nuclear Regulatory Commission and the U.S. Environmental Protection Agency. Other agencies becoming active users of risk assessment methods are the U.S. Department of Energy, U.S. Department of Agriculture, U.S. Department of Defense, U.S. Food and Drug Administration, the National Aeronautics and Space Administration, and the U.S. Department of Transportation. The most active practitioners in the private sector are the nuclear, chemical, and petroleum industries, although the scopes of application vary widely—the nuclear industry being the most consistent user of complex and sophisticated quantitative methods.

Risk assessment has many buzzwords (e.g., Monte Carlo analyses, influence diagrams, multiple attributes, common-cause failures, realizations, minimum cut sets, sensitivity analyses, fault trees, event trees, etc.), but the basic principles are few. The principles focus on the development of scenarios describing how the system under study is supposed to work and scenarios indicating how the system can be made to fail, catastrophically or otherwise. The likelihood of events in the scenario must be linked to the supporting evidence. Events are propagated to an end-state that terminates the scenario (i.e., the consequence). Other principles may be applied to aggregate the various end-states into the desired set of consequences.

The results of risk assessments are easy to interpret, including corrective actions having the biggest payoff in terms of risk reduction. Although the literature suggests many different risk assessment methodologies, in fact the differences are primarily in scope, application, boundary conditions, the degree of quantification, figures-of-merit, and quality. Like many other scientifically based methodologies, quantitative risk assessment is founded on relatively few basic principles.

A.4 Application to Nuclear Power: A Success Story

A.4.1 Why Risk Assessment?

The simple answer to "why risk assessment?" for nuclear power plants is that nations and the world have to make decisions about the best energy mix for the future of planet Earth. Risk to people and the environment is a fundamental attribute of societal decision-making. But there is more to it than just decision-making. Early in the development of nuclear power, it became clear that the large inventories of radiation in nuclear reactors contemplated for generating electricity and the stigma of the dangers of the fission process carried over from the Hiroshima and Nagasaki atomic bombs required a level of safety analysis beyond standard practices. The nuclear power industry was forced to seek new methods of safety analysis of nuclear electric power plants to overcome the "fear anything nuclear" syndrome that prevailed in the minds of some members of the public. New methods were needed to provide answers to the questions, what can go wrong with a nuclear power plant, how likely is it, and what are the consequences. The traditional methods of safety analysis, while somewhat effective in answering questions about what can go wrong and what are the consequences, profoundly failed to adequately answer the question having to do with the likelihood of accidents. The likelihood question held the key for being able to quantify nuclear electric power plant risk. In short, for society to have access to nuclear energy systems that have the potential to provide safe, reliable, and relatively inexpensive electric power, the industry was forced to come up with a more convincing safety case than was possible with past methods of analysis.

The nuclear electric power industry, stimulated by the *Reactor Safety Study* and the early industry studies on the Zion and Indian Point nuclear power plants, has been the leader in the development and widespread use of quantitative risk assessment (QRA). The U.S. nuclear electric power industry gave birth to the term "probabilistic risk assessment" (PRA). The term "probabilistic safety assessment" (PSA) is sometimes used in the international nuclear community as equivalent. The label that appears to be best received across different industries is quantitative risk assessment. In this discussion, the terms quantitative risk assessment, probabilistic risk assessment, and just plain "risk assessment" are used interchangeably.

Risk assessment has survived and flourished in the U.S. nuclear electric power industry because it became an exceptional tool to make better decisions. QRA was able to satisfy the desire of nuclear plant owners to have a decision tool that quantitatively allowed the evaluation of various options that had multiple input variables. The most important variables to the nuclear plant owners were cost, generation, and risk (public health, worker health, and economic). While QRA started out as a tool to address the public health risk, it facilitated evaluating an entire spectrum of variables. The industry's recovery from the Three Mile Island Unit 2 accident in 1979 was greatly aided by the use of quantitative risk assessment because of the ability to better focus on the real safety issues. In fact, the industry has enjoyed an impeccable safety record since embracing contemporary methods of quantitative risk assessment. And safety is not the only benefit that has resulted from the widespread use of risk assessment in the nuclear power industry. Risk assessment provides the ability for plant personnel to balance cost, generation, and risk. While there is no U.S. Nuclear

Regulatory Commission (USNRC) requirement for an existing nuclear power plant to maintain a risk assessment, the plants do so following general industry guidelines. The USNRC does have a requirement for a prospective licensee to submit a PRA with the application for any proposed new nuclear electric power unit in the United States.

A.4.2 Legacy of Nuclear Safety

Today, nuclear power plant safety analysis employs the most advanced methods available for assessing the health and safety of the public. The only significant impact on public health risk comes from the release of fission products from the reactor core following accidents during power operation. The release of fission products from the reactor core is heavily influenced by operator actions both before and after the initiating event that leads to the accident. The PRA includes potential failures of equipment and operators, both before and after the initiating event. Each PRA in the United States is specific for each nuclear power plant because of the differences between the plants.

Many of the methods used for nuclear power plants have been adopted by such high-tech industries as space flight, defense systems, chemical plants, refineries, offshore platforms, and transportation systems. While the probabilistic concepts currently spearhead the level of sophistication of the analyses, there are basic tenets and themes that have guided the safety management of nuclear electric power plants from the beginning. One of the most fundamental of these basic tenets is the concept of multiple barriers.

Multiple barriers are a concept of providing enough barriers between radiation and the environment to provide assurance that the likelihood of simultaneous breach of all barriers is remote. Examples of barriers in a nuclear power plant are high containment capacity fuel with cladding, an isolated reactor coolant system, primary reactor building containment, secondary building containment, and exclusion distance. Effective defense mechanisms developed for nuclear plant safety include improved operator training methods and symptom-based operator procedures. Other defense mechanisms include automatic control systems, single failure criteria (no single failure threatens fuel integrity), and recovery capabilities from equipment malfunctions. QRA provides the ability to determine what risk levels are achieved by each barrier and at what cost. The value of each barrier is placed in the context of the overall risk.

A.4.3 Historical Development of Nuclear Power Plant Safety

Nuclear power plant safety has two major fronts—the physical system itself and the analysis of the physical system. On the physical system front, improvements in safety design included the advent of secondary containment systems (~1953), the inclusion of backup safety systems known as engineered safety features, especially with respect to emergency core cooling systems and electric power (~late 1950s and early 1960s), and the introduction of separate and independent safety trains (~1970s). In the 1980s and 1990s, the nuclear power plants initiated programs for scram reduction based on a complete review and analysis of operating transients. As scrams were reduced, public health risk was also reduced because there were fewer departures from normal steady state operation. Also, in the 1980s and 1990s, each nuclear power plant implemented the concept of "symptom-based procedures" for accident control and installed improved simulators for operator training.

On the analysis front, many events took place leading to a greatly improved understanding of the safety of nuclear power plants. It was demonstrated that the consequences of accidents had little meaning without a better understanding of their likelihood. It became clear that it was not enough to do worst case and maximum credible accident analysis. Everyday transients followed by multiple failures of equipment and mistakes by operators were more likely to result in reactor core damage than previously defined so-called design basis accidents.

The need for probabilistic analysis was recognized as early as the mid-1950s. However, detailed investigations of the probability of reactor accidents did not begin until about 1965. The first major reactor safety study to highlight the need for PRA of reactor accidents was WASH-740, "Theoretical Possibilities and Consequences of Major Accidents in Large Nuclear Power Plants."[19] Speculative estimates were made in WASH-740 that a major reactor accident could occur with a frequency of about one chance in a million during the life of a reactor. The report went on to observe that the complexity of the problem of establishing such a probability, in the absence of operating experience, made these estimates subjective and open to considerable error and criticism. While not offering many specifics, this study did stir interest in probabilistic approaches and many studies were soon to follow. These included British and Canadian efforts, probabilistic analyses of military reactors, and several studies sponsored by the then U.S. Atomic Energy Commission. The thesis referred to earlier by Garrick[12] was written about the same time, advocating a probabilistic approach to assessing nuclear power plant safety. But as also noted earlier, it was the *Reactor Safety Study*[13] that spearheaded the movement toward the application of probabilistic risk assessment.

By the 1980s the question was no longer "why," but how soon could a QRA be developed for every nuclear power plant in the United States That goal has essentially been reached. The benefits of QRA for U.S. nuclear power plants have been demonstrated in terms of the reduction in frequency of core damage events (one reactor core lost in approximately the first 450 reactor years of experience versus zero reactor cores lost in over 2000 actual reactor years of experience since the TMI accident) and improved generation with a reduction in the cost of electricity. The most important benefit is nuclear power plants with reduced public health risk. QRA has not only been effective in calibrating the risk of nuclear power, but has provided better knowledge of the worth of safety systems and allowed the allocation of safety engineering resources to the most important contributors. Effective risk management of nuclear electric power plants in the United States has become a reality, not just a goal.

A.4.4 Nuclear Power Accident Experience

There have only been two accidents worldwide that have resulted in severe core damage of a nuclear power plant designed to generate electricity. The accidents involved the Three Mile Island, Unit 2, plant near Harrisburg, PA, in the United States and the Chernobyl Nuclear Power Station in the Ukraine of the former Soviet Union. Both accidents permanently damaged the nuclear reactors involved, but only the Chernobyl accident resulted in known fatalities and injuries. The on-site consequences of the Chernobyl accident were very serious, as an estimated 30 people are believed to have died from acute doses of radiation and some 300 people required hospital treatment for radiation and burn injuries. No off-site fatalities or injuries have yet been attributed to the Chernobyl accident, although the latent effects are yet to be quantified.

It is important to put these two very serious accidents in context with the safety experience of the nuclear power industry. There are approximately 440 nuclear power plants in the world. Nuclear energy is just over 5% of the world primary energy production and about 17% of its electrical production. In the United States there are some 103 nuclear power plants operating providing approximately 20% of the nation's electricity. The worldwide experience base is approaching 10,000 in-service reactor-years of which about 3000 reactor-years is U.S. experience. The experience base is likely beyond 10,000 reactor-years if all types of reactors are included such as research, test, weapons, and propulsion reactors. Some 70% of the nuclear power plant experience worldwide involves light water reactors for which only one accident has occurred, Three Mile Island. This safety record is most impressive. The challenge is to keep it that way.

Three Mile Island, Unit 2 (TMI-2) Accident

The TMI-2 nuclear power plant, located near Harrisburg, PA, went into commercial operation in December 1978. The plant was designed to generate approximately 800 MW of electricity and used a pressurized water reactor supplied by the Babcock and Wilcox Company. The accident occurred on March 28, 1979.

Routine mechanical malfunctions with the plant resulted in an automatic shutdown ("feedwater trip") of the main feedwater pumps, followed by a trip of the steam turbine and the dumping of steam to the condenser. The loss of heat removal from the primary system resulted in a rise of reactor system pressure and the opening of its power-operated relief valve. This action did not provide sufficient immediate pressure relief, and the control rods were automatically driven into the core to stop the fission process.

These events would have been manageable had it not been for some later problems with such systems as emergency feedwater. Perhaps the turning point of the accident was that the opened pressure relief valve failed to close and the operators did not recognize such. The result was the initiation of the well-studied small loss of coolant accident, known as the small LOCA. The stuck-open valve, together with some other valve closures that had not been corrected from previous maintenance activities, created a shortage of places to put the decay heat loads of the plant. The response of the plant was the initiation of high-pressure emergency cooling. Reactor coolant pump high vibration and concern for pump seal failure resulted in the operators eventually shutting down all of the main reactor coolant pumps and relying on natural circulation in the reactor coolant system. It was during the time that the main reactor coolant pumps were off, some 1–3 h, that the severe damage to the core took place. At about 2 h and 20 min into the accident, the backup valve known as a block valve to the stuck-open relief valve was closed. This action terminated the small LOCA effect of the stuck-open relief valve. While the accident was then under some level of control, it was almost 1 month before complete control was established over the reactor fuel temperature when adequate cooling was provided by natural circulation.

The consequences of the accident were minimal in terms of the threat to public health and safety, but the damage to the reactor was too severe to recover the plant. The accident did confirm the effectiveness of the containment system to contain the fission products escaping from the reactor vessel.

Chernobyl Nuclear Power Station Accident

The Chernobyl nuclear power plant involved a 1000-MW (electrical) boiling water, graphite-moderated, direct cycle reactor of the former Soviet Union. The Chernobyl accident occurred on April 26, 1986, and was initiated during a test of reactor coolant pump operability from the reactor's own turbine generators. The purpose of the test was to determine how long the reactor coolant pumps could be operated, using electric power from the reactor's own turbine generator under the condition of turbine coast down and no steam supply from the reactor. However, the experimenters wanted a continuous steam supply so they decided to conduct the experiment with the reactor running—a serious mistake. The test resulted in a coolant flow reduction in the core and extensive boiling. Because of the inherent properties of this particular reactor design (on boiling, the fission chain reaction increases, rather than decreases as in U.S. plants), a nuclear transient occurred that could not be counteracted by any control system. The result was a power excursion that caused the fuel to overheat, melt, and disintegrate. Fuel fragments were ejected into the coolant, causing steam explosions and rupturing fuel channels with such force that the cover of the reactor was blown off.

This accident resulted in approximately 30 immediate fatalities from acute doses of radiation and the treatment of some 300 people for radiation burn injuries. The off-site consequences are still under investigation. Latent effects are expected, but they have not been quantified.

In summary, nuclear power suffered a severe setback from both of the above accidents, although public support for nuclear power was already beginning to decline before these accidents occurred. Some nuclear plants under construction were cancelled and no new U.S. nuclear plants were ordered between 1979 and 2007.[a] The facts that the TMI-2 accident did not result in any radiation injuries or fatalities and the Chernobyl reactor type is no longer in the mix of viable power reactors has not removed the fear that some segments of the public have of nuclear power. However, the superior performance and safety record in the United States since these two accidents has allowed the USNRC to approve power upgrades and license extensions for numerous U.S. nuclear power plants.

A.5 An Example of Nuclear Power Plant Quantitative Risk Assessment Results

It is appropriate to provide an example of risk assessment results deriving directly from the full application of the methodology of Chapter 2. In particular, the focus in this example is on results only as opposed to the details of how they came about. The emphasis of

[a] Renewed interest in nuclear power has resulted in the first order in 2007 of a new nuclear power plant in the United States with high expectations for several orders in 2008 and 2009

the limited scope case studies of Chapters 3–6 is on the process of quantitative risk assessment, not necessarily the results.

The best application examples are the many comprehensive risk assessments prepared for nuclear power plants worldwide since the 1970s. The example discussed here takes the form of some actual results of such risk assessments.[20] We have chosen to withhold naming specific plants as changes have been made in the plants to improve safety and the risk results no longer apply, but this does not take away the value of the analysis. In general the risks have become less because of the availability of more operating experience and improved analyses and systems.

The example is given in terms of: (1) the bottom line results, (2) importance ranking of contributors to risk, and (3) the use of risk assessment as a design and risk management tool. A consequence of such results is the answer to the important risk management questions, "What in fact is the risk, including the uncertainties involved?" "What scenarios, operator actions, and system failures are driving the risk?" "How should risk assessment be used during design to guide the design of the plant, especially with respect to accident mitigating systems?" and "What corrective actions will have the greatest return for reducing the risk?"

A.5.1 Bottom Line Results

Figure A.1 contains the risk curves from a comprehensive risk assessment performed on a U.S. nuclear power plant during the 1980s.

Figure A.1 represents the integration and assembly of an extensive amount of modeling and analysis covering thousands of pages. The principal elements are initiating events, scenarios, consequences, and likelihoods. The data processing is Bayesian based and involves searches on numerous national databases on human, equipment, and system performance. The results have the forms of Figures 2.9 and 2.10 of Chapter 2. A first impression is that to tell a reasonably complete risk story requires much more than just a number. As a start, it requires a probability curve to communicate the uncertainty in the number. But even a probability curve is not enough to tell the risk story. As can be seen, the risk story told by Figure A.1 involves curves, families of curves, and different representations of families of curves. To make it clearer just what the risk story is, each of the subfigures of Figure A.1 is briefly discussed.

Core Damage Frequency

Core damage frequency is currently the most often used measure of risk in contemporary nuclear power plant risk assessments. Simply put, core damage means that the core coolant is not removing heat as fast as it is being generated, and the core begins to degrade through overheating. Core damage includes a range of degrees of damage to the cladding and fuel elements with the most severe being core melt through the reactor vessel. A partially or totally uncovered core as a result of a loss of coolant accident would be an example of a mechanism for creating a core damage event. Figure A.1(a) is a *probability of frequency* (POF) curve for a risk assessment of a specific U.S. nuclear power plant performed in the 1980s timeframe using the methodology of Chapter 2. This curve tells us that the mean frequency of core damage for this particular plant is about once in some 4300 years. It also tells us that there is uncertainty in the core damage frequency. In particular, it tells us that the core damage frequency is uncertain by a factor of approximately 8.5 between the 5th and 95th percentile. As important as it is to know the core damage frequency and its uncertainty, more information is needed before a basis exists to recommend actions that might reduce that frequency. We begin

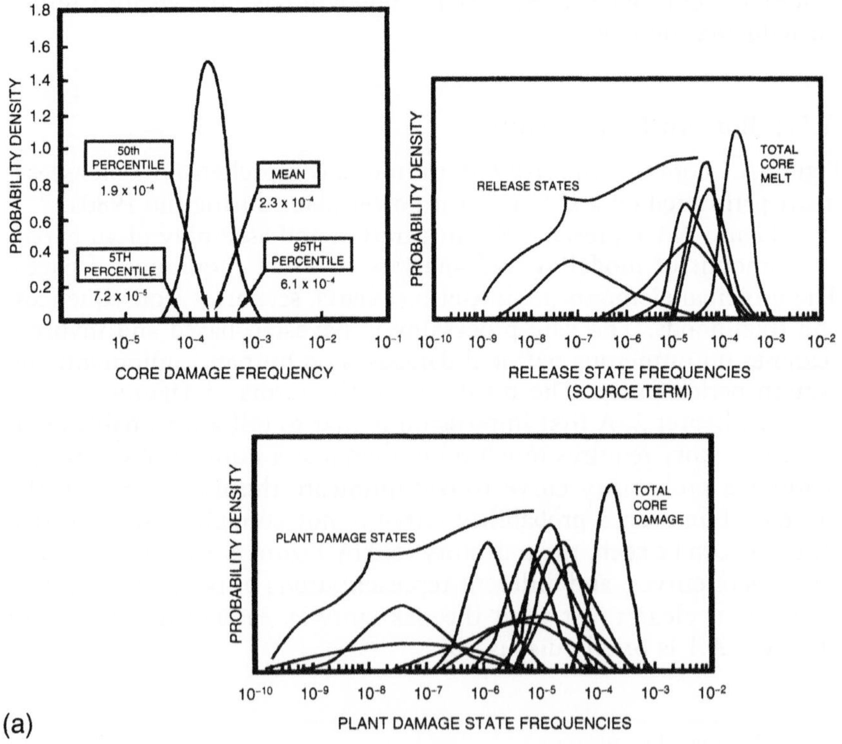

FIGURE A.1. (a) Nuclear power plant risk assessment results.

(Continued)

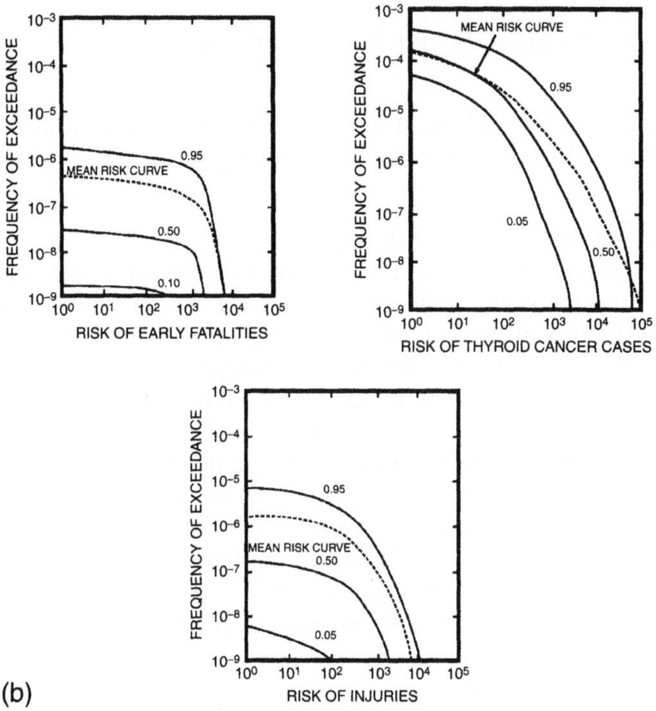

(b)

FIGURE A.1. *Cont'd.* (b) Nuclear power plant risk assessment results.

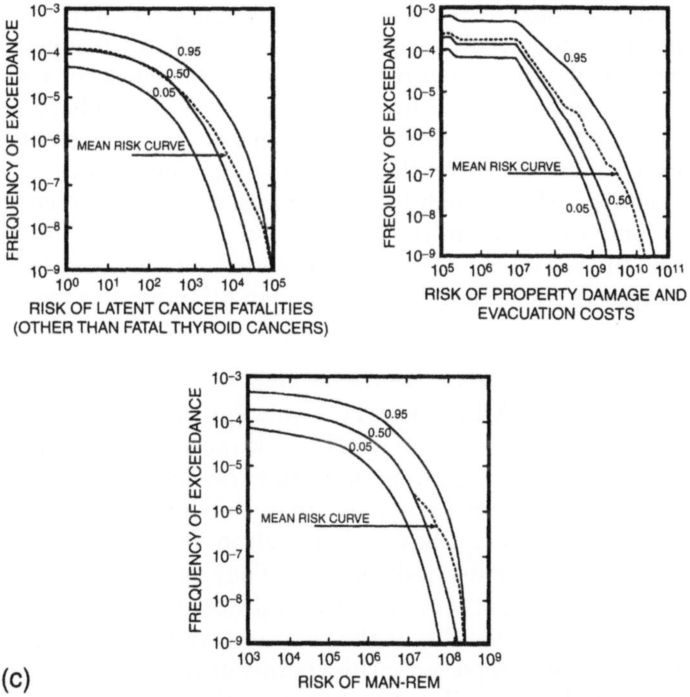

(c)

FIGURE A.1. *Cont'd.* (c) Nuclear power plant risk assessment results.

that process by probing into what kind of damage and radionuclide release might take place as a result of a nuclear power plant accident.

Plant Damage State Frequencies

Plant damage states describe the different possible states (conditions) of the reactor core at the time of reactor vessel failure. Figure A.1(a) shows in probability of frequency format the various plant damage states that were defined for this particular nuclear power plant. Each plant damage state is characterized by such conditions as timing, pressure, temperature, available coolant, and the status of containment systems. For this risk assessment, eight plant damage states were defined. An example of a plant damage state is an overpressure event with failure of pressure suppression systems. The ninth curve in the figure showing the plant damage state frequencies is the core damage frequency curve. A comprehensive risk assessment of a nuclear power plant consists of three models—a plant model, a containment model, and a site model. The plant damage states are the output of the plant model and the input to the containment model as they represent the threats to the containment system.

Release State Frequencies

The release state frequencies, also referred to as release categories, define the source term for the site model and in particular are representations of types, quantities, timings, and elevations of radioisotopes released to the atmosphere. Figure A.1(a) indicates that five release states were defined for this particular assessment. Release states are determined by the state of the containment at the time of core damage, the availability of containment engineered safety features, and the availability of filtration and other mechanisms for radionuclide removal. The release states, or release categories, are the output of the containment model and the input to the site model.

Risk Curves for Various Risk Measures

Figures A.1(b) and A.1(c) answer the question of what is the risk in terms of six different risk measures. The risk measures are early fatalities, injuries, cancer cases, latent fatalities, radiation dose, and property damage. These curves known as "frequency of exceedance" curves or "complementary cumulative distribution functions" are becoming increasingly known as "risk curves." They too are in the "probability of frequency" format, but with the additional feature of consequence as a parameter. Probability is the parameter of the model for displaying uncertainty and is represented by a family of curves, in this case the 5th, 50th, and 95th percentiles. Also shown is the "mean" risk curve. Some of the important observations about the curves are: (1) their comprehensiveness in terms of the quantification of a variety of risk measures, (2) the extremely low levels of risk involved when compared with almost any other natural or technological risk, and (3) a clear display of the uncertainties involved.

A.5.2 Importance Ranking of Contributors to Risk

While the results illustrated in Figure A.1 are extremely important in answering the basic questions of what is the risk and what are the uncertainties involved, they only scratch the surface of what is learned from a comprehensive quantitative risk assessment. In the process of evolving to the bottom line results of Figure A.1, information is developed to answer many more questions such as what is contributing to the risk and what actions can be taken to reduce the risk and assure that the risk is reasonably managed. Also, there is the issue of how to use risk assessment during the design process of a nuclear power plant, or any facility, to evolve to a design that is balanced in terms of actions to control the risk. That facet is covered in Section A.5.3.

Figure A.2 is taken from an actual probabilistic risk assessment of a U.S. nuclear power plant. It ranks the scenarios contributing to the different risk measures. The scenario descriptions are greatly abbreviated to simplify the figure, but the most important events in the scenarios are identified.

Besides the fact that the risks are very small, the most important point made by Figure A.2 is that there are different measures of risk, and each is driven by different scenarios. This is an extremely important result of a quantitative risk assessment. This means that when doing risk assessment and ranking the importance of contributors to risk, it is essential to be very clear about what is being used to measure the risk. It is also important to realize that different measures of risk are often necessary to answer the question, what is the risk? Consider the different risk measures of Figure A.2, core damage frequency, frequency of early deaths, and frequency of latent effects. While Scenario 1 is the most important contributor to core damage frequency, it ranks as 15th in importance to the risk measure of frequency of early deaths and 7th in importance to the frequency of latent effects (injuries and deaths). On the other hand, the most important scenario leading to deaths is not the most important contributor leading to core damage and ranks 7th in the list of scenarios important to latent effects.

The above brings up an important issue in making decisions about how to measure risk. That issue is, "which measure is best?" In the case of a nuclear power plant it is clear that if the reactor core is not severely damaged, then there is no risk of radiation exposure

Scenario	Rank with Respect to Core Melt Frequency	Mean Annual Frequency (Contribution to Core Melt)	Relative Rank with Respect to Latent Effects Release Frequency	Mean Annual Frequency of Latent Effects Release	Relative Rank with Respect to Early Deaths Release Frequency	Mean Annual Frequency of Early Deaths Release
Small LOCA; Failure of High-Pressure Recirculation Cooling	1	8.2–5	7	8.2–9	2	8.2–9
Large LOCA; Failure of Low-Pressure Recirculation Cooling	2	1.1–5	8	1.1–9	4	1.1–9
Medium LOCA; Failure of Low-Pressure Recirculation Cooling	3	1.1–5	9	1.1–9	5	1.1–9
Fire; Other Fire Areas Such as the Cable Spreading Room, Auxiliary Feedwater Pump Room, etc.	4	6.7–6	10	6.7–10	7	6.7–10
Large LOCA; Failure of Safety Injection	5	6.4–6	11	6.4–10	8	6.4–10
Fire; Specific Fires in Switchgear Room and Cable Spreading Room Causing RCP Seal LOCA and Failure of Power Cables to the Safety Injection Pumps, the Containment Spray Pumps, and Fan Coolers	6	5.7–6	1	5.7–6	3	1.1–9
Seismic; Loss of Control or AC Power	7	4.7–6	2	4.7–6	6	9.4–10
Interfacing System LOCA	15	5.7–7	5	5.7–7	1	5.7–7

FIGURE A.2. Comparison of core damage and release frequency contributions.

to members of the public. On first impression, this would suggest that core damage frequency would be a good measure of risk as it is a surrogate for radiation safety to the public. Furthermore, core damage frequency is much easier to determine with respect to nuclear plant accidents than is the frequency of health effects to the public. That is, there is more confidence (less uncertainty) in the calculation of the likelihood of a core damage event than there is in the calculation of the likelihood of radiation health effects to the public. This is one of the important reasons why core damage frequency has been favored as a measure of nuclear power plant risk. Otherwise, the obvious choice for safety risk would be injuries and fatalities. But there is a need to be cautious about using surrogates to such indicators as health and safety risk as is suggested by Figure A.2. The caution comes from the fact that the risk measure is a frequency of something, not just core damage, injuries, or fatalities. These frequencies are not linearly related. Taking actions to change the frequency of a precursor risk measure has to be examined for how it impacts downstream risk measures. The important fact is the impact could be negative or positive, i.e., it is not always clear whether the impact decreases or increases the value of the downstream risk measure. For example, it would not be good to decrease the core damage frequency and find that the fix to do so results in increasing the frequency of deaths. How could that happen? Suppose we design a plant or make changes to the plant that increases the pressure capacity of the primary system as a way to reduce the core damage frequency. Now we have a situation where we have lowered the core damage frequency for a certain class of core damage scenarios and, most likely, this will result in a lower total core damage frequency. Under these conditions fewer high-pressure transients will fail the primary system, but when they do there are now higher pressure transients per transient seen by the containment system. The result is a greater threat to the containment per transient. In other words, depending on the mismatch of the frequencies of the risk measures, we may have increased the failure frequency of the secondary containment and possibly the frequency of injuries and fatalities.

What this all means is that it is not enough to design just for lower core damage frequencies in the case of nuclear power plants. It is important to take a total systems approach and to understand the coupled processes between events that lead to core damage and events that lead to containment failure and off-site consequences.

Other important results from a comprehensive quantitative risk assessment are illustrated in Figures A.3–A.5.

Because of radioactive decay of fission products and the heat that they produce, there remains the need to maintain cooling of a reactor core for some time following termination of the chain reaction, i.e.,

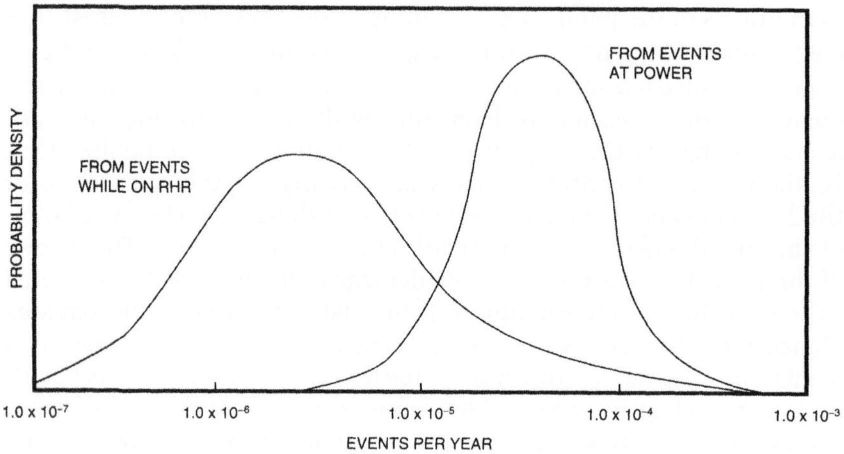

FIGURE A.3. Probability distribution for core damage.

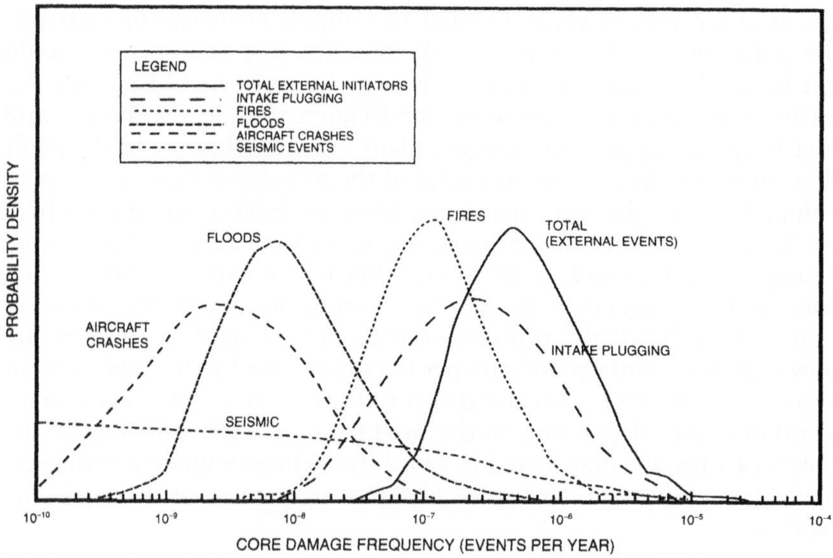

FIGURE A.4. Contributions of specific external events to CDF.

following shutdown of the reactor. Consequently, there remains the risk of fuel damage and the possible release of radiation and radionuclides during times when the reactor is not operating at power. This contribution to the overall risk of exposing workers and the public to radiation must also be a part of a comprehensive risk assessment.

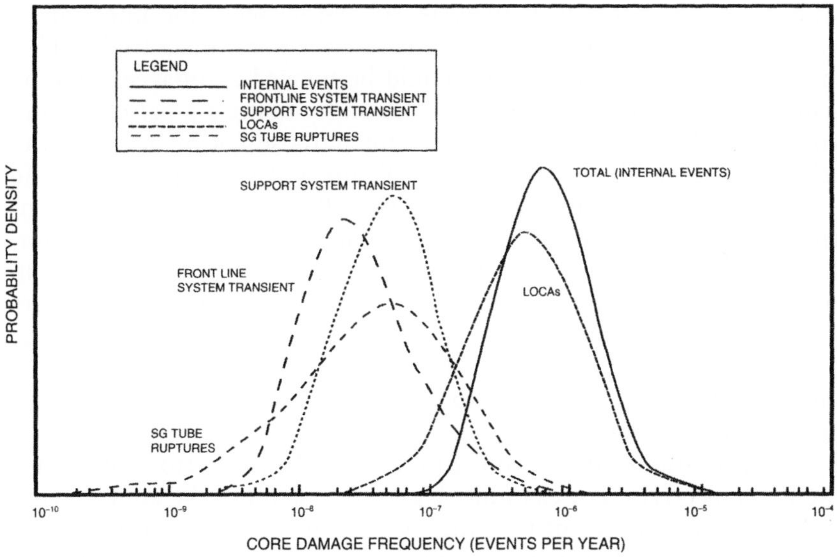

FIGURE A.5. Contributions of major initiating event classes to CDF from internal events.

Figure A.3 not only presents the core damage frequency of a particular nuclear power plant during off-power conditions, but also provides a direct comparison with the risk of core damage frequency during at-power conditions. More than just comparing the two risks, Figure A.3 quantifies the uncertainties associated with the two contributions to overall risk. As can be seen from Figure A.3, the risk from off-power conditions is very small compared to at-power conditions, but it is certainly not zero. Mean values of the risk of several plants indicate that the contribution to overall core damage frequency of off-power conditions varies approximately from 10% to 30% of the total risk.

In the process of "importance ranking" contributors to risk, it is also essential to quantify any external threats to the plant, i.e., threats that are not due to inherent plant operations but are as a result of such phenomena as external fires, severe weather, aircraft impacts, and earthquakes. Figure A.4 is an example of such an analysis. The most significant external event for this particular plant was the risk of a severe storm creating river debris and other conditions that could lead to plugging the intake lines for secondary cooling. Other visible contributors were external fires, floods, aircraft crashes, and earthquakes.

Figure A.4 makes a very important point about just how much analysis of a contribution should be performed to reduce the uncertainty in the risk measure. It also provides strong evidence for the value of doing uncertainty analysis. As can be seen, the uncertainty range of seismic events covers many orders of magnitude. This amount of uncertainty by itself could be viewed as unacceptable by many analysts, and the push would be on for doing a great deal more work; more work than is necessary for purposes of quantifying the risks. Even with all the uncertainty, Figure A.4 tells us that seismic is not a significant contributor to the overall risk considering the contribution from other sources. If it turned out to be a major contributor, then efforts to reduce the uncertainty could be justified. Summing up the point, if the uncertainty of a contributor varies over many orders of magnitude, say five between the 5th and 95th percentiles, and the risk over that interval varies from 10^{-12} to 10^{-7}, then a risk that is already in the 10^{-4} range due to other contributors isn't going to be significantly impacted.

Presentations similar to Figure A.4 can be made from many different perspectives. For example, Figure A.5 displays the contribution of different categories of initiating events to the risk. An initiating event is an event that disturbs an otherwise normally operating system and triggers a possible accident scenario. Figure A.5 indicates that for this particular plant, the initiating event category of loss of coolant accidents was the major contributor to core damage frequency. Other important contributors were transients of support systems (emergency power, equipment cooling systems, ventilation systems, etc.), front-line system transients (turbines, feedwater pumps, primary coolant, etc.), and steam generator tube ruptures. Once the scenarios have been ranked in terms of importance to risk, it is usually straightforward to identify the events and equipment items within that scenario most important to risk.

A.5.3 Risk Assessment as a Design Tool

Figure A.6 illustrates how risk assessment can be used to achieve maximum benefit from the performance of safety systems during the evolution of a nuclear power plant design. As with Figures A.1–A.5, Figure A.6 shows the actual results taken from a nuclear power plant design project.

The first iteration of the risk assessment determined a core damage frequency and identified the major system and operator activities contributing to the risk. Design flaws were identified with several systems, including the service water system and the safeguards chilled water system—systems critical to cooling safety and support systems during emergency conditions. As a result of the risk assessment, several design changes were made to reduce their contribution to risk. Meanwhile, the core damage frequency was reduced by a factor of over three. The second iteration several months later identified a different

System(s) or Operator Action	Percent Reduction in Core Damage Frequency if the Individual System (or Operator Action) Failure Frequency Could Be Reduced to Zero			
	First Iteration	Second Iteration	Third Iteration	Fourth Iteration
1. Electric Power	11	65	43	52
2. Auxiliary Feedwater	9	11	11	31
3. Two Trains of Electric Power Recovered				21
4. Low-Pressure Injection / Decay Heat Removal	4	3	8	19
5. Failure to Reclose PORV / PSVs		5	20	17
6. ESFAS / ECCAS			14	15
7. High-Pressure Injection Systems	3	9	15	14
8. Operator Recovery of Electric Power during Station Blackout		50	14	14
9. Sump Recirculation Water Source				11
10. Component Cooling Water			3	8
11. Throttle HPI Flow (Operator Action)			1	4
12. Failure of Main Steam Safety Valve to Reclose			1	4
13. Service Water	32	<1	10	4
14. Safeguards Chilled Water	20	8	13	1
15. BWST Suction Valve				1
16. Containment Isolation			1	
Relative Core Melt Frequency	1.00	0.30	0.10	0.06

FIGURE A.6. Contributors to core damage for four phases of risk management.

set of contributors, but against a much lower risk baseline due to the improvements in the design. Again, there were some standout contributors, most notably the systems for the recovery of the loss of off-site electric power during station blackout.

The third iteration occurring later in the design process began to manifest more balance in the performance of safety systems and less opportunity for any single fix having a major contribution to reducing the risk. In this case it was necessary to consider several design changes to have a significant impact on the risk. However, by the third iteration the mean core damage frequency had been reduced by a factor of 10. By the fourth iteration and the near completion of the design, the calculated core damage frequency had been reduced by a factor of almost 17 and the safety systems were much more balanced in their contribution to protecting the plant. There are some things worth pointing out in the fourth iteration. The 52% contribution from electric power may appear to be a source for further reducing the risk. It is not a good source for corrective action because of its pervasiveness throughout the plant and its association with thousands of subsystems and components. No single or few component, or subsystem, fixes would materially improve the core damage frequency.

An analogy that sometimes clarifies the above process is the mental exercise that comes from thinking of a lake with rocks rising above the surface of the water and viewing the rocks as contributors to risk. What happens when the big rocks are removed from the lake? First, they are no longer there; second, the lake level drops as a result of their removal; and third, new rocks appear because of the lower lake level. Repeating this process is very analogous to what is taking place with the risk assessment and design process portrayed in Figure A.6. As the core damage frequency is reduced by removing the important contributors, other contributors begin to appear that become candidates for removal or modification. Eventually you get to a point of diminishing risk benefit.

The point of this example is to illustrate the robust amount of information that comes from a comprehensive risk assessment and the options it provides for cost-effective risk management. The case studies of Chapters 3–6 of this book involve much more limited scope assessments and were primarily for the purpose of demonstrating the six-step process of quantitative risk assessment.

References

[1] Jeffreys, H., *Scientific Inference*, 2nd edition, University Press, Cambridge, United Kingdom, Cambridge, 1957.

[2] Cox, R. T., "Probability, Frequency, and Reasonable Expectation," *Am. J. Phys.* 1946, *14*, 1–13.

[3] Shannon, C. E., "A Mathematical Theory of Communication," *Bell Syst. Tech. J.* 1948, *27*, 379–423 see also pp. 623–656, July and October.

[4] Pólya, G., *How to Solve It, Mathematics and Plausible Reasoning*, Princeton University Press, Vols. I and II, 1954.

[5] Raiffa, H., *Decision Analysis*, McGraw-Hill Primis Custom Publishing, Columbus, Ohio, 1996.

[6] Jaynes, E. T., *Probability Theory: The Logic of Science*, Cambridge University Press, United Kingdom, Cambridge, 2003.

[7] Green, A. E., Bourne, A. J., *Reliability Technology*, John Wiley, London, United Kingdom, 1972.

[8] Kumamoto, H., Henley, E. J., *Probabilistic Risk Assessment and Management for Engineers and Scientists*, 2nd edition, IEEE Press, New York, 1996, pp. 1–54.

[9] Neufeld, J., *The Development of Ballistic Missiles in the United States Air Force 1945–1960*, Office of Air Force History, U.S. Air Force, Washington, D.C., 1990, pp. 169–215.

[10] Farmer, F. R., The Growth of Reactor Safety Criteria in the United Kingdom, In: *Proceedings of the Anglo-Spanish Nuclear Power Symposium*, Madrid, Spain, 1964.

[11] Holmes and Narver, Inc., *Reliability Analysis of Nuclear Power Plant Protective Systems*, Prepared for U.S. Atomic Energy Commission, Washington, D.C., 1967, HN-190.

[12] Garrick, B. J., *Unified Systems Safety Analysis for Nuclear Power Plants,* Ph.D. Thesis, University of California, Los Angeles, 1968.
[13] U.S. NRC (U.S. Nuclear Regulatory Commission), Reactor Safety Study: An Assessment of Accident Risks in U.S. Commercial Nuclear Power Plants, WASH-1400 (NUREG-75/014), 1975.
[14] PLG (Pickard, Lowe, and Garrick, Inc.), Westinghouse Electric Corporation, and Fauske and Associates, Inc. "Zion Probabilistic Safety Study," prepared for Commonwealth Edison Company, Chicago, Illinois, 1981.
[15] PLG (Pickard, Lowe, and Garrick, Inc.), Westinghouse Electric Corporation, and Fauske and Associates, Inc. "Indian Point Probabilistic Safety Study," prepared for Consolidated Edison Company of New York, Inc., and the New York Power Authority, 1982.
[16] Kaplan, S., Garrick, B. J., "On the Quantitative Definition of Risk," *Risk Anal.* 1981, *1*(1), 11–27.
[17] (NASA) National Aeronautics and Space Administration, Probabilistic Risk Assessment Procedures Guide for NASA Managers and Practitioners, prepared for Office of Safety and Mission Assurance, National Aeronautics and Space Administration, Washington, D.C., 2002.
[18] Boyd, G. J., St. Pierre, G., "Risk Management Program for the Disposal of Chemical Agents and Munitions," presented at the Society for Risk Analysis: Special Symposium on Quantitative Risk Assessment, sponsored by the Family Foundations of Chauncey Starr and B. John Garrick; B. John Garrick Foundation for the Advancement of the Risk Sciences, Laguna Beach, CA, May 31–June 2 2001.
[19] AEC (U.S. Atomic Energy Commission)., Theoretical Possibilities and Consequences of Major Accidents in Large Nuclear Power Plants, WASH-740, March, 1957.
[20] Garrick, B. J., "Lessons Learned from 21 Nuclear Plant Probabilistic Risk Assessments," *Nucl. Technol.* 1989, *84*, 319–330.

Appendix B: Supporting Evidence for the Case Study of the Hurricane Risk in New Orleans, LA

B.1 Hurricane Risk Assessment for the Period 1900–2004

Chapter 3, Section 3.2.4, describes the overall quantification of the human fatality risk of a major hurricane impacting New Orleans based on hurricane data for the period 1900–2004. The data used in the quantification of the event trees in Chapter 3 is contained in Figures B.1–B.42 of this Appendix, where Figures B.1–B.15 are based on Table 3.4 of Chapter 3 and the basis for Figures B.37–B.42 is discussed in Section 3.2. These figures show (1) the original uncertainty distribution (prior) used for the initiating events and various branch points in the event

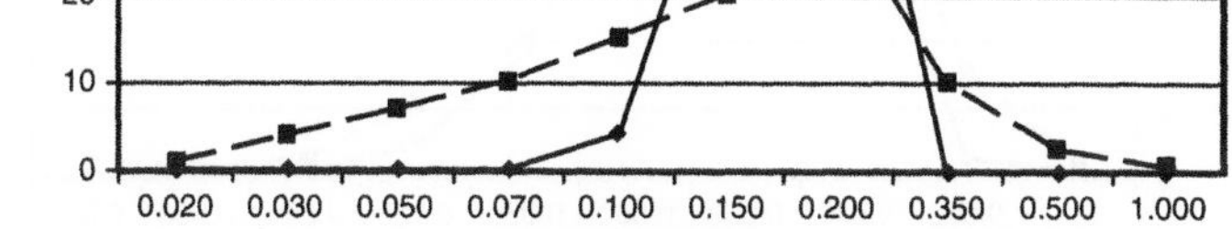

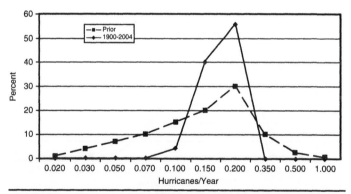

Based on hurricane tracking records, 18 of the 41 hurricanes were categorized as being in the Gulf of Mexico for 48 hours or less before landfall. These hurricanes were placed in Initiating Event 1. There is a wide variation in the period of time between successive hurricanes making up Initiating Event 1, with a range of periodicity of approximately one per year to one per 20 years. The uncertainty about the frequency of Initiating Event 1 is shown above. A subjective "Prior" uncertainty distribution was generated and updated using Bayesian techniques with the actual experience of 18 hurricanes in 105 years. The mean value for the frequency of Initiating Event 1 was calculated to be equal to 0.176 hurricanes per year or approximately one Initiating Event 1 every 5.7 years.

FIGURE B.1. Uncertainty distribution, Initiating Event 1 frequency (hurricanes from 1900–2004).

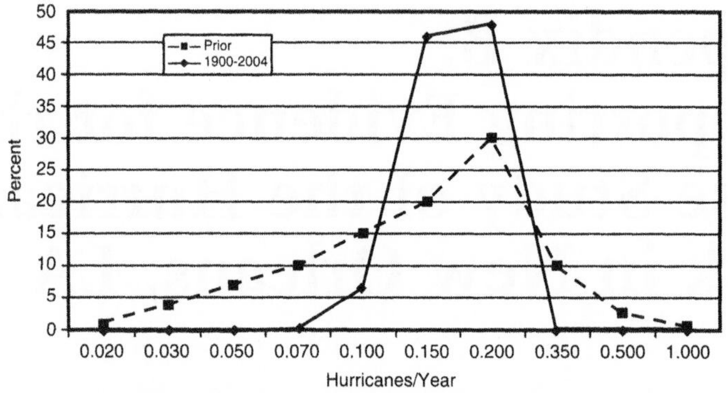

Based on hurricane tracking records, 17 of the 41 hurricanes were categorized as being in the Gulf of Mexico for between 48 hours and 72 hours before landfall. These hurricanes were placed in Initiating Event 2. There is a wide variation in the period of time between successive hurricanes making up Initiating Event 2, with a range of periodicity of approximately two per year to one per 20 years. The uncertainty about the frequency of Initiating Event 2 is shown above. A subjective "Prior" uncertainty distribution was generated and updated using Bayesian techniques with the actual experience of 17 hurricanes in 105 years. The mean value for the frequency of Initiating Event 2 was calculated to be equal to 0.170 hurricanes per year or approximately one Initiating Event 1 every 5.8 years.

FIGURE B.2. Uncertainty distribution, Initiating Event 2 frequency (hurricanes from 1900–2004).

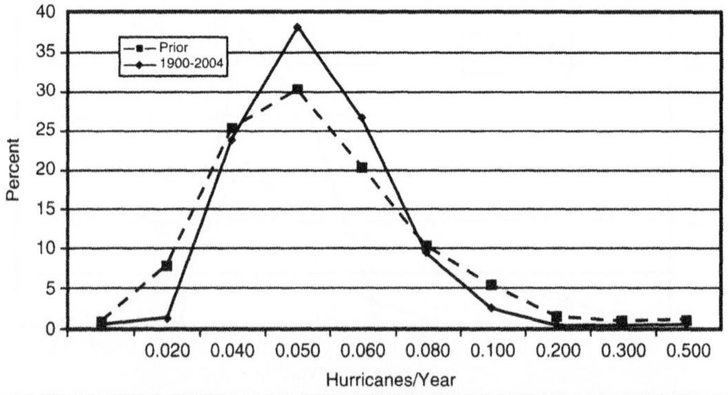

Based on hurricane tracking records, 6 of the 41 hurricanes were categorized as being in the Gulf of Mexico for greater than 72 hours before landfall. These hurricanes were placed in Initiating Event 3. There is a wide variation in the period of time between successive hurricanes making up Initiating Event 3, with a range of periodicity of approximately one per 3 years to one per 30 years. The uncertainty about the frequency of Initiating Event 2 is shown above. A subjective "Prior" uncertainty distribution was generated and updated using Bayesian techniques with the actual experience of 6 hurricanes in 105 years. The mean value for the frequency of Initiating Event 3 was calculated to be equal to 0.0538 hurricanes per year or approximately one Initiating Event 3 every 18 years.

FIGURE B.3. Uncertainty distribution, Initiating Event 3 frequency (hurricanes from 1900–2004).

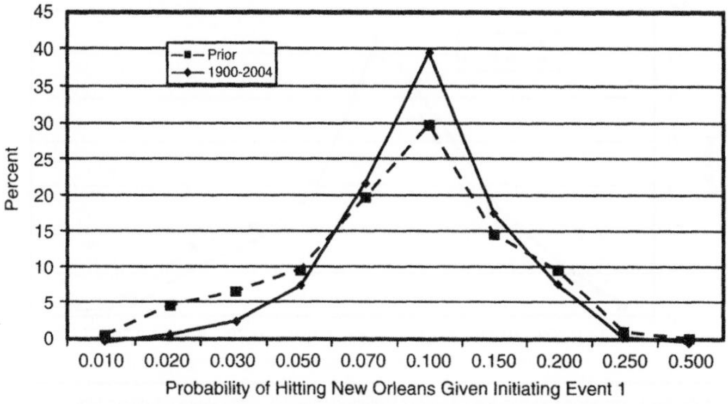

Of the 18 hurricanes listed in the category Initiating Event 1, 2 hurricanes were considered to directly impact New Orleans, unnamed hurricane of 1915 and Hurricane Betsy in 1965. Of the 18 hurricanes listed in the category Initiating Event 1, 2 hurricanes were considered to be close calls, unnamed hurricane of 1947 and Hurricane Andrew in 1992. The uncertainty about the fraction of Initiating Event 1 hurricanes that directly impact New Orleans is shown above. A subjective "Prior" uncertainty distribution was generated and updated using Bayesian techniques with the actual experience of 2 hurricanes out of 18 hitting New Orleans. The mean value for the fraction of Initiating Event 1 hurricanes hitting New Orleans was calculated to be equal to 0.105, or approximately one of every ten hurricanes.

FIGURE B.4. Uncertainty distribution, Initiating Event 1 hits New Orleans (hurricanes from 1900–2004).

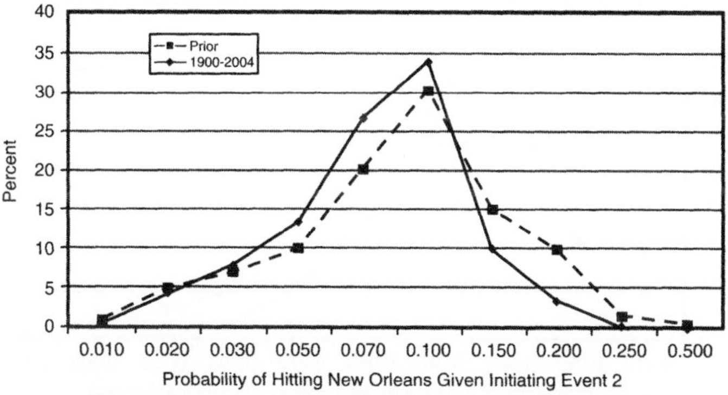

Of the 17 hurricanes listed in the category Initiating Event 2, 1 hurricane was considered to directly impact New Orleans, unnamed hurricane of 1909 (Grand Isle, LA). Of the 17 hurricanes listed in the category Initiating Event 2, 1 hurricane was considered to be a close call, Hurricane Camille in 1969. The uncertainty about the fraction of Initiating Event 2 hurricanes that directly impact New Orleans is shown above. A subjective "Prior" uncertainty distribution was generated and updated using Bayesian techniques with the actual experience of 1 hurricane out of 17 hitting New Orleans. The mean value for the fraction of Initiating Event 2 hurricanes hitting New Orleans was calculated to be equal to 0.0847, or approximately 1 of every 13 hurricanes.

FIGURE B.5. Uncertainty distribution, Initiating Event 2 hits New Orleans (hurricanes from 1900–2004).

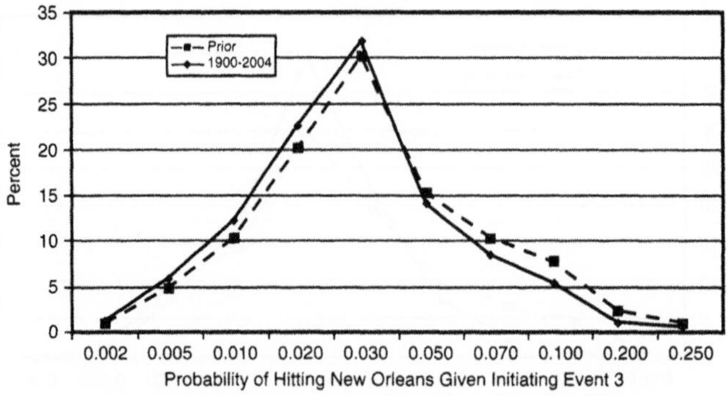

Of the six hurricanes listed in the category Initiating Event 3, there were no direct hits or close calls. The uncertainty about the fraction of Initiating Event 3 hurricanes that directly impact New Orleans is shown above. A subjective "Prior" uncertainty distribution was generated and updated using Bayesian techniques with the actual experience of zero hurricanes out of six hitting New Orleans. The mean value for the fraction of Initiating Event 3 hurricanes hitting New Orleans was calculated to be equal to 0.0348, or approximately 1 of every 29 hurricanes.

FIGURE B.6. Uncertainty distribution, Initiating Event 3 hits New Orleans (hurricanes from 1900–2004).

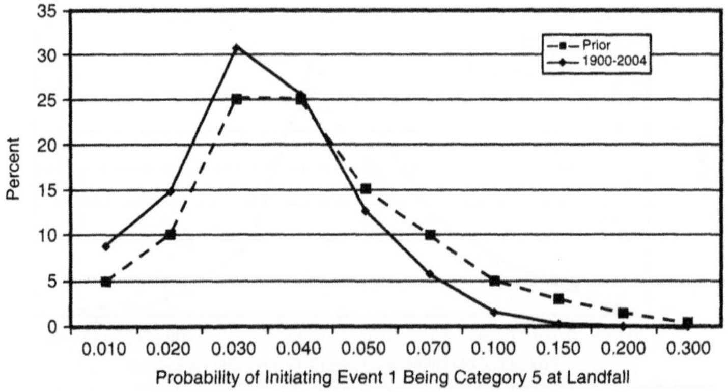

Of the 18 hurricanes listed in the category Initiating Event 1, there were no hurricanes that were Category 5 at landfall. There was one hurricane (Allen, 1980) that weakened from Category 5 to Category 4 before landfall. The uncertainty about the fraction of Initiating Event 1 hurricane making landfall as Category 5 is shown above. A subjective "Prior" uncertainty distribution was generated and updated using Bayesian techniques with the actual experience of 0 hurricanes out of 18 hitting landfall as Category 5. The mean value for the fraction of Initiating Event 1 hurricanes hitting landfall as Category 5 was calculated to be equal to 0.0357, or approximately 1 of every 28 hurricanes.

FIGURE B.7. Uncertainty distribution, Initiating Event 1 hits as Category 5 (hurricanes from 1900–2004).

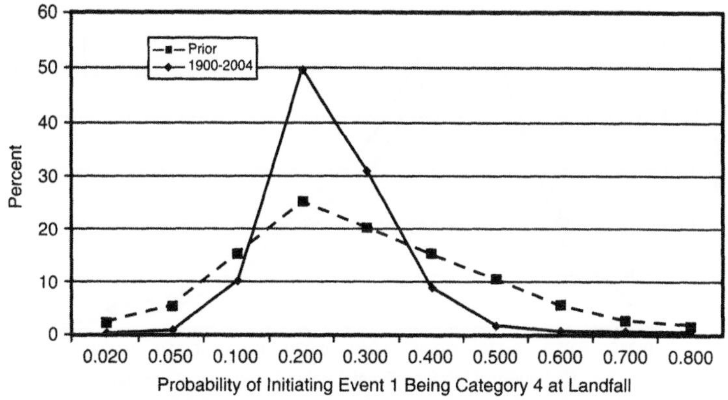

Of the 18 hurricanes listed in the category Initiating Event 1, there were 4 hurricanes that were Category 4 at landfall. There were also four hurricanes that weakened from Category 4 to Category 3 before landfall. The uncertainty about the fraction of Initiating Event 1 hurricanes making landfall as Category 4 is shown above. A subjective "Prior" uncertainty distribution was generated and updated using Bayesian techniques with the actual experience of 4 hurricanes out of 18 hitting landfall as Category 4. The mean value for the fraction of Initiating Event 1 hurricanes hitting landfall as Category 4 was calculated to be equal to 0.2410, or approximately one of every four hurricanes.

FIGURE B.8. Uncertainty distribution, Initiating Event 1 hits as Category 4 (hurricanes from 1900–2004).

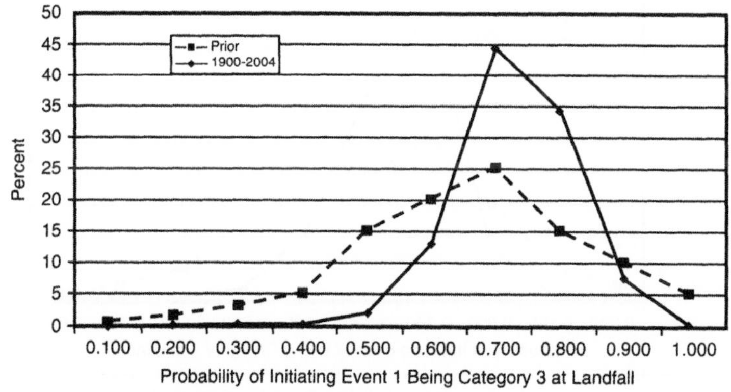

Of the 18 hurricanes listed in the category Initiating Event 1, there were 14 hurricanes that were Category 3 at landfall. The uncertainty about the fraction of Initiating Event 1 hurricanes making landfall as Category 3 is shown above. A subjective "Prior" uncertainty distribution was generated and updated using Bayesian techniques with the actual experience of 14 hurricanes out of 18 hitting landfall as Category 3. The mean value for the fraction of Initiating Event 1 hurricanes hitting landfall as Category 3 was calculated to be equal to 0.7320, or approximately three of every four hurricanes.

FIGURE B.9. Uncertainty distribution, Initiating Event 1 hits as Category 3 (hurricanes from 1900–2004).

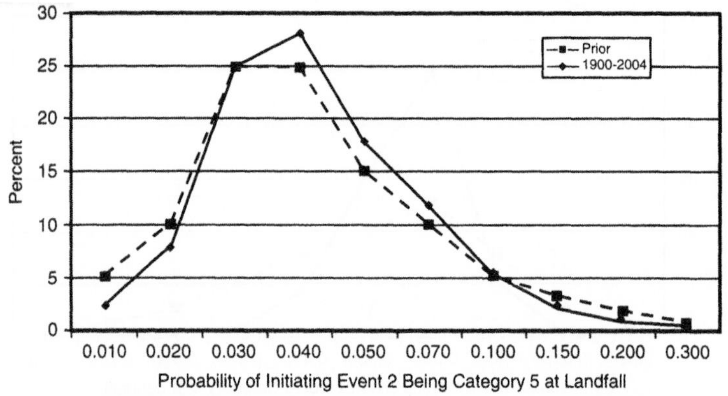

Of the 17 hurricanes listed in the category Initiating Event 2, there was 1 hurricane that was Category 5 at landfall (Camille, 1969). There were two hurricanes that weakened from Category 5 to Category 3 before landfall. The uncertainty about the fraction of Initiating Event 2 hurricanes making landfall as Category 5 is shown above. A subjective "Prior" uncertainty distribution was generated and updated using Bayesian techniques with the actual experience of 1 hurricane out of 17 hitting landfall as Category 5. The mean value for the fraction of Initiating Event 2 hurricanes hitting landfall as Category 5 was calculated to be equal to 0.046, or approximately 1 of every 22 hurricanes.

FIGURE B.10. Uncertainty distribution, Initiating Event 2 hits as Category 5 (hurricanes from 1900–2004).

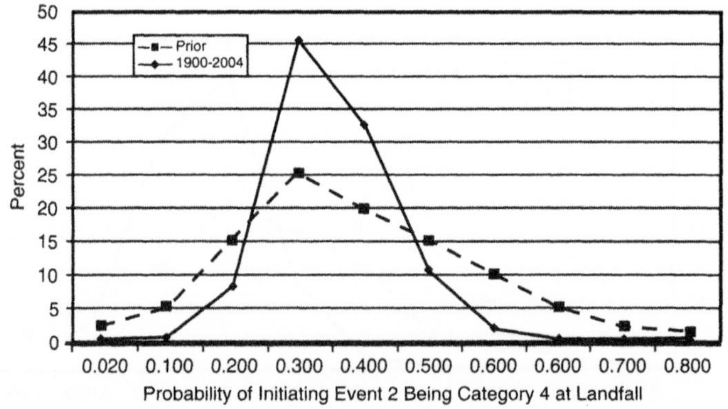

Of the 17 hurricanes listed in the category Initiating Event 2, there were 4 hurricanes that were Category 4 at landfall. There were also three hurricanes that weakened from Category 4 to Category 3 before landfall. The uncertainty about the fraction of Initiating Event 2 hurricanes making landfall as Category 4 is shown above. A subjective "Prior" uncertainty distribution was generated and updated using Bayesian techniques with the actual experience of 4 hurricanes out of 17 hitting landfall as Category 4. The mean value for the fraction of Initiating Event 2 hurricanes hitting landfall as Category 4 was calculated to be equal to 0.251, or approximately one of every four hurricanes.

FIGURE B.11. Uncertainty distribution, Initiating Event 2 hits as Category 4 (hurricanes from 1900–2004).

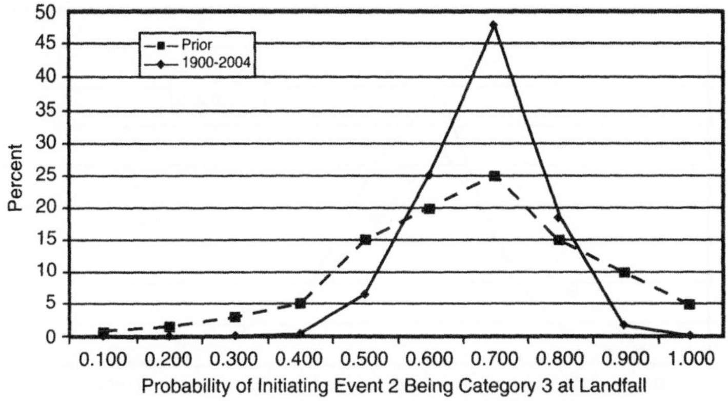

Of the 17 hurricanes listed in the category Initiating Event 2, there were 12 hurricanes that were Category 3 at landfall. The uncertainty about the fraction of Initiating Event 2 hurricanes making landfall as Category 3 is shown above. A subjective "Prior" uncertainty distribution was generated and updated using Bayesian techniques with the actual experience of 12 hurricanes out of 17 hitting landfall as Category 3. The mean value for the fraction of Initiating Event 2 hurricanes hitting landfall as Category 3 was calculated to be equal to 0.682, or approximately two of every three hurricanes.

FIGURE B.12. Uncertainty distribution, Initiating Event 2 hits as Category 3 (hurricanes from 1900–2004).

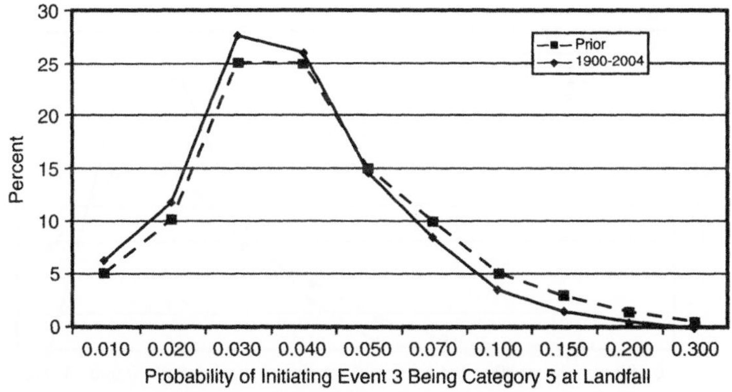

Of the six hurricanes listed in the category Initiating Event 3, there was no hurricane that was Category 5 at landfall. The uncertainty about the fraction of Initiating Event #3 hurricanes making landfall as Category 5 is shown above. A subjective "Prior" uncertainty distribution was generated and updated using Bayesian techniques with the actual experience of zero hurricanes out of six hitting landfall as Category 5. The mean value for the fraction of Initiating Event 3 hurricanes hitting landfall as Category 5 was calculated to be equal to 0.042, or approximately 1 of every 24 hurricanes.

FIGURE B.13. Uncertainty distribution, Initiating Event 3 hits as Category 5 (hurricanes from 1900–2004).

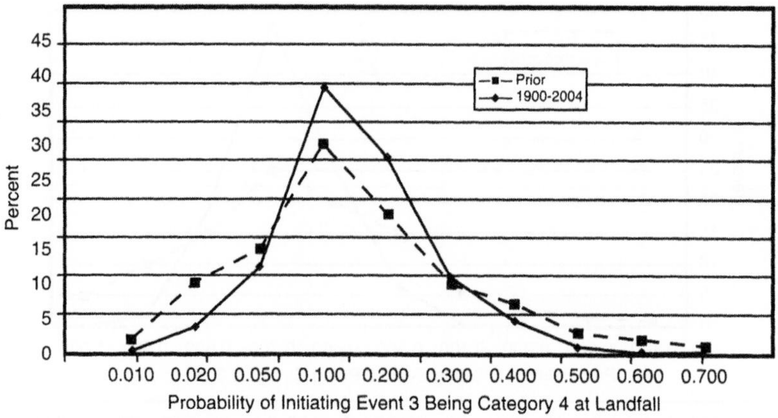

Of the six hurricanes listed in the category Initiating Event 3, there was one hurricane that was Category 4 at landfall. There were also two hurricanes that weakened from Category 4 to Category 3 before landfall. The uncertainty about the fraction of Initiating Event 3 hurricanes making landfall as Category 4 is shown above. A subjective "Prior" uncertainty distribution was generated and updated using Bayesian techniques with the actual experience of one hurricane out of six hitting landfall as Category 4. The mean value for the fraction of Initiating Event 3 hurricanes hitting landfall as Category 4 was calculated to be equal to 0.160, or approximately one of every six hurricanes.

FIGURE B.14. Uncertainty distribution, Initiating Event 3 hits as Category 4 (hurricanes from 1900–2004).

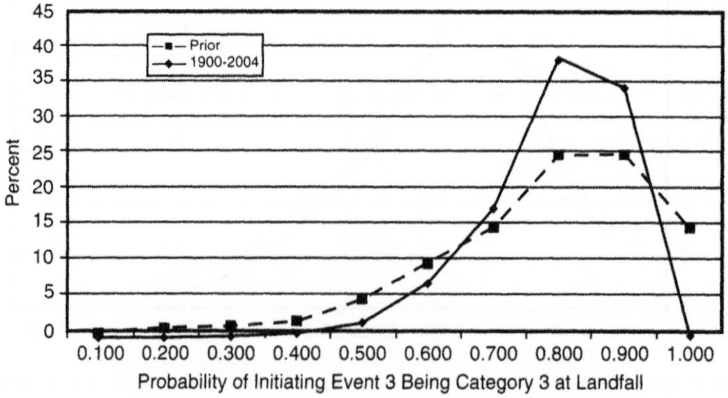

Of the six hurricanes listed in the category Initiating Event 3, there were five hurricanes that were Category 3 at landfall. The uncertainty about the fraction of Initiating Event 3 hurricanes makes landfall as Category 3 is shown above. A subjective "Prior" uncertainty distribution was generated and updated using Bayesian techniques with the actual experience of five hurricanes out of six hitting landfall as Category 3. The mean value for the fraction of Initiating Event 3 hurricanes hitting landfall as Category 3 was calculated to be equal to 0.795, or approximately eight of every ten hurricanes.

FIGURE B.15. Uncertainty distribution, Initiating Event 3 hits as Category 3 (hurricanes from 1900–2004).

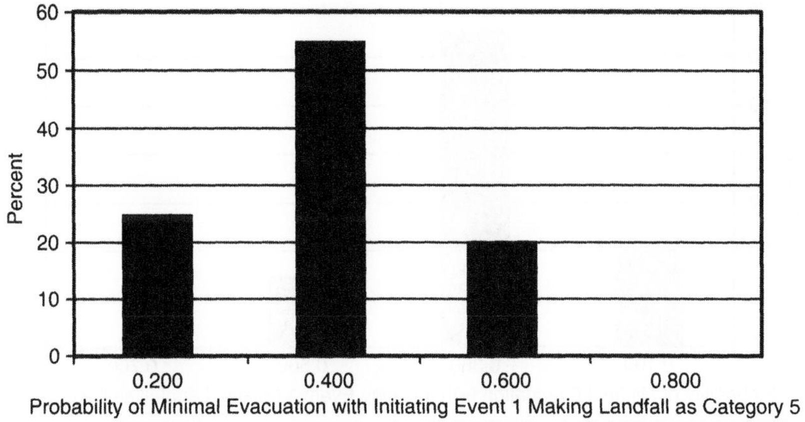

For Initiating Event 1 making landfall as Category 5 hurricane, the uncertainty about the percentage of these hurricanes that will involve "minimal" evacuation is shown above. The uncertainty distribution is subjective. There was no Bayesian updating since there is no data to support an update. The mean percent of minimal evacuations was calculated from the uncertainty distribution to be 39.

FIGURE B.16. Uncertainty distribution, Initiating Event 1 hits as Category 5 following minimal evacuation (hurricanes from 1900–2004 and 1900–1950).

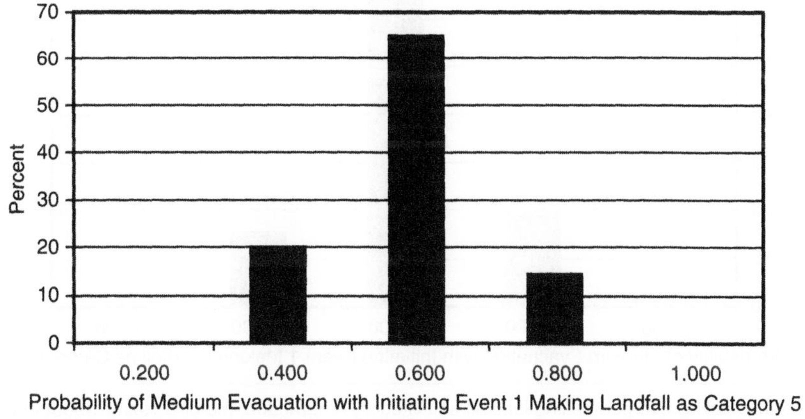

For an Initiating Event 1 making landfall as Category 5 hurricane, the uncertainty about the percentage of these hurricanes that will involve "medium" evacuation is shown above. This uncertainty distribution is subjective. There was no Bayesian updating since there is no data to support an update. The mean percent of "medium" evacuations was calculated from the uncertainty distribution to be 59.

FIGURE B.17. Uncertainty distribution, Initiating Event 1 hits as Category 5 following medium evacuation (hurricanes from 1900–2004 and 1900–1950).

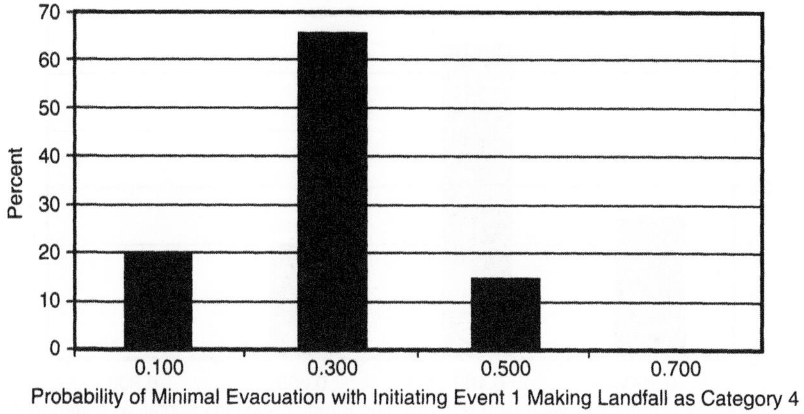

For an Initiating Event 1 making landfall as Category 4 hurricane, the uncertainty about the percentage of these hurricanes that will involve "minimal" evacuation is shown above. This uncertainty is subjective. There was no Bayesian updating since there is no data to support an update. The mean percent of "minimal" evacuations was calculated from the uncertainty distribution to be 29.

FIGURE B.18. Uncertainty distribution, Initiating Event 1 hits as Category 4 following minimal evacuation (hurricanes from 1900–2004 and 1900–1950).

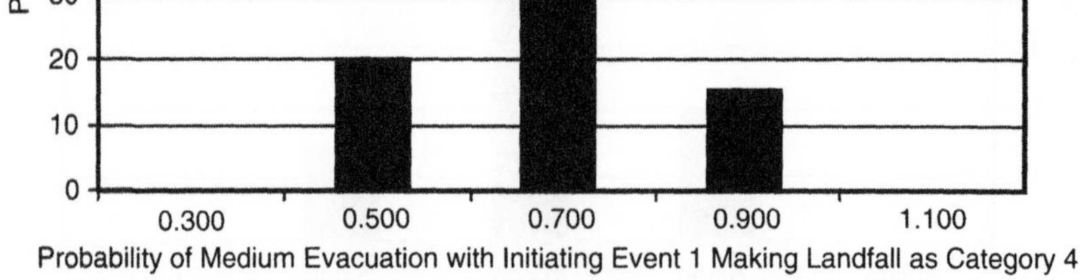

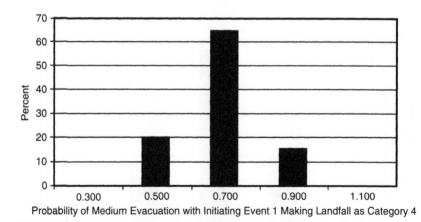

For an Initiating Event 1 making landfall as Category 4 hurricane, the uncertainty about the percentage of these hurricanes that will involve "medium" evacuation is shown above. This uncertainty is subjective. There was no Bayesian updating since there is no data to support an update. The mean percent of "medium" evacuations was calculated from the uncertainty distribution to be 69.

FIGURE B.19. Uncertainty distribution, Initiating Event 1 hits as Category 4 following medium evacuation (hurricanes from 1900–2004 and 1900–1950).

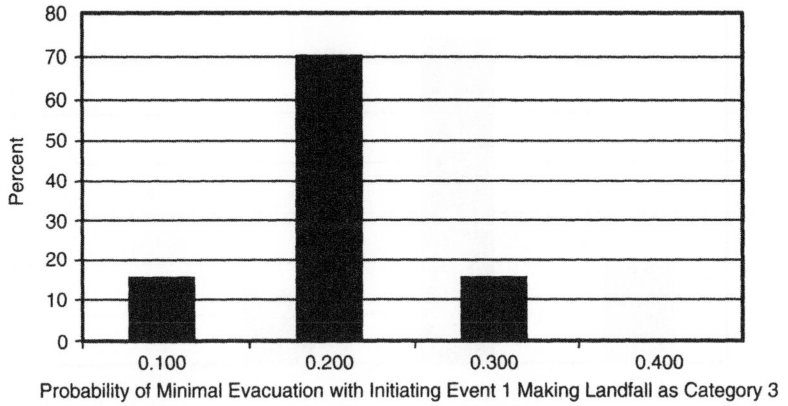

For an Initiating Event 1 making landfall as Category 3 hurricane, the uncertainty about the percentage of these hurricanes that will involve a "minimal" evacuation is shown above. This uncertainty is subjective. There was no Bayesian updating since there is no data to support an update. The mean percent of "minimal" evacuations was calculated from the uncertainty distribution to be 20.

FIGURE B.20. Uncertainty distribution, Initiating Event 1 hits as Category 3 following minimal evacuation (hurricanes from 1900–2004 and 1900–1950).

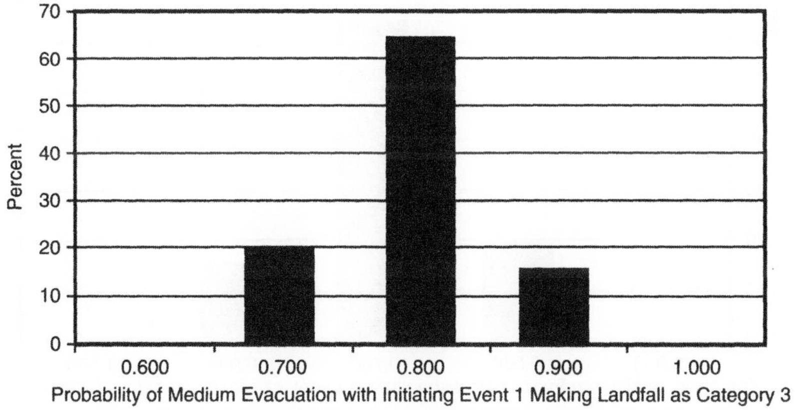

For an Initiating Event 1 making landfall as Category 3 hurricane, the uncertainty about the percentage of these hurricanes that will involve "medium" evacuation is shown above. This uncertainty is subjective. There was no Bayesian updating since there is no data to support an update. The mean percent of "medium" evacuations was calculated from the uncertainty distribution to be 79.5.

FIGURE B.21. Uncertainty distribution, Initiating Event 1 hits as Category 3 following medium evacuation (hurricanes from 1900–2004 and 1900–1950).

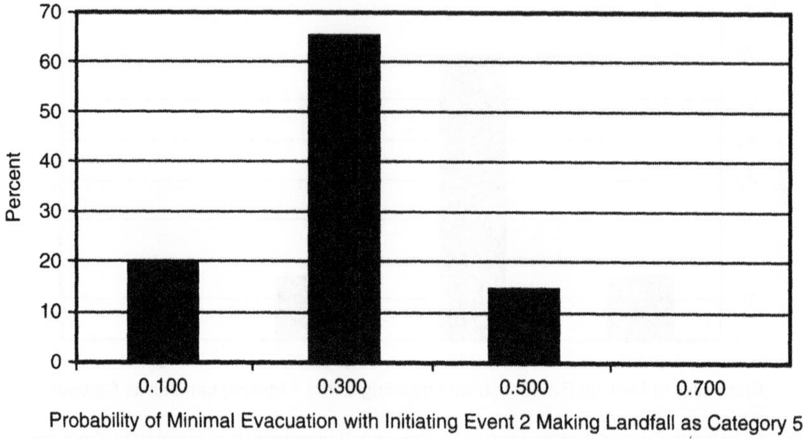

For an Initiating Event 2 making landfall as Category 5 hurricane, the uncertainty about the percentage of these hurricanes that will involve "minimal" evacuation is shown above. This uncertainty is subjective. There was no Bayesian updating since there is no data to support an update. The mean percent of "minimal" evacuations was calculated from the uncertainty distribution to be 29.

FIGURE B.22. Uncertainty distribution, Initiating Event 2 hits as Category 5 following minimal evacuation (hurricanes from 1900–2004 and 1900–1950).

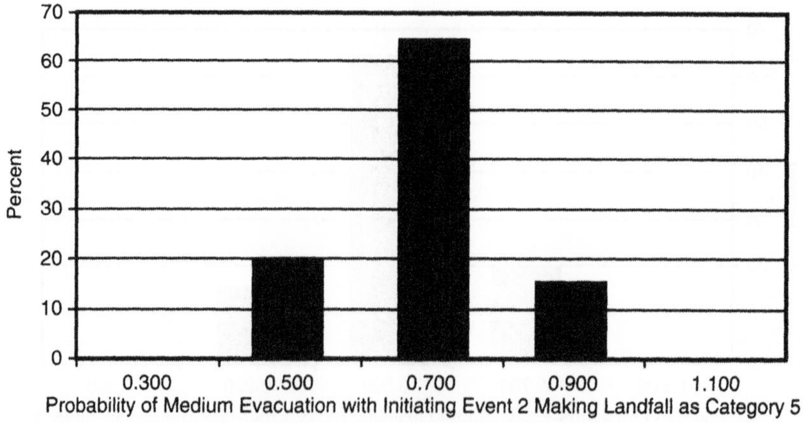

For an Initiating Event 2 making landfall as Category 5 hurricane, the uncertainty about the percentage of these hurricanes that will involve "medium" evacuation is shown above. This uncertainty is subjective. There was no Bayesian updating since there is no data to support an update. The mean percent of "medium" evacuation was calculated from the uncertainty distribution to be 69.

FIGURE B.23. Uncertainty distribution, Initiating Event 2 hits as Category 5 following medium evacuation (hurricanes from 1900–2004 and 1900–1950).

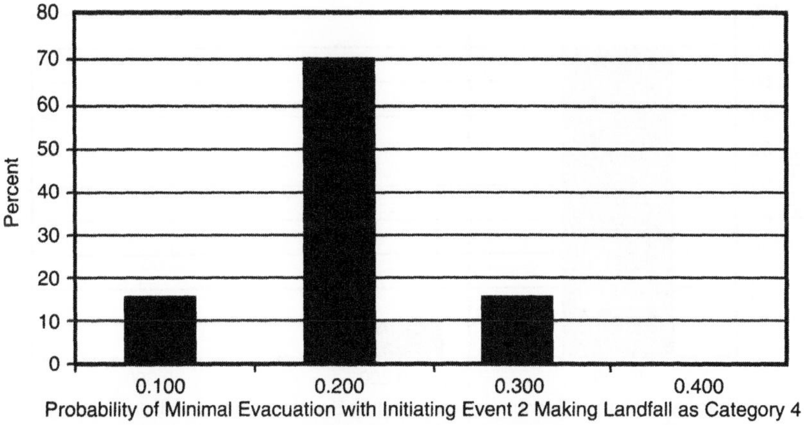

For an Initiating Event 2 making landfall as Category 4 hurricane, the uncertainty about the percentage of these hurricanes that will involve "minimal" evacuation is shown above. This uncertainty is subjective. There was no Bayesian updating since there is no data to support an update. The mean percent of "minimal" evacuation was calculated from the uncertainty distribution to be 20.

FIGURE B.24. Uncertainty distribution, Initiating Event 2 hits as Category 4 following minimal evacuation (hurricanes from 1900–2004 and 1900–1950).

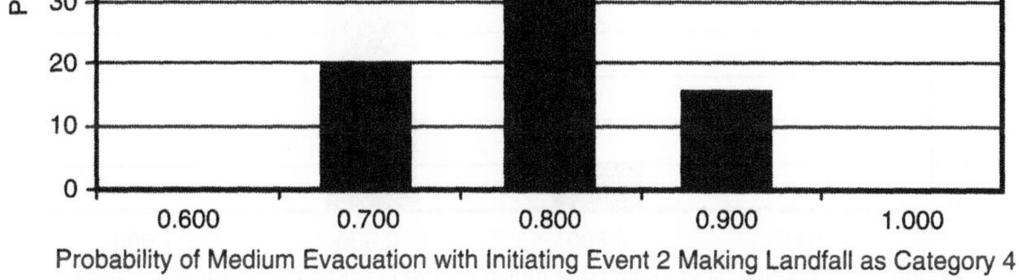

For an Initiating Event 2 making landfall as Category 4 hurricane, the uncertainty about the percentage of these hurricanes that will involve "medium" evacuation is shown above. This uncertainty is subjective. There was no Bayesian updating since there is no data to support an update. The mean percent of "medium" evacuation was calculated from the uncertainty distribution to be 79.5.

FIGURE B.25. Uncertainty distribution, Initiating Event 2 hits as Category 4 following medium evacuation (hurricanes from 1900–2004 and 1900–1950).

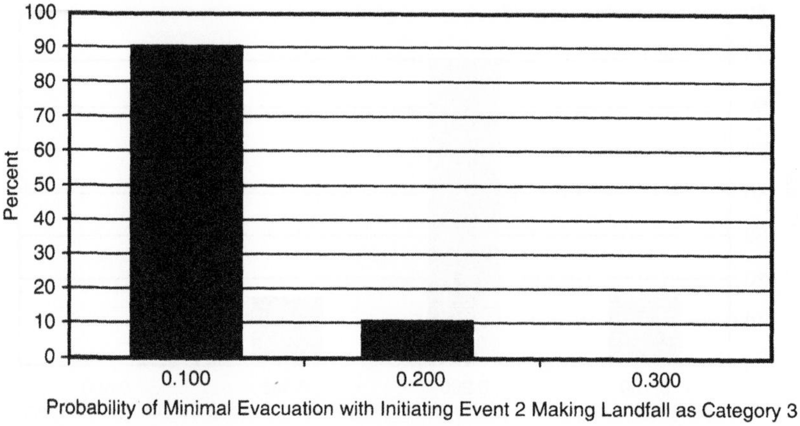

For an Initiating Event 2 making landfall as Category 3 hurricane, the uncertainty about the percentage of these hurricanes that will involve "minimal" evacuation is shown above. This uncertainty is subjective. There was no Bayesian updating since there is no data to support an update. The mean percent of "minimal" evacuation was calculated from the uncertainty distribution to be 11.

FIGURE B.26. Uncertainty distribution, Initiating Event 2 hits as Category 3 following minimal evacuation (hurricanes from 1900–2004 and 1900–1950).

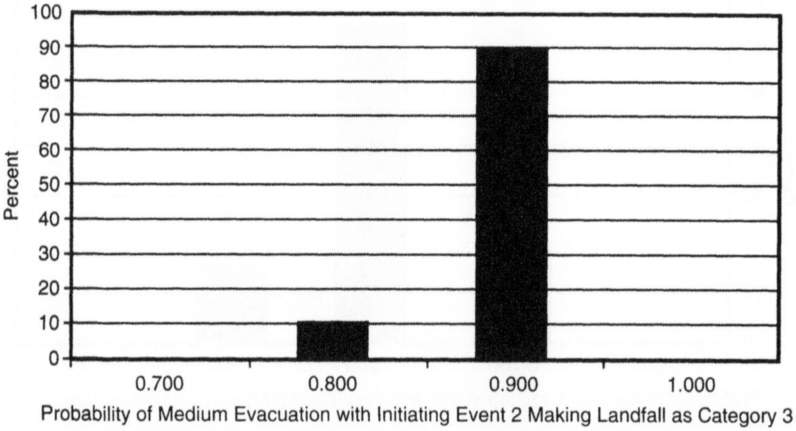

For an Initiating Event 2 making landfall as Category 3 hurricane, the uncertainty about the percentage of these hurricanes that will involve "medium" evacuation is shown above. This uncertainty is subjective. There was no Bayesian updating since there is no data to support an update. The mean percent of "medium" evacuation was calculated from the uncertainty distribution to be 89.

FIGURE B.27. Uncertainty distribution, Initiating Event 2 hits as Category 3 following medium evacuation (hurricanes from 1900–2004 and 1900–1950).

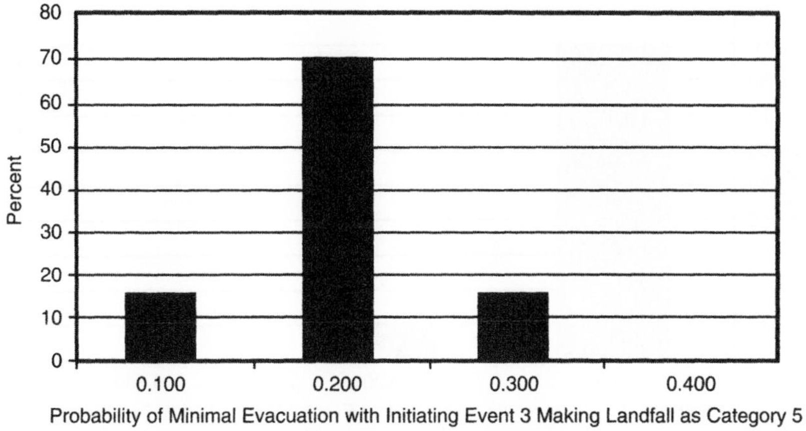

For an Initiating Event 3 making landfall as Category 5 hurricane, the uncertainty about the percentage of these hurricanes that will involve "minimal" evacuation is shown above. This uncertainty is subjective. There was no Bayesian updating since there is no data to support an update. The mean percent of "minimal" evacuation was calculated from the uncertainty distribution to be 20.

FIGURE B.28. Uncertainty distribution, Initiating Event 3 hits as Category 5 following minimal evacuation (hurricanes from 1900–2004 and 1900–1950).

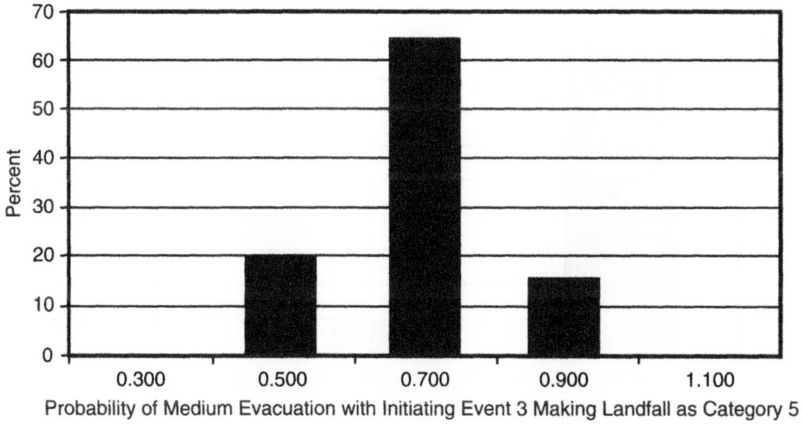

For an Initiating Event 3 making landfall as Category 5 hurricane, the uncertainty about the percentage of these hurricanes that will involve "medium" evacuation is shown above. This uncertainty is subjective. There was no Bayesian updating since there is no data to support an update. The mean percent of "medium" evacuation was calculated from the uncertainty distribution to be 69.

FIGURE B.29. Uncertainty distribution, Initiating Event 3 hits as Category 5 following medium evacuation (hurricanes from 1900–2004 and 1900–1950).

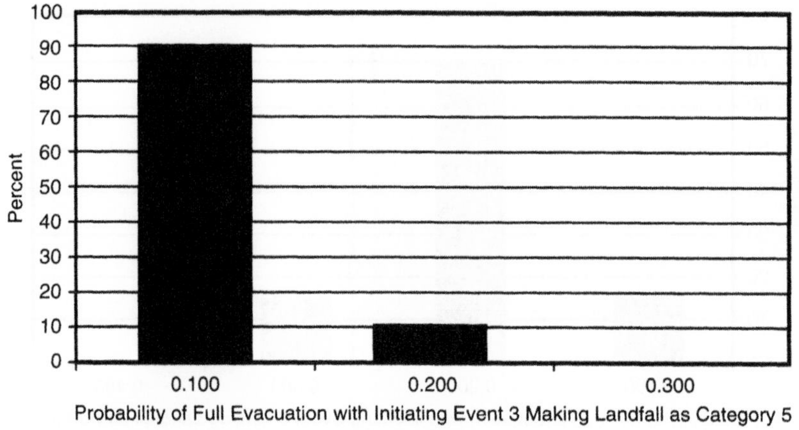

For an Initiating Event 3 making landfall as Category 5 hurricane, the uncertainty about the percentage of these hurricanes that will involve "full" evacuation is shown above. This uncertainty is subjective. There was no Bayesian updating since there is no data to support an update. The mean percent of "full" evacuation was calculated from the uncertainty distribution to be 11.

FIGURE B.30. Uncertainty distribution, Initiating Event 3 hits as Category 5 following full evacuation (hurricanes from 1900–2004 and 1900–1950).

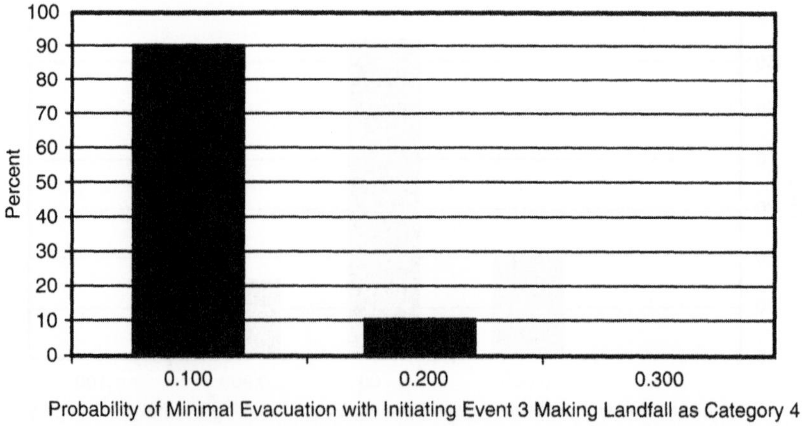

For an Initiating Event 3 making landfall as Category 4 hurricane, the uncertainty about the percentage of these hurricanes that will involve "minimal" evacuation is shown above. This uncertainty is subjective. There was no Bayesian updating since there is no data to support an update. The mean percent of "minimal" evacuation was calculated from the uncertainty distribution to be 11.

FIGURE B.31. Uncertainty distribution, Initiating Event 3 hits as Category 4 following minimal evacuation (hurricanes from 1900–2004 and 1900–1950).

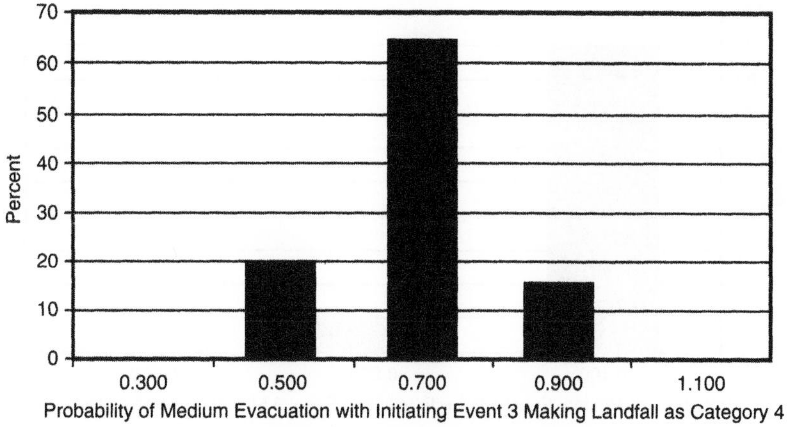

For an Initiating Event 3 making landfall as Category 4 hurricane, the uncertainty about the percentage of these hurricanes that will involve "medium" evacuation is shown above. This uncertainty is subjective. There was no Bayesian updating since there is no data to support an update. The mean percent of "medium" evacuation was calculated from the uncertainty distribution to be 69.

FIGURE B.32. Uncertainty distribution, Initiating Event 3 hits as Category 4 following medium evacuation (hurricanes from 1900–2004 and 1900–1950).

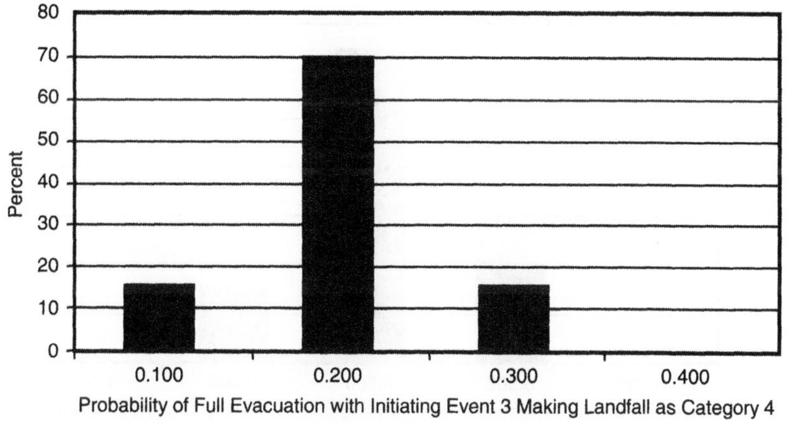

For an Initiating Event 3 making landfall as Category 4 hurricane, the uncertainty about the percentage of these hurricanes that will involve "full" evacuation is shown above. This uncertainty is subjective. There was no Bayesian updating since there is no data to support an update. The mean percent of "full" evacuation was calculated from the uncertainty distribution to be 20.

FIGURE B.33. Uncertainty distribution, Initiating Event 3 hits as Category 4 following full evacuation (hurricanes from 1900–2004 and 1900–1950).

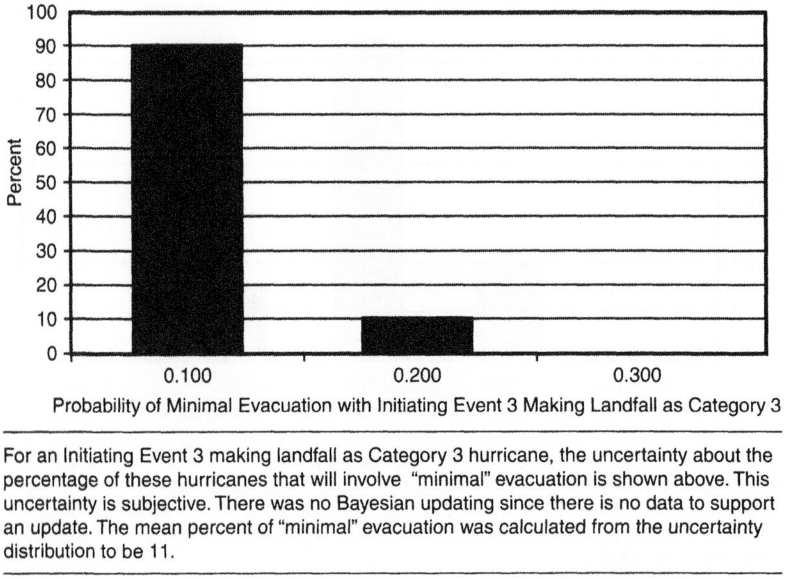
100
90
80
70
60
50
40
30
20
10
0
Percent
0.100
0.200
0.300
Probability of Minimal Evacuation with Initiating Event 3 Making Landfall as Category 3
For an Initiating Event 3 making landfall as Category 3 hurricane, the uncertainty about the percentage of these hurricanes that will involve "minimal" evacuation is shown above. This uncertainty is subjective. There was no Bayesian updating since there is no data to support an update. The mean percent of "minimal" evacuation was calculated from the uncertainty distribution to be 11.

FIGURE B.34. Uncertainty distribution, Initiating Event 3 hits as Category 3 following minimal evacuation (hurricanes from 1900–2004 and 1900–1950).

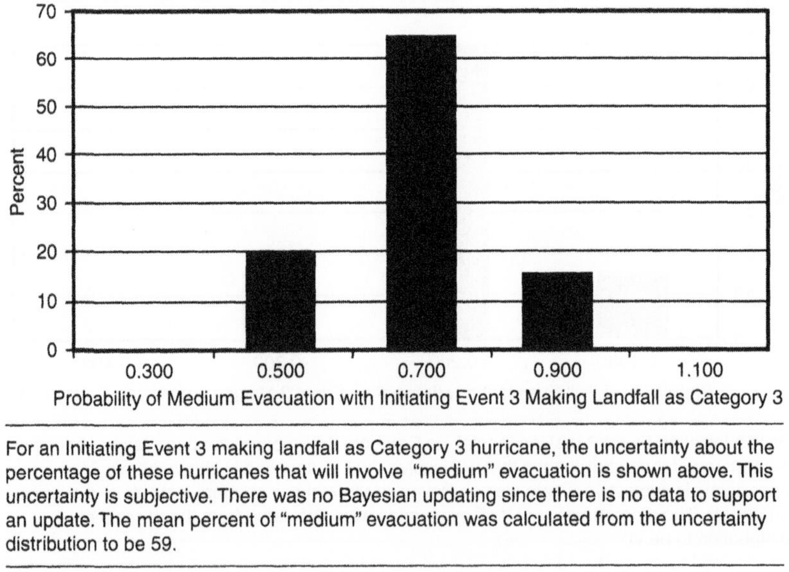
70
60
50
40
30
20
10
0
Percent
0.300
0.500
0.700
0.900
1.100
Probability of Medium Evacuation with Initiating Event 3 Making Landfall as Category 3
For an Initiating Event 3 making landfall as Category 3 hurricane, the uncertainty about the percentage of these hurricanes that will involve "medium" evacuation is shown above. This uncertainty is subjective. There was no Bayesian updating since there is no data to support an update. The mean percent of "medium" evacuation was calculated from the uncertainty distribution to be 59.

FIGURE B.35. Uncertainty distribution, Initiating Event 3 hits as Category 3 following medium evacuation (hurricanes from 1900–2004 and 1900–1950).

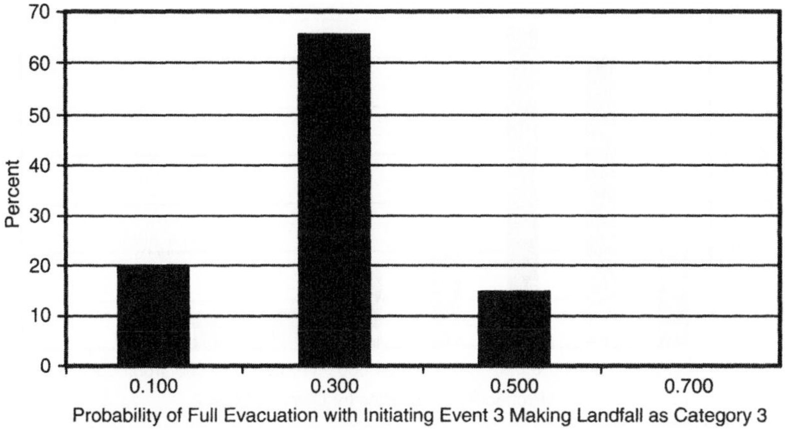

For an Initiating Event 3 making landfall as Category 3 hurricane, the uncertainty about the percentage of these hurricanes that will involve "full" evacuation is shown above. This uncertainty is subjective. There was no Bayesian updating since there is no data to support an update. The mean percent of "full" evacuation was calculated from the uncertainty distribution to be 29.

FIGURE B.36. Uncertainty distribution, Initiating Event 3 hits as Category 3 following full evacuation (hurricanes from 1900–2004 and 1900–1950).

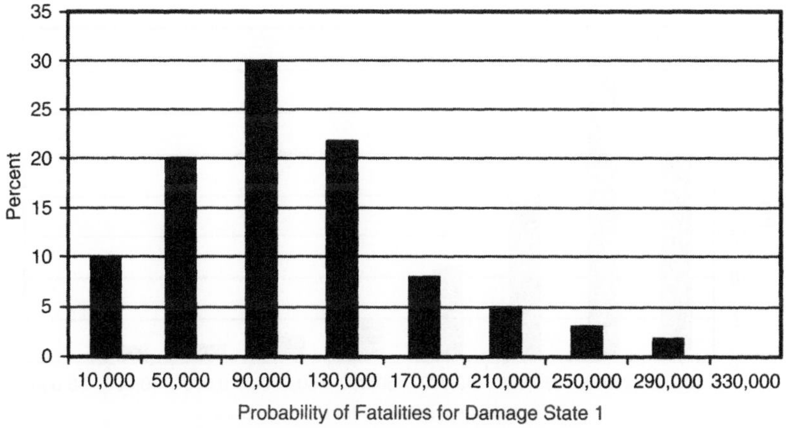

Damage State 1 occurs if "minimal" evacuation is achieved during a Category 5 hurricane. The uncertainty about the number of fatalities for Damage State 1 is shown above. This uncertainty is subjective. There was no Bayesian updating since there is no data to support an update. The mean number of fatalities was calculated from the uncertainty distribution to be 104,000.

FIGURE B.37. Uncertainty distribution, Damage State 1 fatalities (hurricanes from 1900–2004).

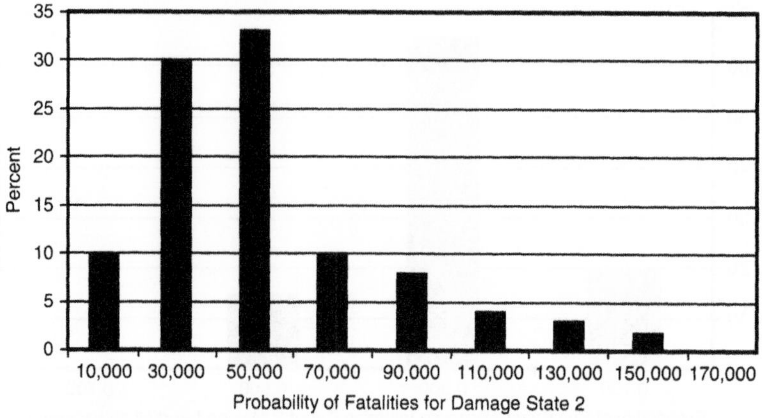

Damage State 2 occurs if "medium" evacuation is achieved during a Category 5 hurricane or if "minimal" evacuation is achieved during a Category 4 hurricane. The uncertainty about the number of fatalities for Damage State 2 is shown above. This uncertainty is subjective. There was no Bayesian updating since there is no data to support an update. The mean number of fatalities was calculated from the uncertainty distribution to be 52,000.

FIGURE B.38. Uncertainty distribution, Damage State 2 fatalities (hurricanes from 1900–2004).

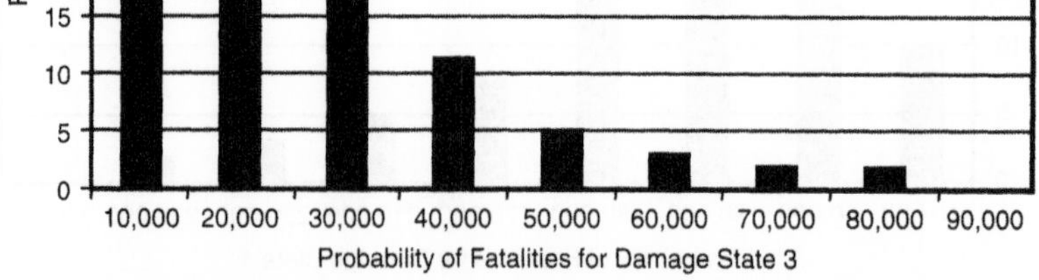

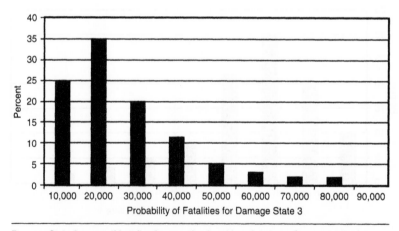

Damage State 3 occurs if "medium" evacuation is achieved during a Category 4 hurricane. The uncertainty about the number of fatalities for Damage State 3 is shown above. This uncertainty is subjective. There was no Bayesian updating since there is no data to support an update. The mean number of fatalities was calculated from the uncertainty distribution to be 26,000.

FIGURE B.39. Uncertainty distribution, Damage State 3 fatalities (hurricanes from 1900–2004).

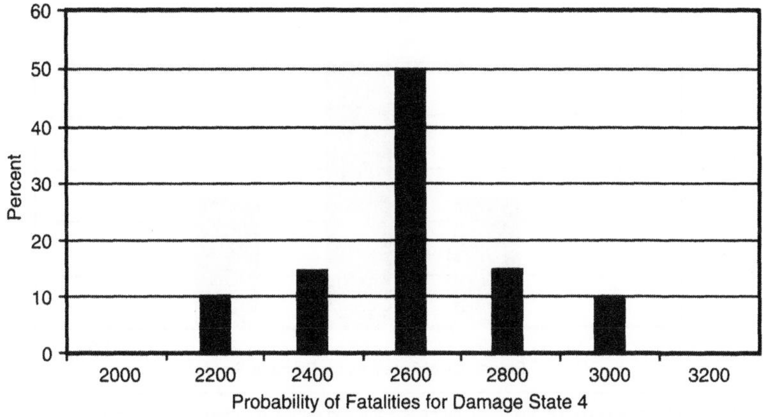

Damage State 4 occurs if "full" evacuation is achieved during a Category 5 hurricane or if "minimal" evacuation is achieved during a Category 3 hurricane. The uncertainty about the number of fatalities for Damage State 4 is shown above. This uncertainty is subjective. There was no Bayesian updating since there is no data to support an update. The mean number of fatalities was calculated from the uncertainty distribution to be 2600.

FIGURE B.40. Uncertainty distribution, Damage State 4 fatalities (hurricanes from 1900–2004).

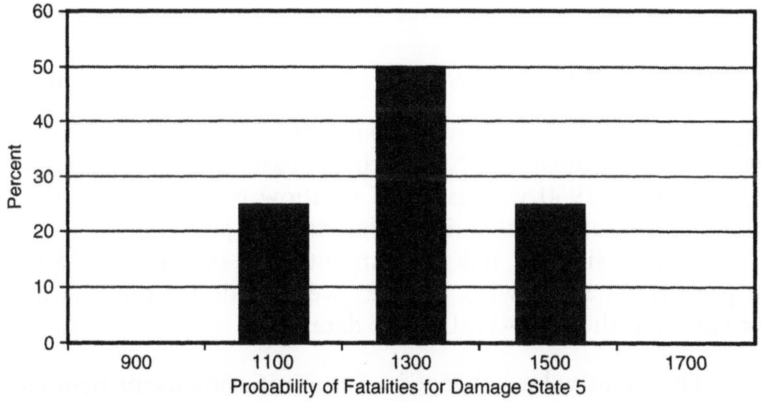

Damage State 5 occurs if "full" evacuation is achieved during a Category 4 hurricane or if "medium" evacuation is achieved during a Category 3 hurricane. The uncertainty about the number of fatalities for Damage State 5 is shown above. This uncertainty is subjective. There was no Bayesian updating since there is no data to support an update. The mean number of fatalities was calculated from the uncertainty distribution to be 1300.

FIGURE B.41. Uncertainty distribution, Damage State 5 fatalities (hurricanes from 1900–2004).

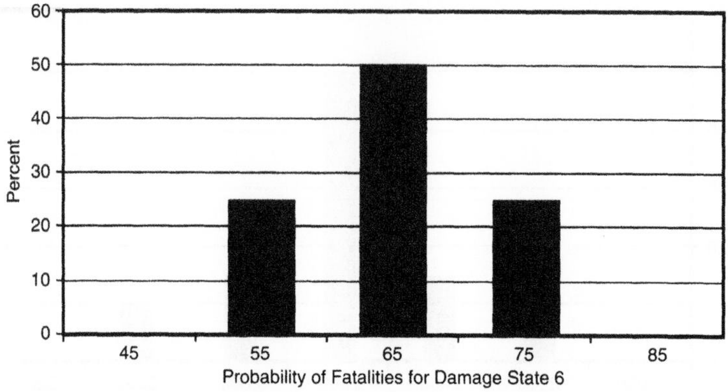

Damage State 6 occurs if "full" evacuation is achieved during a Category 3 hurricane. The uncertainty about the number of fatalities for Damage State 6 is shown above. This uncertainty is subjective. There was no Bayesian updating since there is no data to support an update. The mean number of fatalities was calculated from the uncertainty distribution to be 65.

FIGURE B.42. Uncertainty distribution, Damage State 6 fatalities (hurricanes from 1900–2004).

trees, (2) the data from 1900–2004 used to perform a Bayesian update on the prior distribution, and (3) the updated uncertainty distribution. The mean values used for the initiating events and branch points are described in Chapter 3, Section 3.2.4, and are shown in Table 3.2.

B.2 Hurricane Risk Assessment for the Period 1900–1950

Using the same methodology as above, the human fatality risk of a major hurricane impacting New Orleans based on hurricane data for the period 1900–1950 was assessed to allow comparison of the risk results using databases covering different time periods.

The same six-step risk assessment process was used for both time periods. The following changes were made for the risk assessment based on the 1900–1950 input data.

1. The input values changed for the initiating event frequencies. See Figures B.43–B.45, which are based on Table 3.8 of Chapter 3.
2. The input values changed for the probability of a hurricane impacting New Orleans. See Figures B.46–B.48, which are based on Table 3.8 of Chapter 3.
3. The input values changed for the probability of an initiating event making landfall as a major hurricane. See Figures B.49–B.57, which are based on Table 3.8 of Chapter 3.
4. The input values changed for the probability of fatalities for damage states. See Figures B.58–B.63, which are discussed in Section 3.4.

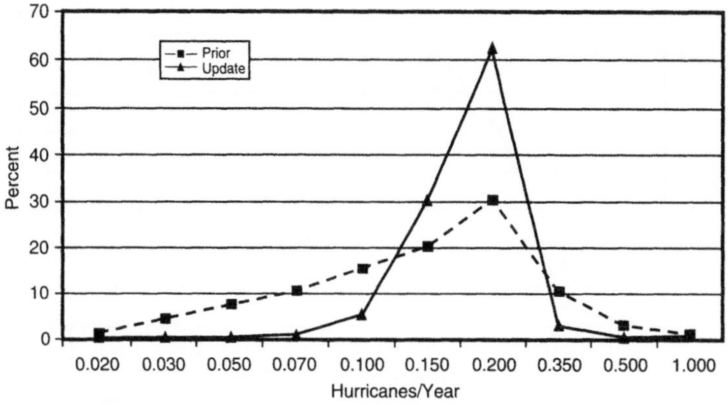

Based on hurricane tracking records, 10 of the 23 hurricanes were categorized as being in the Gulf of Mexico for 48 hours or less before landfall. These hurricanes were placed in Initiating Event 1. As seen above, the same "Prior" uncertainty distribution for the years 1900-2004 was used and updated using Bayesian techniques with the actual experience of ten hurricanes in 51 years. The mean value for the frequency of Initiating Event 1 was calculated to be equal to 0.184 hurricanes per year or approximately one Initiating Event 1 every 5.4 years.

FIGURE B.43. Uncertainty distribution, Initiating Event 1 frequency (hurricanes from 1900–1950).

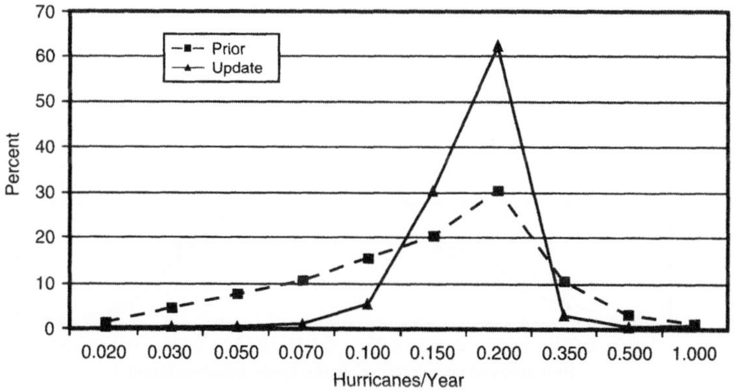

Based on hurricane tracking records, 10 of the 23 hurricanes were categorized as being in the Gulf of Mexico for between 48 hours and 72 hours before landfall. These hurricanes were placed in Initiating Event#2. As seen above, the same "Prior" uncertainty distribution for the years 1900-2004 was generated and updated using Bayesian techniques with the actual experience of ten hurricanes in 51 years. The mean value for the frequency of Initiating Event 2 was calculated to be equal to 0.184 hurricanes per year or approximately one Initiating Event 2 every 5.4 years.

FIGURE B.44. Uncertainty distribution, Initiating Event 2 frequency (hurricanes from 1900–1950).

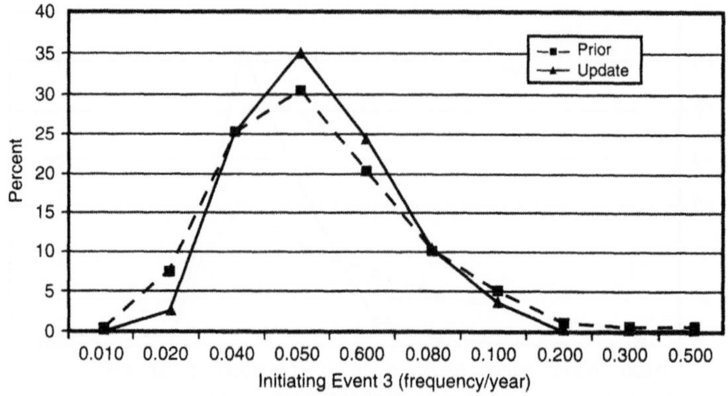

Based on hurricane tracking records, 3 of the 23 hurricanes were categorized as being in the Gulf of Mexico for greater than 72 hours before landfall. These hurricanes were placed in Initiating Event 3. As seen above, the same "Prior" uncertainty distribution for the years 1900-2004 was generated and updated using Bayesian techniques with the actual experience of three hurricanes in 51 years. The mean value for the frequency of Initiating Event 3 was calculated to be equal to 0.0541 hurricanes per year or approximately one Initiating Event 3 every 18 years.

FIGURE B.45. Uncertainty distribution, Initiating Event 3 frequency (hurricanes from 1900–1950).

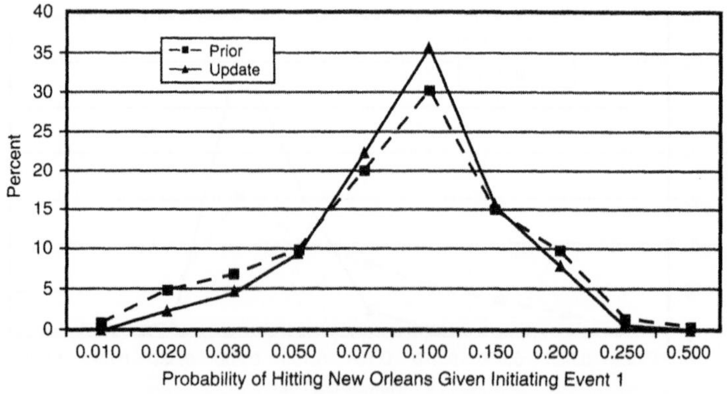

Of the ten hurricanes listed in the category Initiating Event 1, one hurricane was considered to directly impact New Orleans, unnamed hurricane of 1915. As seen above, the same "Prior" uncertainty distribution for the years 1900-2004 was generated and updated using Bayesian techniques with the actual experience of one hurricane out of ten hitting New Orleans. The mean value for the fraction of Initiating Event 1 hurricanes hitting New Orleans was calculated to be equal to 0.100, or approximately one of every ten hurricanes.

FIGURE B.46. Uncertainty distribution, Initiating Event 1 hits New Orleans (hurricanes from 1900–1950).

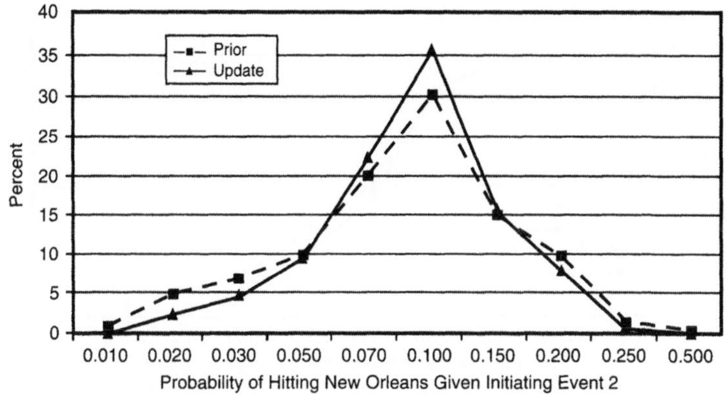

Of the ten hurricanes listed in the category Initiating Event 2, one hurricane was considered to directly impact New Orleans, unnamed hurricane of 1909 (Grand Isle, LA). As seen above, the same "Prior" uncertainty distribution for the years 1900-2004 was generated and updated using Bayesian techniques with the actual experience of one hurricane out of ten hitting New Orleans. The mean value for the fraction of Initiating Event 2 hurricanes hitting New Orleans was calculated to be equal to 0.100, or approximately one of every ten hurricanes.

FIGURE B.47. Uncertainty distribution, Initiating Event 2 hits New Orleans (hurricanes from 1900–1950).

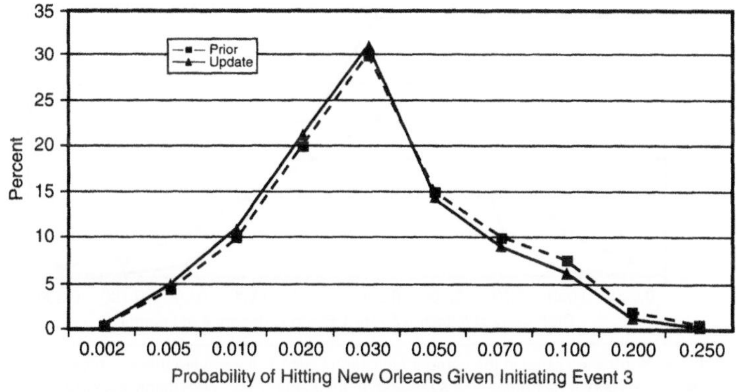

Of the three hurricanes listed in the category Initiating Event 3, there were no direct hits or close calls. As seen above, the same "Prior" uncertainty distribution for the years 1900-2004 was generated and updated using Bayesian techniques with the actual experience of zero hurricanes out of three hitting New Orleans. The mean value for the fraction of Initiating Event 3 hurricanes hitting New Orleans was calculated to be equal to 0.0376, or approximately 1 of every 27 hurricanes.

FIGURE B.48. Uncertainty distribution, Initiating Event 3 hits New Orleans (hurricanes from 1900–1950).

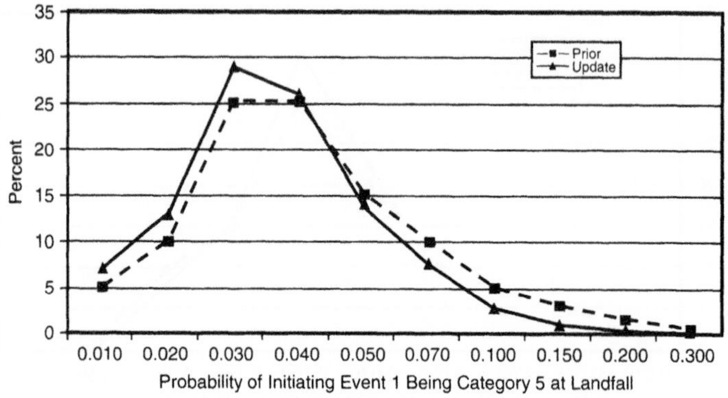

Of the ten hurricanes listed in the category Initiating Event 1, there were no hurricanes that were Category 5 at landfall. As seen above, the same "Prior" uncertainty distribution for the years 1900-2004 was generated and updated using Bayesian techniques with the actual experience of zero hurricanes out of ten hitting landfall as Category 5. The mean value for the fraction of Initiating Event 1 hurricanes hitting landfall as Category 5 was calculated to be equal to 0.0392, or approximately 1 of every 26 hurricanes.

FIGURE B.49. Uncertainty distribution, Initiating Event 1 hits as Category 5 (hurricanes from 1900–1950).

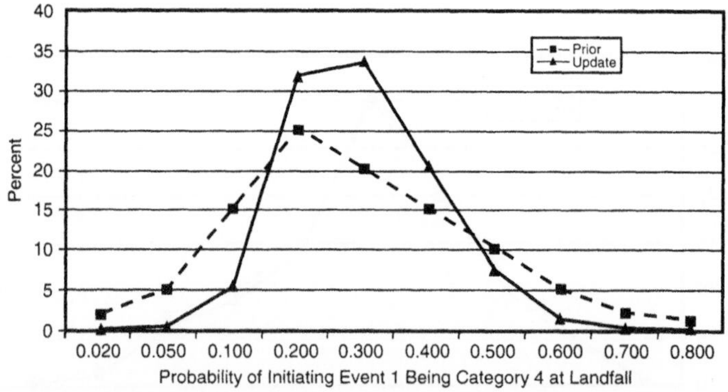

Of the ten hurricanes listed in the category Initiating Event 1, there were three hurricanes that were Category 4 at landfall. As seen above, the same "Prior" uncertainty distribution for the years 1900-2004 was generated and updated using Bayesian techniques with the actual experience of three hurricanes out of ten hitting landfall as Category 4. The mean value for the fraction of Initiating Event 1 hurricanes hitting landfall as Category 4 was calculated to be equal to 0.2960, or approximately two of every seven hurricanes.

FIGURE B.50. Uncertainty distribution, Initiating Event 1 hits as Category 4 (hurricanes from 1900–1950).

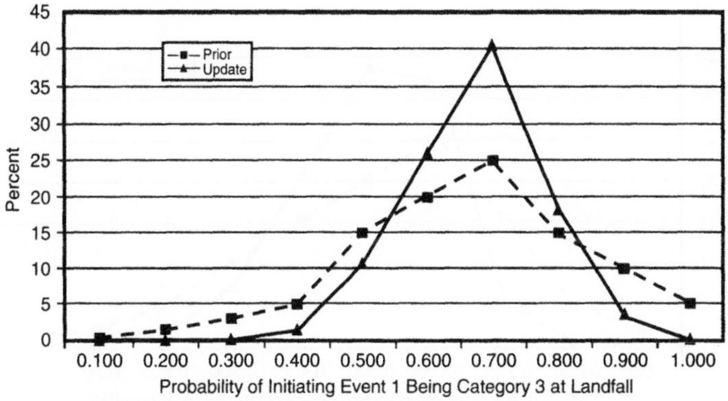

Of the ten hurricanes listed in the category Initiating Event 1, there were seven hurricanes that were Category 3 at landfall. As seen above, the same "Prior" uncertainty distribution for the years 1900-2004 was generated and updated using Bayesian techniques with the actual experience of seven hurricanes out of ten hitting landfall as Category 3. The mean value for the fraction of Initiating Event 1 hurricanes hitting landfall as Category 3 was calculated to be equal to 0.6740, or approximately two of every three hurricanes.

FIGURE B.51. Uncertainty distribution, Initiating Event 1 hits as Category 3 (hurricanes from 1900–1950).

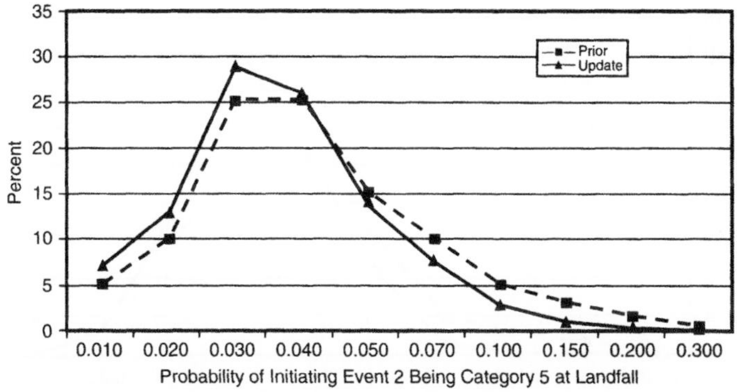

Of the ten hurricanes listed in the category Initiating Event 2, there were no hurricanes that were Category 5 at landfall. As seen above, the same "Prior" uncertainty distribution for the years 1900-2004 was generated and updated using Bayesian techniques with the actual experience of zero hurricanes out of ten hitting landfall as Category 5. The mean value for the fraction of Initiating Event 2 hurricanes hitting landfall as Category 5 was calculated to be equal to 0.039, or approximately 1 of every 26 hurricanes.

FIGURE B.52. Uncertainty distribution, Initiating Event 2 hits as Category 5 (hurricanes from 1900–1950).

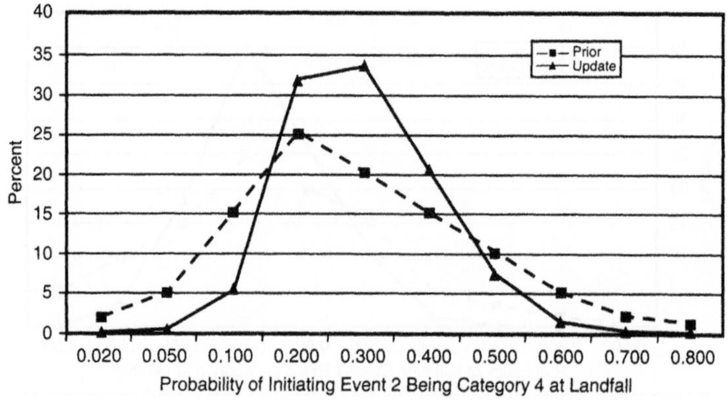

Of the ten hurricanes listed in the category Initiating Event 2, there were three hurricanes that were Category 4 at landfall. As seen above, the same "Prior" uncertainty distribution for the years 1900-2004 was generated and updated using Bayesian techniques with the actual experience of three hurricanes out of ten hitting landfall as Category 4. The mean value for the fraction of Initiating Event 2 hurricanes hitting landfall as Category 4 was calculated to be equal to 0.296, or approximately two of every seven hurricanes.

FIGURE B.53. Uncertainty distribution, Initiating Event 2 hits as Category 4 (hurricanes from 1900–1950).

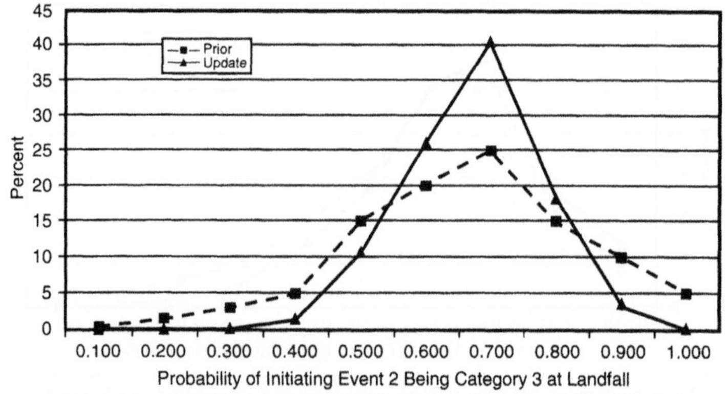

Of the ten hurricanes listed in the category Initiating Event 2, there were seven hurricanes that were Category 3 at landfall. As seen above, the same "Prior" uncertainty distribution for the years 1900-2004 was generated and updated using Bayesian techniques with the actual experience of seven hurricanes out of ten hitting landfall as Category 3. The mean value for the fraction of Initiating Event 2 hurricanes hitting landfall as Category 3 was calculated to be equal to 0.674, or approximately two of every three hurricanes.

FIGURE B.54. Uncertainty distribution, Initiating Event 2 hits as Category 3 (hurricanes from 1900–1950).

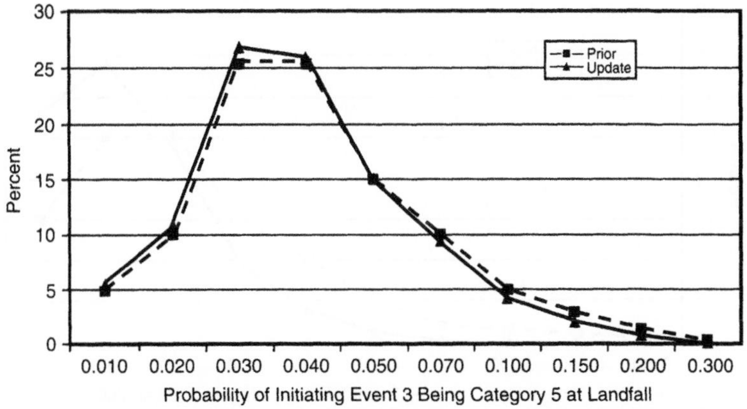

Of the three hurricanes listed in the category Initiating Event 3, there was no hurricane that was Category 5 at landfall. As seen above, the same "Prior" uncertainty distribution for the years 1900-2004 was generated and updated using Bayesian techniques with the actual experience of zero hurricanes out of three hitting landfall as Category 5. The mean value for the fraction of Initiating Event 3 hurricanes hitting landfall as Category 5 was calculated to be equal to 0.045, or approximately 1 of every 22 hurricanes.

FIGURE B.55. Uncertainty distribution, Initiating Event 3 hits as Category 5 (hurricanes from 1900–1950).

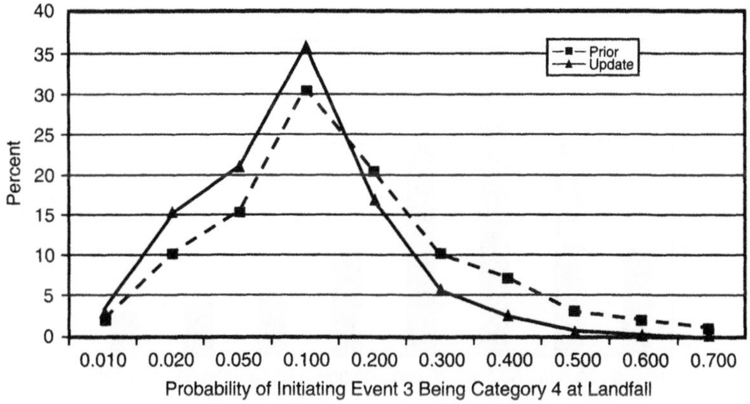

Of the three hurricanes listed in the category Initiating Event 3, there were no hurricanes that were Category 4 at landfall. As seen above, the same "Prior" uncertainty distribution for the years 1900-2004 was generated and updated using Bayesian techniques with the actual experience of zero hurricanes out of three hitting landfall as Category 4. The mean value for the fraction of Initiating Event 3 hurricanes hitting landfall as Category 4 was calculated to be equal to 0.113, or approximately one of every nine hurricanes.

FIGURE B.56. Uncertainty distribution, Initiating Event 3 hits as Category 4 (hurricanes from 1900–1950).

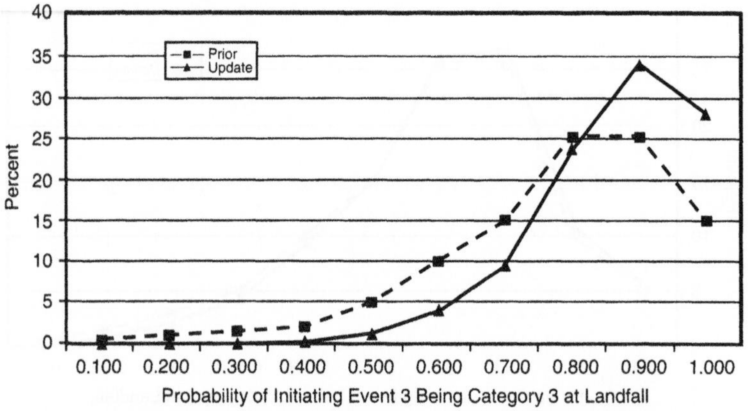

Of the three hurricanes listed in the category Initiating Event 3, there were three hurricanes that were Category 3 at landfall. As seen above, the same "Prior" uncertainty distribution for the years 1900-2004 was generated and updated using Bayesian techniques with the actual experience of three hurricanes out of three hitting landfall as Category 3. The mean value for the fraction of Initiating Event 3 hurricanes hitting landfall as Category 3 was calculated to be equal to 0.866, or approximately nine of every ten hurricanes.

FIGURE B.57. Uncertainty distribution, Initiating Event 3 hits as Category 3 (hurricanes from 1900–1950).

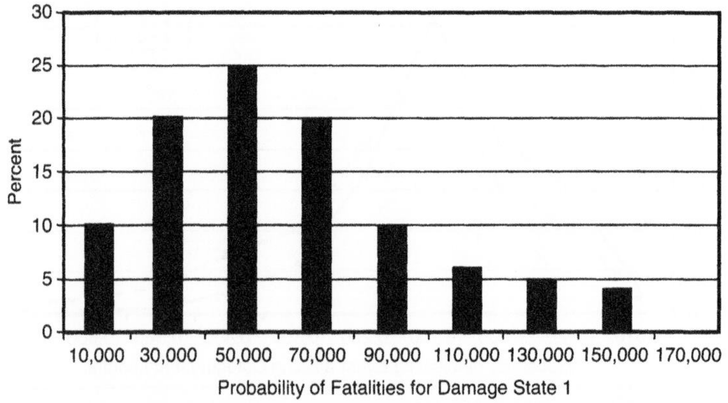

Damage State 1 occurs if "minimal" evacuation is achieved during a Category 5 hurricane. The uncertainty about the number of fatalities for Damage State 1 is shown above. This uncertainty is subjective. There was no Bayesian updating since there is no data to support an update. The mean number of fatalities was calculated from the uncertainty distribution to be 61,600.

FIGURE B.58. Uncertainty distribution, Damage State 1 fatalities (hurricanes from 1900–1950).

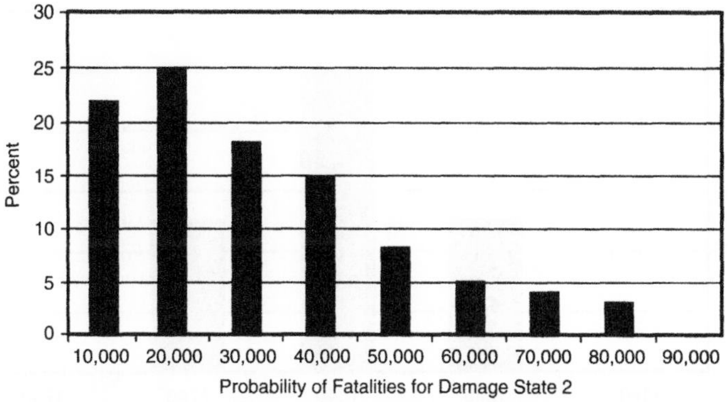

Damage State 2 occurs if "medium" evacuation is achieved during a Category 5 hurricane or if "minimal" evacuation is achieved during a Category 4 hurricane. The uncertainty about the number of fatalities for Damage State 2 is shown above. This uncertainty is subjective. There was no Bayesian updating since there is no data to support an update. The mean number of fatalities was calculated from the uncertainty distribution to be 30,800.

FIGURE B.59. Uncertainty distribution, Damage State 2 fatalities (hurricanes from 1900–1950).

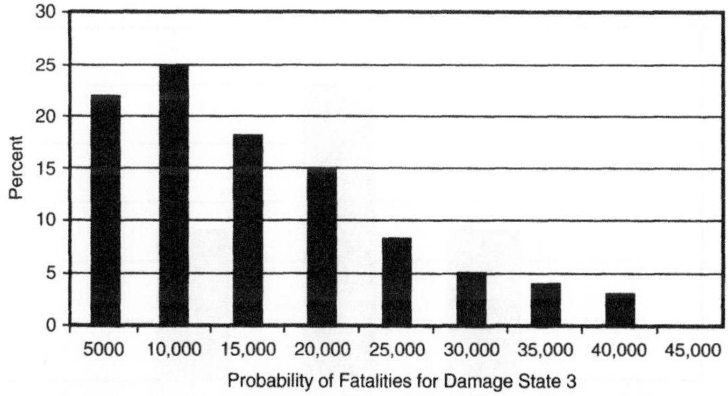

Damage State 3 occurs if "medium" evacuation is achieved during a Category 4 hurricane. The uncertainty about the number of fatalities for Damage State 3 is shown above. This uncertainty is subjective. There was no Bayesian updating since there is no data to support an update. The mean number of fatalities was calculated from the uncertainty distribution to be 15,400.

FIGURE B.60. Uncertainty distribution, Damage State 3 fatalities (hurricanes from 1900–1950).

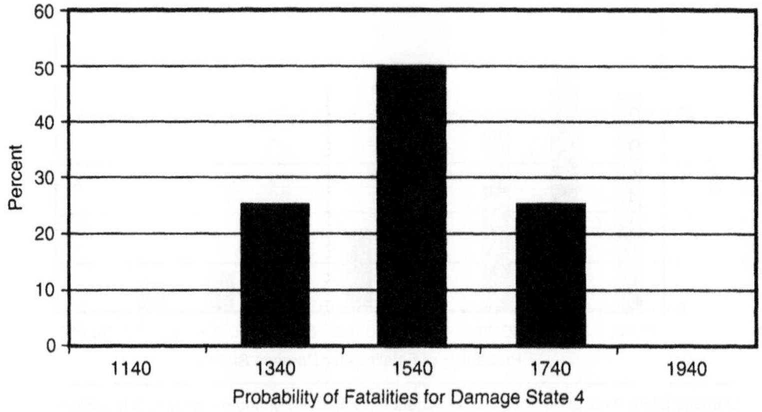

Damage State 4 occurs if "full" evacuation is achieved during a Category 5 hurricane or if "minimal" evacuation is achieved during a Category 3 hurricane. The uncertainty about the number of fatalities for Damage State 4 is shown above. This uncertainty is subjective. There was no Bayesian updating since there is no data to support an update. The mean number of fatalities was calculated from the uncertainty distribution to be 1540.

FIGURE B.61. Uncertainty distribution, Damage State 4 fatalities (hurricanes from 1900–1950).

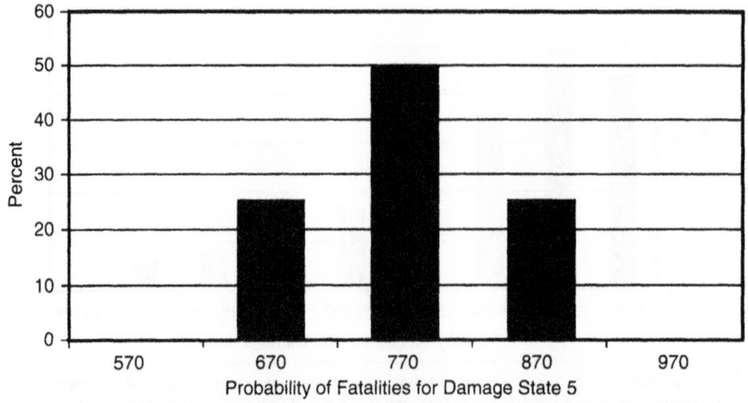

Damage State 5 occurs if "full" evacuation is achieved during a Category 4 hurricane or if "medium" evacuation is achieved during a Category 3 hurricane. The uncertainty about the number of fatalities for Damage State 5 is shown above. This uncertainty is subjective. There was no Bayesian updating since there is no data to support an update. The mean number of fatalities was calculated from the uncertainty distribution to be 770.

FIGURE B.62. Uncertainty distribution, Damage State 5 fatalities (hurricanes from 1900–1950).

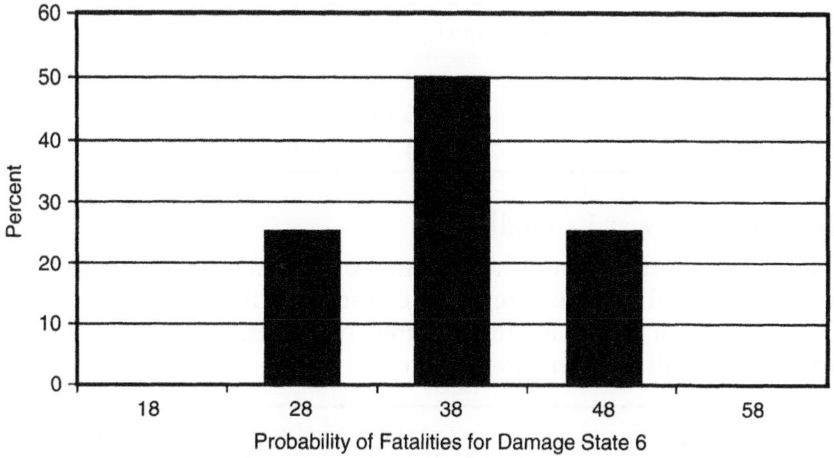

Damage State 6 occurs if "full" evacuation is achieved during a Category 3 hurricane. The uncertainty about the number of fatalities for Damage State 6 is shown above. This uncertainty is subjective. There was no Bayesian updating since there is no data to support an update. The mean number of fatalities was calculated from the uncertainty distribution to be 38.

FIGURE B.63. Uncertainty distribution, Damage State 6 fatalities (hurricanes from 1900–1950).

B.2.1 Initiating Events

An evaluation of the 23 major hurricanes making landfall from 1900 to 1950 resulted in the initiating event divisions shown in Table 3.8 of Chapter 3.

The uncertainties about the frequency of initiating events are shown in Figures B.43–B.45.

The mean values for the frequencies of the initiating events are:

Initiating Event	*Frequency*
1	1 in 5.4 years
2	1 in 5.4 years
3	1 in 18 years

B.2.2 Impact on New Orleans

The uncertainties about the fraction of initiating event hurricanes that directly impact New Orleans are shown in Figures B.46–B.48.

The mean values for the fractions of each initiating event hurricanes impacting New Orleans are:

Initiating Event	*Fraction Impacting New Orleans*
1	1 of 10 hurricanes
2	1 of 10 hurricanes
3	1 of 27 hurricanes

B.2.3 Category at Landfall

The uncertainties about the fraction of initiating event hurricane making landfall as each category are described in Figures B.49–B.57.

The mean values for the fractions of initiating event hurricanes making landfall as each category are:

Initiating Event	*Category*	*Fraction of Hurricanes Making Landfall by Initiating Event and Category*
1	5	1 of 26
1	4	2 of 7
1	3	2 of 3
2	5	1 of 26
2	4	2 of 7
2	3	2 of 3
3	5	1 of 22
3	4	1 of 9
3	3	9 of 10

B.2.4 Type of Evacuation

No changes were made to the distributions for evacuations for the time period 1900–1950.

These distributions are shown in Figures B.16–B.36.

B.2.5 Damage States

The damage state distributions were changed to reflect the smaller population in 1950 (770,000 vs 1.3 million).

Damage states are discussed in the "Damage States" section under Section 3.2.3. The uncertainty about the number of fatalities for the damage states is described in Figures B.58–B.63.

Appendix C: Supporting Evidence for the Case Study on Asteroid Risk

C.1 Asteroid Risk Assessment for the 48 Contiguous States of the United States of America

C.1.1 Initiating Events

As described in the "Initiating events" section under Section 4.2.3, there are six initiating events used in the quantitative risk assessment as shown in the table below:

Initiating Event	*Impact Energy (Megatons of TNTE)*	*Recurrence Interval (years)*	*Frequency (Per Year)*
1	0.1–10	8	1.34×10^{-1}
2	10–10,000	260	3.84×10^{-3}
3	10,000–100,000	81,000	1.24×10^{-5}
4	100,000–1,000,000	470,000	2.15×10^{-6}
5	1,000,000–10,000,000	3.8 Million	2.66×10^{-7}
Global		50 Million	2.00×10^{-8}

The asteroid impact frequencies and energies forming the basis for the initiating events defined in the above table were based on the processing of data developed by NASA[1] and Toon.[2] The NASA and Toon data are "best estimate" median values, that is, point estimates. In keeping with a probabilistic approach, it was necessary to cast the Toon and NASA data into a form that accounts for uncertainty. Once the recasting of the data was performed, then based on data quality considerations the two databases were combined through a weighting process into a single set of curves that provides the necessary uncertainty distributions of asteroid impact frequency as a function of energy. These are the impact frequency versus impact energy curves forming the basis for Initiating Events 1–5. The global initiating event was treated as a special case as discussed in Chapter 4.

The approach to accounting for the uncertainties in Initiating Events 1–5 is consistent with recommended practice for assigning uncertainties derived primarily from expert judgment. The uncertainty distributions were not taken directly from the NASA and Toon data, but from the authors' processing of their data. The "best estimate" values of NASA and Toon were assigned a 60% probability that they are correct. Uncertainties about the best estimate values were then assessed to derive "high" and "low" estimate values. Each of the "high" and "low" estimate values was assigned a 20% probability of being correct. Error factors were subjectively assigned to impact energy intervals, where the error factor is defined as the ratio of the square roots of the 95th and 5th percentile values. The assigned values are shown in Table C.1.

Sources of uncertainty include: (1) the population of asteroids within each size range, (2) the fraction of the population that impacts the Earth, and (3) the velocity of the asteroid at impact. Moderate uncertainty was assigned to the estimated frequencies at the low end of the impact energy range. This accounts for the fact that measurable impact frequency data is available for small objects with the lowest energies. Somewhat higher uncertainty was assigned to the estimated frequencies in the middle of the impact energy range. This accounts for the fact that these impacts are very rare events, and less effort has been made to identify and characterize the populations of these moderate-size objects. Lower uncertainty was assigned to the estimated frequencies at the high end of the impact energy range to take into consideration the extensive efforts that have been made to identify and classify very large asteroids having the potential to

TABLE C.1
Impact frequency uncertainty factors

Impact Energy (MT)	*Uncertainty Error Factor*
10^{-2}–10^{-1}	5
10^{-1}–1	5
1–10	5
10–10^{2}	6
10^{2}–10^{3}	6
10^{3}–10^{4}	7
10^{4}–10^{5}	6
10^{5}–10^{6}	5
10^{6}–10^{7}	4
$>10^{7}$	3

impact the Earth. The "best estimate" frequency value was multiplied by the error factor to derive the "high estimate" frequency and was divided by the error factor to derive the "low estimate" frequency.

The NASA data did not cover the full range of the Toon impact energies, especially at the low end of the energy range. Extrapolations were made of the NASA data to cover the same range as the Toon data. The result of this exercise is shown in Figure C.1.

The next task was to combine the Toon and NASA data into a single data set of impact frequencies versus impact energies. The NASA impact frequency estimates were assigned a weight of 0.70. The NASA estimates were given greater weight because they were derived from more recent asteroid population data. The NASA estimates assume an impact velocity of 20 kilometers per second (km/sec). The Toon impact frequency estimates, derived from older asteroid population data, are assigned a weight of 0.30. The weighting is judgmental. The Toon estimates assume an impact velocity of 15 km/sec. The weighting process was performed by taking slices at discrete energies of Figure C.1 to obtain a distribution of the impact frequencies for that energy. The median and mean frequency values from each slice were preserved and the resulting 5th percentile and 95th percentile values were calculated, extrapolating and interpolating from the cumulative probability distribution for the slice. The resulting percentiles were plotted as Figure C.2.

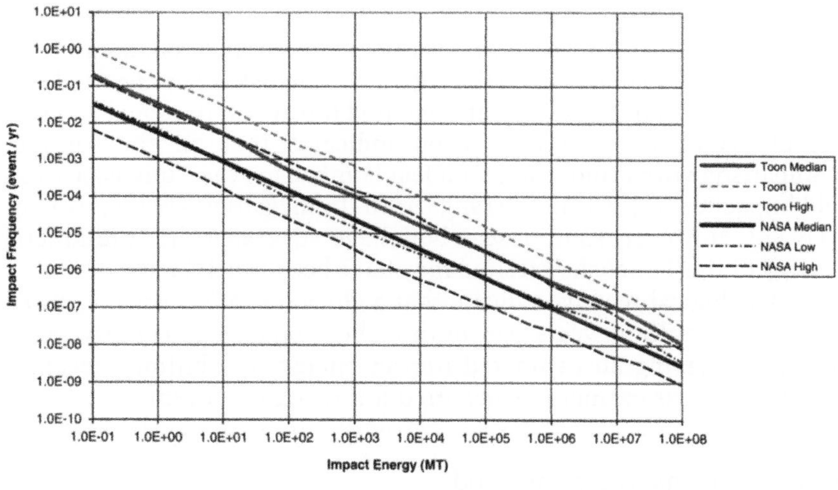

FIGURE C.1. Comparison of NASA and Toon asteroid impact frequency and energy estimates (NASA estimates extrapolated to cover full range of Toon impact energies).

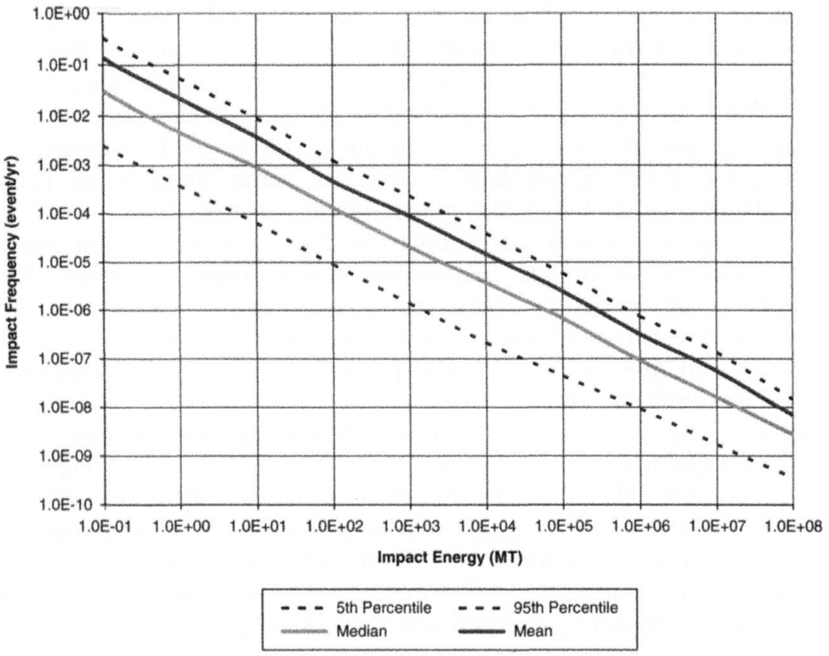

FIGURE C.2. Combined NASA and Toon asteroid impact frequency and energy estimates (NASA weight: 0.70, Toon weight: 0.30).

Finally, the initiating event frequencies for the five asteroid impact energy ranges were calculated from the curves of Figure C.2.

With respect to data, two events are noted. In 1908, an asteroid impacted Siberia in Russia in the region of Tunguska. This asteroid produced an "air burst" over an uninhabited forest and flattened approximately 2000 km^2 of forest. Toon indicates that on the basis of comparison with atmospheric nuclear bomb tests, the Tunguska asteroid impact energy was in the range of 10–15 megatons TNT equivalent (TNTE). However, some researchers have suggested that the kinetic energy of the Tunguska asteroid may have been much larger.

Geological records indicate that a global event occurred approximately 65 million years ago during the Cretaceous–Tertiary (K–T) boundary having an estimated impact energy of a billion megatons of TNTE. Mass extinctions occurred following this event.

C.1.2 Impacts Water or Land

The Earth's total water surface area is approximately 357 million km^2, some 70% of the Earth's total surface area. The Earth's total land surface area is approximately 153 million km^2, i.e., about 30% of the total.

Impacts 48 States—Gulf of Mexico

The coastal distance of the states (Florida, Alabama, Mississippi, Louisiana, and Texas) bordering the Gulf of Mexico was taken to be 2610 km. The total surface area of the Gulf of Mexico is approximately 1.6 million km^2. Because of the configuration of the Gulf of Mexico, the risk assessment did not include some portion of the Gulf of Mexico that is close to Mexico. The total surface of the Gulf of Mexico included in the risk assessment was 1.2 million km^2. This is approximately 0.34% of the Earth's total water surface area or 1.2 million divided by 357 million km^2.

The ocean depths in the Gulf of Mexico are relatively shallow compared to the Pacific and Atlantic oceans. Approximately 60% of the Gulf of Mexico is composed of water less than 200 m in depth. Approximately another 20% is part of a continental slope that varies from 200 to 3000 m. The last 20%, classified here as deep water, is greater than 3000 m.

The total surface of the Gulf of Mexico included in the risk assessment (1.2 million km^2) was broken into three classifications: shallow water, medium water, and deep water. See Figure 4.6 of Chapter 4. The total area of shallow water was taken to be 640,000 km^2 (640,000 divided by 1.2 million km^2 = 53%) with an average depth of 70 m. The total area of medium depth water was 260,000 km^2 divided by 1.2 million km^2 or 21%, with an average depth of 1500 m. The total area of deep water was 320,000 km^2 divided by 1.2 million km^2 or 26%, with an average depth of 4000 m.

Impacts 48 States—Pacific Coast

The coastal distance of the states (Washington, Oregon, and California) bordering the Pacific Ocean was taken to be 2070 km. The total surface area of the Pacific Ocean considered in the case study was approximately 7.5 million km^2. This surface area is approximately 2.1% of the Earth's total water surface area (7.5 million divided by 357 million km^2). To some degree, this surface area represents the Pacific Ocean area that is within 2500 km of the Pacific coast of the 48 states.

The Pacific Ocean floor is relatively uniform with an average depth of approximately 4000 m. The Pacific Ocean floor bordering the United States drops off rather quickly. The risk assessment assumed that there is no continental shelf and that the average depth (4000 m) applied to all the regions in the Pacific Ocean.

The total surface of the Pacific Ocean included in the risk assessment, 7.5 million km^2, was broken into three areas: Region I—less than 100 km from land, Region II—from 100 to 500 km to land, and Region III—from 500 to 2500 km to land. See Figure 4.7 of Chapter 4. The total area of Region I was taken to be 217,000 km^2 divided by 7.5 million km^2 or 2.9%, with an average depth of 4000 m. The total area of Region II was taken to be 1.1 million km^2 divided by 7.5 million km^2 or 14%, with an average depth of 4000 m. The total area of Region III was taken to be 6.2 million km^2 divided by 7.5 million km^2 or 83%, with an average depth of 4000 m.

Impacts 48 States—Atlantic Coast

The coastal distance of the states (Maine, New Hampshire, Rhode Island, Massachusetts, Connecticut, New York, New Jersey, Delaware, Maryland, Virginia, North Carolina, South Carolina, Georgia, and Florida) bordering the Atlantic Ocean was calculated to be 3310 km. The total surface area of the Atlantic Ocean considered in the case study was approximately 10.1 million km^2. This surface area is approximately 2.8% of the Earth's total water surface area, i.e., 10.1 million divided by 357 million km^2. To some degree, this Atlantic Ocean surface area represents the area that is within 2500 km of the Atlantic coast of the 48 states.

The Atlantic Ocean floor is relatively uniform with an average depth of approximately 4000 m. The case study risk assessment assumed that the continental shelf in the Atlantic Ocean floor was limited and that the average depth of the Atlantic Ocean (4000 m) applied to all the regions in the Atlantic Ocean.

The total surface of the Atlantic Ocean included in the case study, 10.1 million km^2, was divided into three regions: Region I—less than 100 km from land, Region II—from 100 to 500 km to land, and Region III—from 500 to 2500 km to land. See Figure 4.8 of Chapter 4. The total area of Region I was taken to be 340,000 km^2 divided by 10.1 million km^2 or 3.4% with an average depth of 4000 m. The total area of Region II was taken to be 1.6 million km^2 divided by 10.1 million km^2 or 15% with an average depth of 4000 m. The total area of Region III was taken to be 8.2 million km^2 divided by 10.1 million km^2 or 81% with an average depth of 4000 m.

Impacts 48 States—Land

The 48 states land surface area is approximately 7.7 million km^2. This is approximately 5% of the Earth's total land surface area (7.7 million divided by 153 million km^2).

A critical parameter with respect to fatalities following an asteroid impact is population density in the impact area. Obviously, the number of fatalities would be directly proportional to the population density for a given impact. In the case study risk assessment, the land area of the 48 states was divided into three categories. These categories were: land areas with high population density (metropolitan areas with greater than 200 people/km^2), medium population density (metropolitan areas with less than 200 people/km^2), and land areas with low population density (non-metropolitan areas).

Using the year 2000 U.S. census data for metropolitan areas, the following table shows the population, land area, population density, and fraction of land by category.

Population Density	*Total Population*	*Total Land Area (km^2)*	*Population Density (km^2)*	*Fraction of Land Area in Category*
High	116,157,107	310,828	374	0.041
Medium	108,546,506	1,244,337	87.2	0.162
Low	54,879,824	6,109,891	9	0.797
Total	279,583,437	7,665,056	36.5	

There were 41 metropolitan areas in the high population density category. The largest land area was approximately 21,000 km^2 (metropolitan New York City).

The average land area of the high-density sites was approximately 7600 km^2.

There were 237 metropolitan areas in the medium population density category. The largest land area was approximately 93,000 km^2 (metropolitan Los Angeles). The average land area of the medium density sites was approximately 5200 km^2.

C.1.3 Water Damage Area

The damage area of a tsunami depends on:

- The coastline impacted by the asteroid.
- The energy impact of the asteroid.
- The amplitude of the tsunami at the edge of the asteroid impact site.
- The amplitude of the tsunami when it reaches the shoreline.
- The distance inland that the tsunami travels.

The damage area of the tsunami is the product of the coastline impacted times the distance inland.

Data developed by Toon is the basis for the calculations in the case study for damage area of an asteroid impacting water.

Coastline Impacted by the Asteroid

The coastline impacted by the asteroid in the Gulf of Mexico is discussed in the "Impacts 48 States—Gulf of Mexico" section and is the entire Gulf of Mexico coastline for all initiating events and all impact sites. The value used was 2610 km.

The coastline impacted by the asteroid in the Pacific Ocean is discussed in the "Impacts 48 States—Pacific Coast" section and is summarized in the table below for all initiating events. The regions are defined in Figure 4.7 of Chapter 4.

Pacific Region	*Coastline Impacted (km)*
I	200
II	1200
III	2070

The coastline impacted by the asteroid in the Atlantic Ocean is discussed in the "Impacts 48 States—Atlantic Coast" section and is summarized in the table below for all initiating events. The regions are defined in Figure 4.8 of Chapter 4.

Atlantic Region	*Coastline Impacted (km)*
I	200
II	1200
III	3310

Energy Impact

The impact energy and frequency of the asteroid is defined by the initiating events and is discussed in the "High Energy Asteroids" section under Section 4.2.2 and the "Initiating Events" section under Section 4.2.3, and Section C.1.1.

Damage Area

The calculation of the damage area is described below.

The amplitude of the tsunami at the impact site, the amplitude of the tsunami at the shoreline, and the maximum distance the tsunami reaches inland are described by Toon. The damage area is defined as the product of the coastline impacted times the distance inland that the tsunami travels.

The amplitude of the tsunami (W) at the impact site in km is the smaller of $(0.66)(0.68)(Y)^{1/4}$ or $(0.3)d_o$ where Y is the impact energy of the asteroid in megatons and d_o is the depth of the ocean at the impact site in km.

The amplitude of the tsunami (A) at the shoreline is defined as the amplitude of the tsunami at the impact site (W) divided by the distance from the impact site to the shoreline in km (D).

The maximum distance the tsunami reaches inland = $(10)^4\,(A)^{4/3}$, where A is the amplitude of the tsunami at the shoreline calculated above. The maximum distance the tsunami reaches inland was reduced to account for factors such as reefs, barrier islands, continental shelves, etc. The uncertainty distributions for the distance inland are shown in Figures C.3–C.12.

The results are summarized in the table below:

Initiating Event	*Impact Depth (km)*	*Wave Amplitude at Impact (km)*	*Distance from Land (km)*	*Wave Amplitude at Shoreline (km)*	*Mean Distance Inland (km)*	*Mean Damage Area (km²)*
2–5—Gulf—shallow	0.07	0.021	100	0.00021	0.0674	176
2–5—Gulf—medium	1.5	0.45	300	0.0015	1.017	2650
2—Gulf—deep	4.0	0.9	400	0.002275	1.592	4160
3–5—Gulf—deep	4.0	1.2	400	0.003	2.325	5022
2—Atlantic or Pacific Region I	4.0	0.91	75	0.0121	14.8	2960
3–5—Atlantic or Pacific Region I	4.0	1.2	75	0.016	23.08	4616
2—Atlantic or Pacific Region II	4.0	0.91	450	0.00205	1.398	1680
3–5—Atlantic or Pacific Region II	4.0	1.2	450	0.0027	2.005	2406
2—Atlantic or Pacific Region III	4.0	0.91	1750	0.00052	0.231	765
3–5—Atlantic or Pacific Region III	4.0	1.2	1750	0.000686	0.327	1080

C.1.4 Land Blast Area

Figures 5 and 19 in the Toon reference were used to determine a correlation between asteroid impact energy and blast area. Using the impact energy, the blast area was then correlated to the initiating event frequencies described in the "Initiating Events" section under Section 4.2.3 of Chapter 4 and Section C.1.1. The following results were developed.

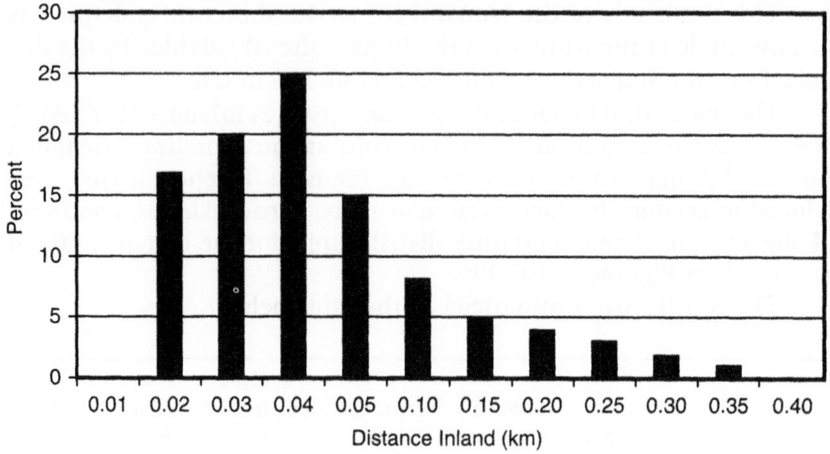

FIGURE C.3. Gulf of Mexico tsunami uncertainty distribution (shallow water, distance inland, all initiating events).

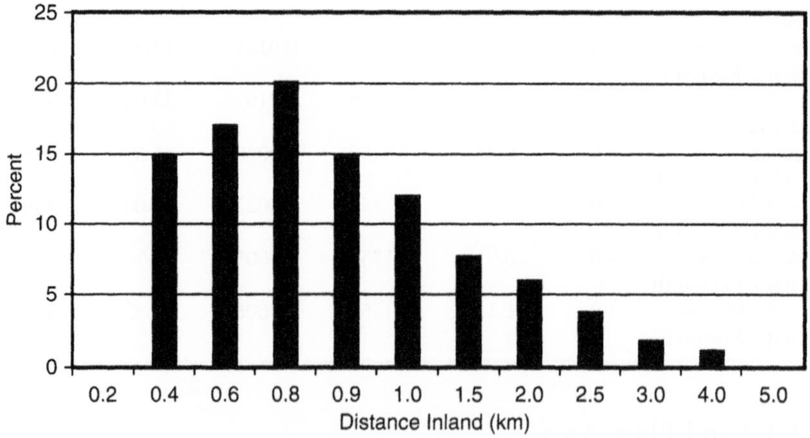

FIGURE C.4. Gulf of Mexico tsunami uncertainty distribution (medium water, distance inland, all initiating events).

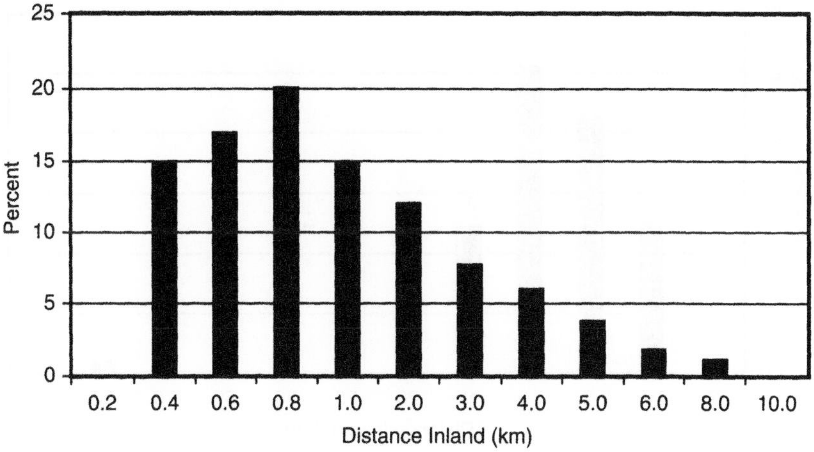

FIGURE C.5. Gulf of Mexico tsunami uncertainty distribution (deep water, distance inland, Initiating Event 2).

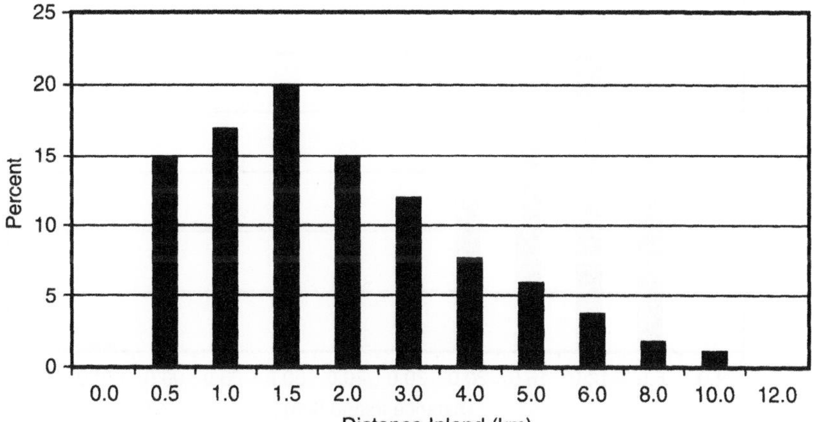

FIGURE C.6. Gulf of Mexico tsunami uncertainty distribution (deep water, distance inland, Initiating Events 3–5).

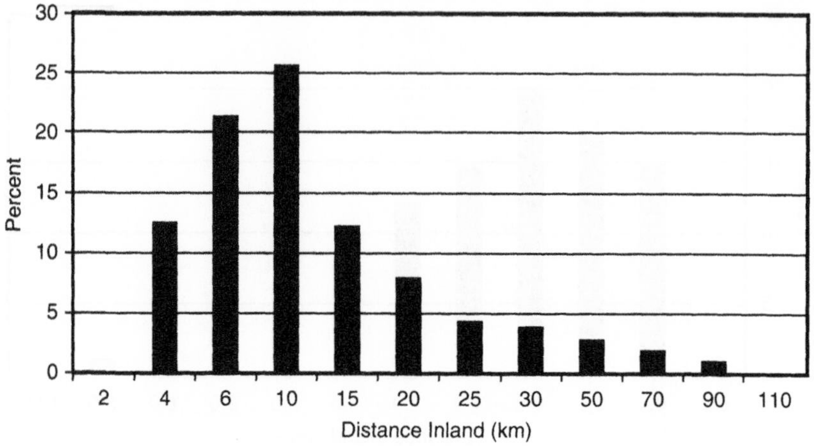

FIGURE C.7. Atlantic or Pacific tsunami uncertainty distribution (Region I, distance inland, Initiating Event 2).

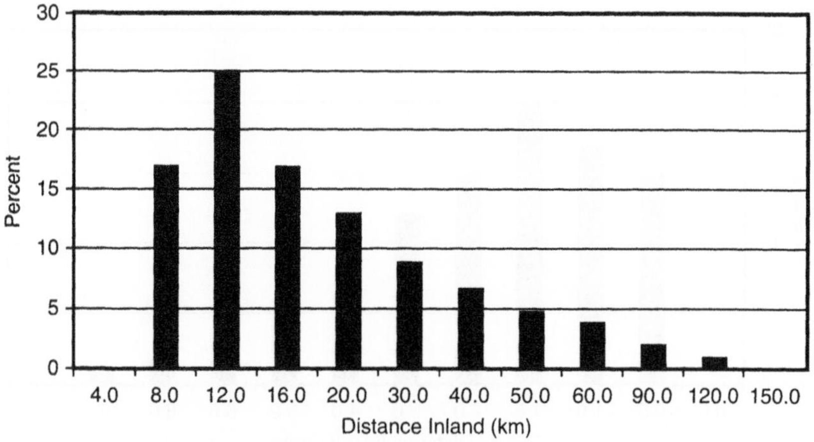

FIGURE C.8. Atlantic or Pacific tsunami uncertainty distribution (Region I, distance inland, Initiating Events 3–5).

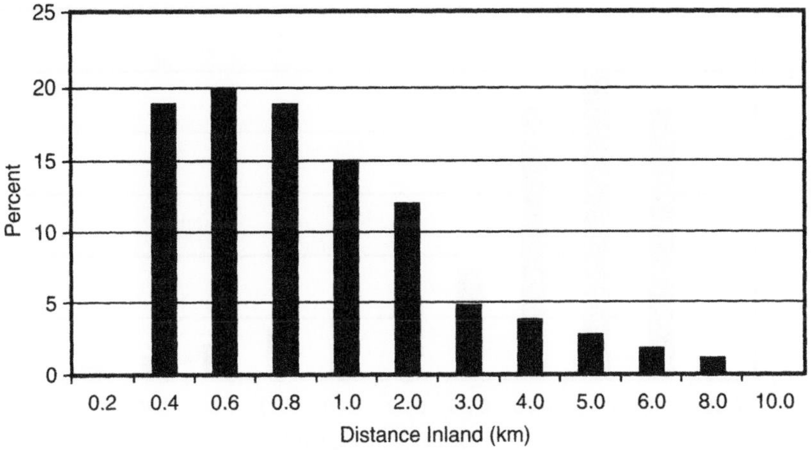

FIGURE C.9. Atlantic or Pacific tsunami uncertainty distribution (Region II, distance inland, Initiating Event 2).

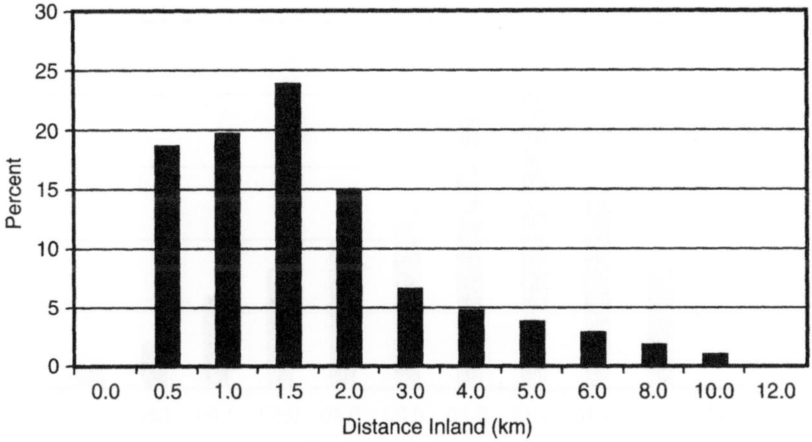

FIGURE C.10. Atlantic or Pacific tsunami uncertainty distribution (Region II, distance inland, Initiating Events 3–5).

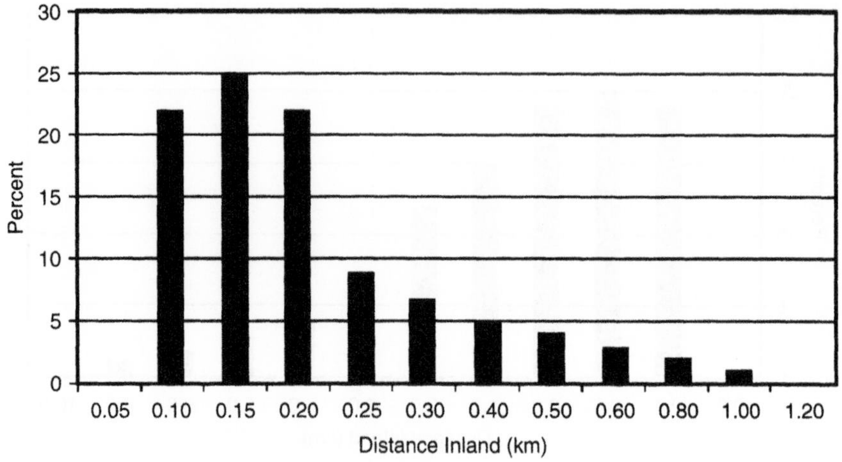

FIGURE C.11. Atlantic or Pacific tsunami uncertainty distribution (Region III, distance inland, Initiating Event 2).

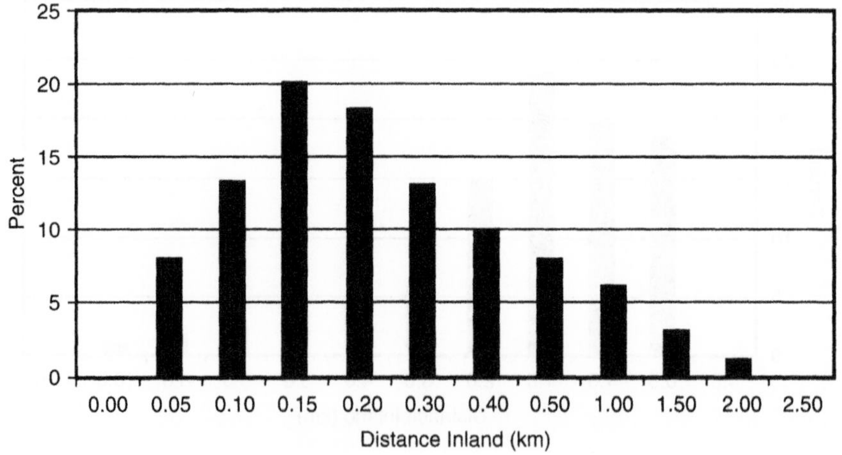

FIGURE C.12. Atlantic or Pacific tsunami uncertainty distribution (Region III, distance inland, Initiating Events 3–5).

Impact Energy (Megatons of TNTE)	*Frequency Per Year (Mean)*	*Blast Area (km^2)*
0.1	1.38E−1[a]	20
0.2	7.00E−2	100
0.4	4.00E−2	250
0.8	3.00E−2	350
1.0	2.26E−2	400
2.0	1.00E−2	550
4.0	8.00E−3	700
8.0	4.00E−3	800
10	3.84E−3	900
20	2.00E−3	1200
40	1.00E−3	1400
80	6.00E−4	1600
100	4.71E−4	1700
200	2.00E−4	2500
400	1.80E−4	3000
800	1.20E−4	4300
1000	9.10E−5	6000
2000	5.00E−5	8000
4000	3.00E−5	10,000
8000	1.80E−5	13,000
10,000	1.49E−5	15,000
20,000	9.00E−5	19,000
40,000	5.00E−6	25,000
80,000	3.00E−6	40,000
100,000	2.47E−6	50,000
200,000	1.10E−6	95,000
400,000	8.00E−7	130,000
800,000	4.00E−7	190,000
1,000,000	3.22E−7	230,000
2,000,000	1.50E−7	480,000
4,000,000	1.00E−7	700,000
8,000,000	7.00E−8	1,000,000
10,000,000	5.59E−8	1,200,000

[a]The notation of 1.38E−1 is the same as 1.38×10^{-1}, which is the same as 0.138.

This table was then modified to the intervals of each initiating event first by weighting the blast area by the frequency of the interval and then expanding the range of blast area for each initiating event to cover the uncertainties. This led to the uncertainty distributions in Figures C.13–C.17.

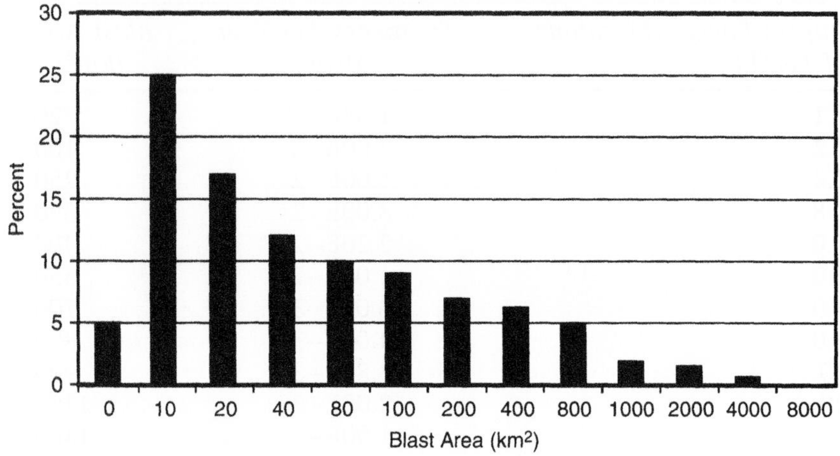

FIGURE C.13. U.S. 48 states land blast area uncertainty distribution (Initiating Event 1).

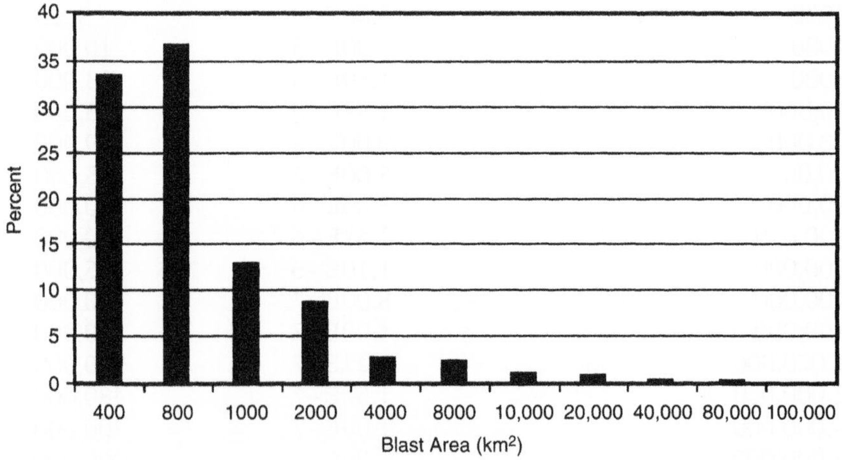

FIGURE C.14. U.S. 48 states land blast area uncertainty distribution (Initiating Event 2).

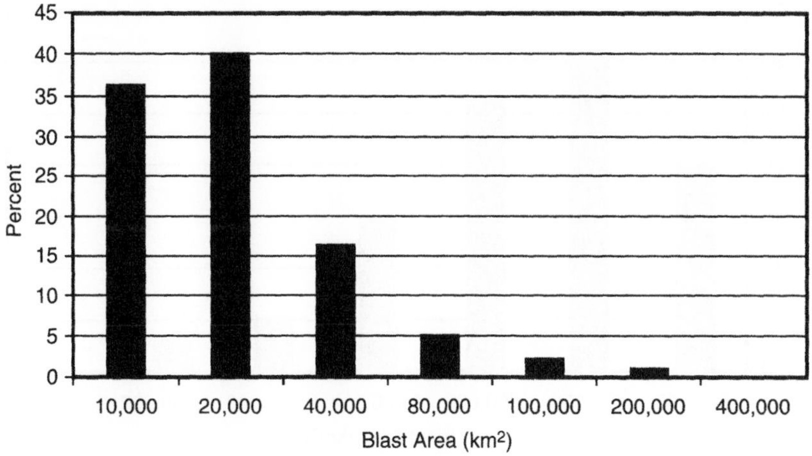

FIGURE C.15. U.S. 48 states land blast area uncertainty distribution (Initiating Event 3).

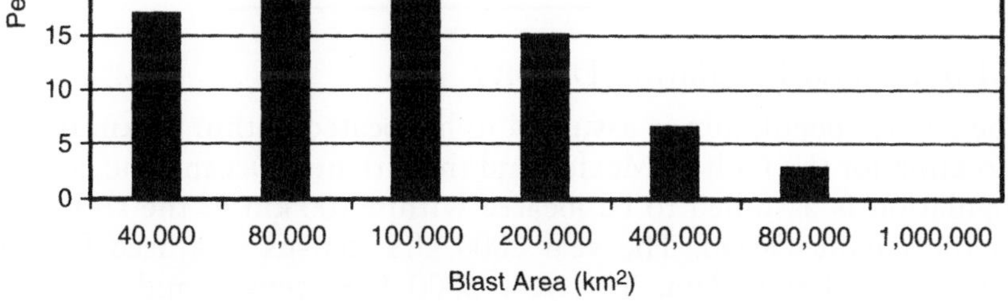

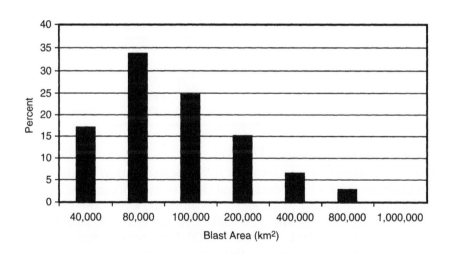

FIGURE C.16. 'U.S. 48 states land blast area uncertainty distribution (Initiating Event 4).

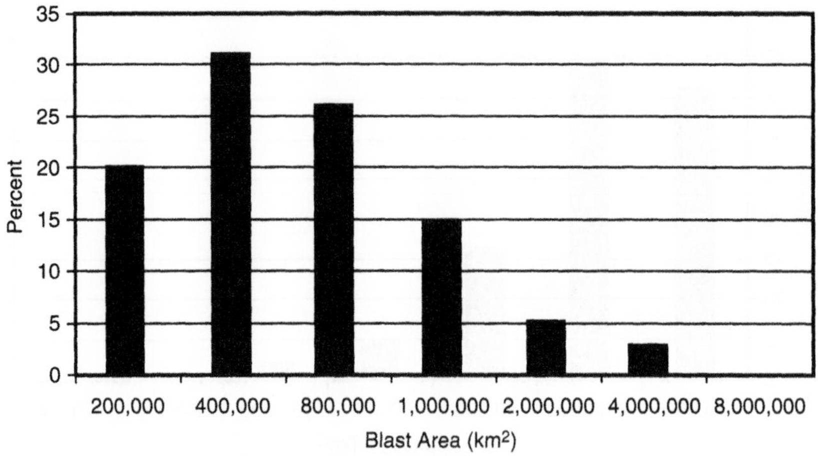

FIGURE C.17. U.S. 48 states land blast area uncertainty distribution (Initiating Event 5).

The mean values are summarized in the table below.

Initiating Event	*Mean Blast Area (km^2)*
1	176
2	1420
3	26,000
4	137,000
5	742,000

C.1.5 Coastal Population Density

The coastal population is assumed to be located within 80 km of the shoreline for the Gulf of Mexico and the Atlantic Ocean. The coastal population is assumed to be located within 100 km of the shoreline for the Pacific Ocean. The year 2000 U.S. census was used for the coastal population data. The year 2000 U.S. census and the year 1990 U.S. census were used for the land area data.

Gulf of Mexico

There were 19 metropolitan areas on the Gulf of Mexico coast between Brownsville, Texas, and Naples, Florida, with a total population of 13,838,934 and a total metropolitan land area of approximately 92,117 km^2, which yields an average population density of approximately 150 people/km^2. Using the coastal length of 2610 km and an inland assumption of 80 km, the total area is 208,880 km^2. On this basis, the non-metropolitan area is 116,683 km^2, and we assume a population density of 9 people/km^2 resulting in a non-metropolitan population of 1,050,000 people. With these assumptions, the average coastal population density for the Gulf of Mexico coast is 71.4 people/km^2. The results are summarized in the table below. The uncertainty distribution for the Gulf states coastal population density is shown in Figure C.18.

Gulf of Mexico	*Metropolitan*	*Non-Metropolitan*	*Total*
Population	13,838,934	1,050,000	14,888,934
Area (km^2)	92,117	116,683	208,880
Population density/km^2	150	9	71.4

Pacific Ocean

There were eight metropolitan areas on the Pacific coast between Bellingham, Washington, and San Diego, California, with a total population of 30,996,204 and a total metropolitan land area of approximately 164,990 km^2, which yields an average population density of approximately 188 people/km^2. Using the coastal length of 2070 km and an inland assumption of 100 km, the total area is 207,000 km^2. On this basis, the non-metropolitan area is 42,010 km^2, and we assume a population density of nine people/km^2 resulting in a non-metropolitan population of 378,090 people. With these assumptions, the average coastal population density for the Pacific coast is 152 people/km^2.

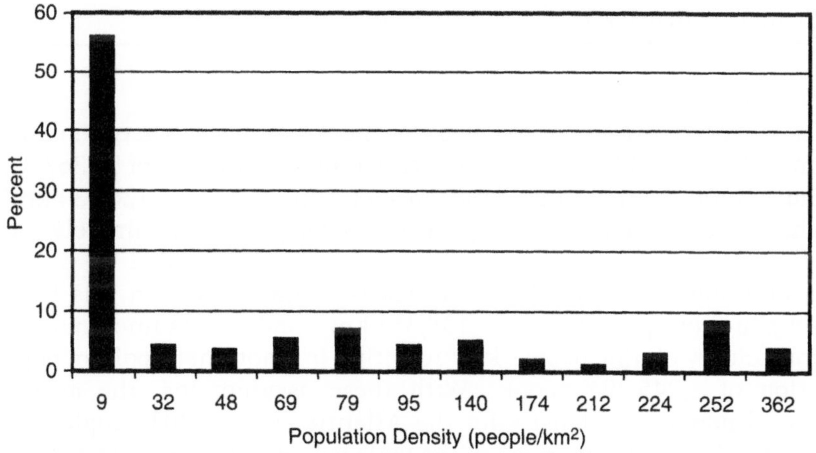

FIGURE C.18. Gulf states coastal population density uncertainty distribution (all initiating events).

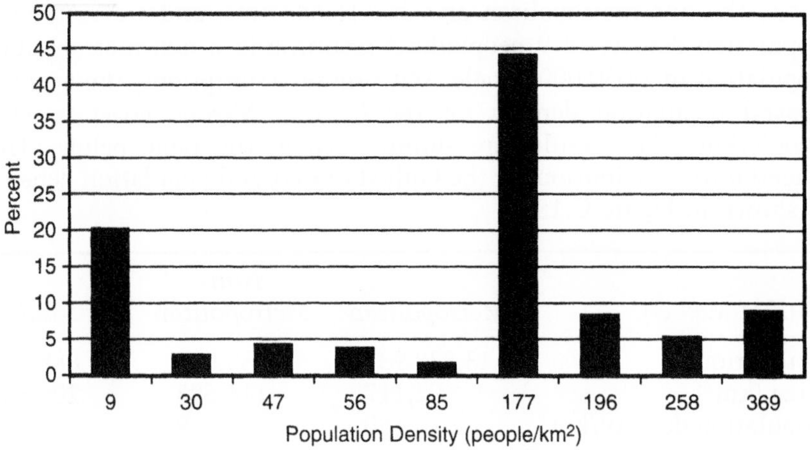

FIGURE C.19. Pacific states coastal population density uncertainty distribution (all initiating events).

The results are summarized in the table below. The uncertainty distribution for the Pacific coast population density is shown in Figure C.19.

Pacific	*Metropolitan*	*Non-Metropolitan*	*Total*
Population	30,996,204	378,090	31,374,294
Area (km^2)	164,990	42,010	207,000
Population density/km^2	188	9	152

Atlantic Ocean

There were 22 metropolitan areas on the Atlantic coast between Portland, Maine, and Miami, Florida, with a total population of 52,367,336 and a total metropolitan land area of approximately 126,434 km^2, which yields an average population density of approximately 414 people/km^2. Using the coastal length of 3310 km and an inland assumption of 80 km, the total area is 264,800 km^2. On this basis, the non-metropolitan area is 138,366 km^2, and we assume a population density of nine people/km^2 resulting in a non-metropolitan population of 1,245,294 people. With these assumptions, the average coastal population density for the Atlantic coast is 203 people/km^2. The results are summarized in the table below. The uncertainty distribution for the Atlantic coast population density is shown in Figure C.20.

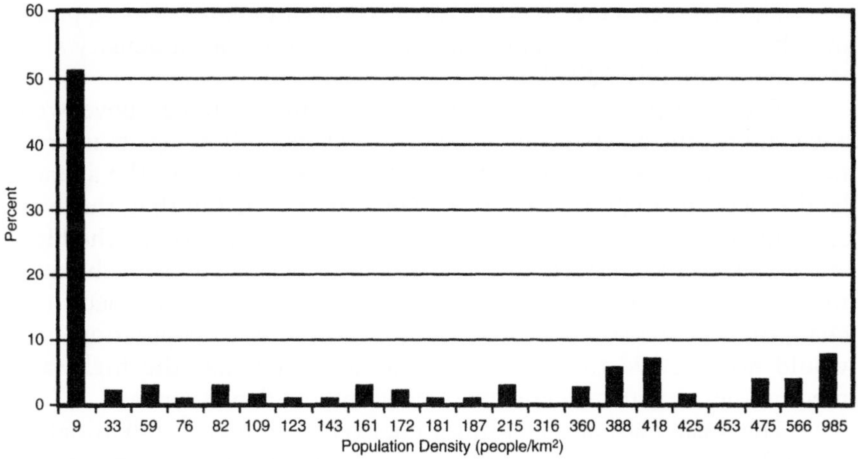

FIGURE C.20. Atlantic states coastal population density uncertainty distribution (all initiating events).

Atlantic	*Metropolitan*	*Non-Metropolitan*	*Total*
Population	52,367,336	1,245,294	53,612,630
Area (km^2)	126,434	138,366	264,800
Population density (km^2)	414	9	203

C.1.6 Land Population Density

As discussed in the "Impacts 48 States—Land" section, using the year 2000 U.S. census data for metropolitan areas, the following table was produced showing the population, land area, population density, and fraction of land area in each category that were used in the risk assessment.

Population Density	*Total Population*	*Total Land Area (km^2)*	*Population Density (km^2)*	*Fraction of Land Area in Category*
High	116,157,107	310,828	374	0.041
Medium	108,546,506	1,244,337	87.2	0.162
Low	54,879,824	6,109,891	9	0.797
Total	279,583,437	7,665,056	36.5	

The average land area of the high-density sites was approximately 7600 km^2. The average land area of the medium-density sites was approximately 5200 km^2.

The land population densities shown in the table above were adequate for the calculation of the people at risk following an asteroid impact on land as long as the blast area was smaller than the average land area of the high-density sites (7600 km^2) or smaller than the average land area of the medium-density sites (5200 km^2). When the blast area was larger than the high or medium average land area for the metropolitan areas, the population density in the blast area needed to be changed because some people from a low-density population area would not survive the impact. It was assumed that the high- and medium-density areas were surrounded by low-density areas.

The mean blast areas for Initiating Event 1 (176 km^2) and Initiating Event 2 (1420 km^2) were lower than average high- and medium-density areas. Thus, the population densities in the table above could be used with no modification in the calculation of the people at risk for Initiating Events 1 and 2. However, the mean blast areas for Initiating Event 3 (26,000 km^2), Initiating Event 4 (137,000 km^2), and Initiating Event 5 (742,000 km^2) were much larger than the average high- or medium-density average land areas and the population densities needed to be modified.

The modifications were completed by:

1. Multiplying the average land area by the appropriate population density (high or medium) to get a total population in the high or medium portion.
2. Subtracting the average land area (high or medium) from the blast area and multiplying the difference by nine people/km^2 to get the total population in the surrounding area.
3. Adding the two populations calculated above to get the total population in the blast area.
4. Dividing the total population in the blast area by the blast area to get the population density for the total blast area.

For land impact, these calculations produced the population densities shown in the table below.

Initiating Event	*High-Density Impact (People/km^2)*	*Medium-Density Site (People/km^2)*	*Low-Density Site (People/km^2)*
1	374	87.2	9
2	374	87.2	9
3	208	42.8	9
4	46	21.5	9
5	15.3	11.3	9

The uncertainty distributions for the population densities for land impact are shown in Figures C.21–C.29.

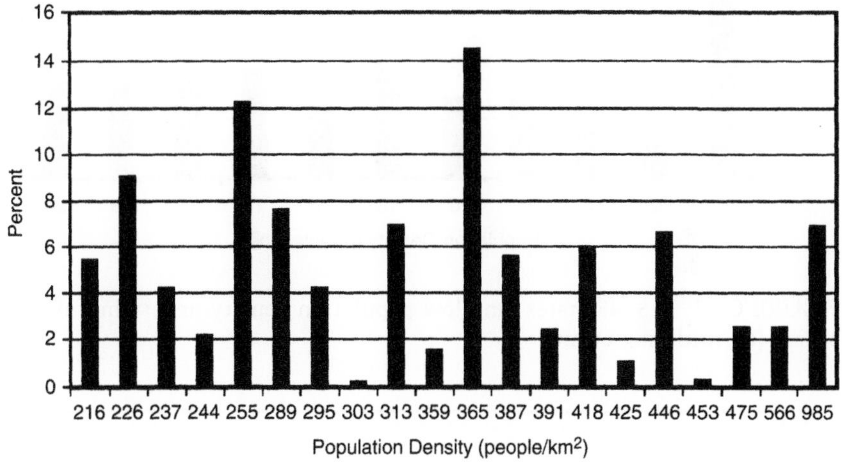

FIGURE C.21. U.S. 48 states land high population density uncertainty distribution (Initiating Events 1 and 2).

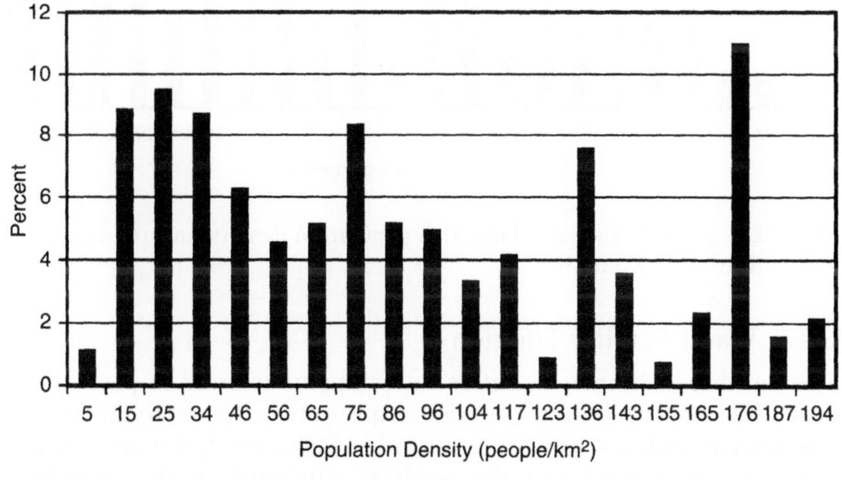

FIGURE C.22. U.S. 48 states land medium population density uncertainty distribution (Initiating Events 1 and 2).

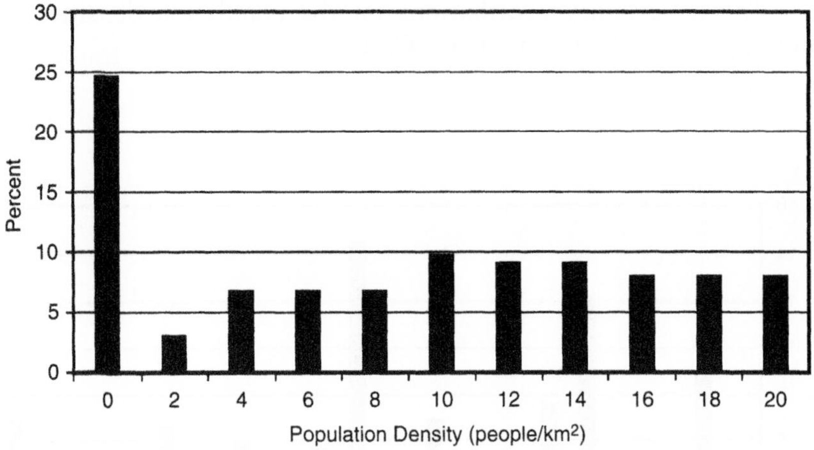

FIGURE C.23. U.S. 48 states land low population density uncertainty distribution (all initiating events).

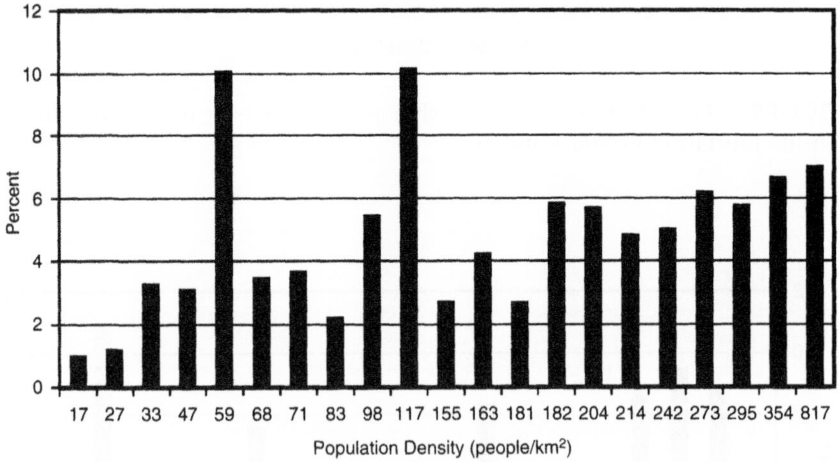

FIGURE C.24. U.S. 48 states land high population density uncertainty distribution (Initiating Event 3).

C.1.7 Water Fatality Fraction

The number of human fatalities following an asteroid impact in the ocean will depend on the tsunami amplitude at the shoreline. In the quantitative risk assessment case study of Chapter 4, the assumption was made that the larger the tsunami amplitude at the shoreline, the larger the fatality fraction. The amplitude of the tsunami at the shoreline is calculated as described in the "Damage Area" section

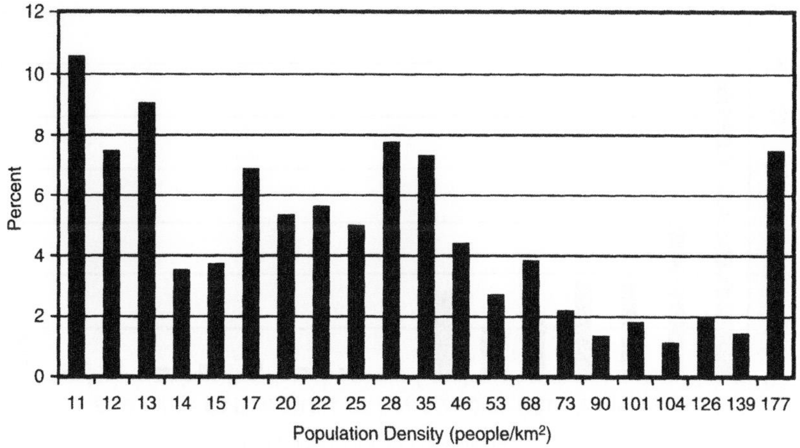

FIGURE C.25. U.S. 48 states land medium population density uncertainty distribution (Initiating Event 3).

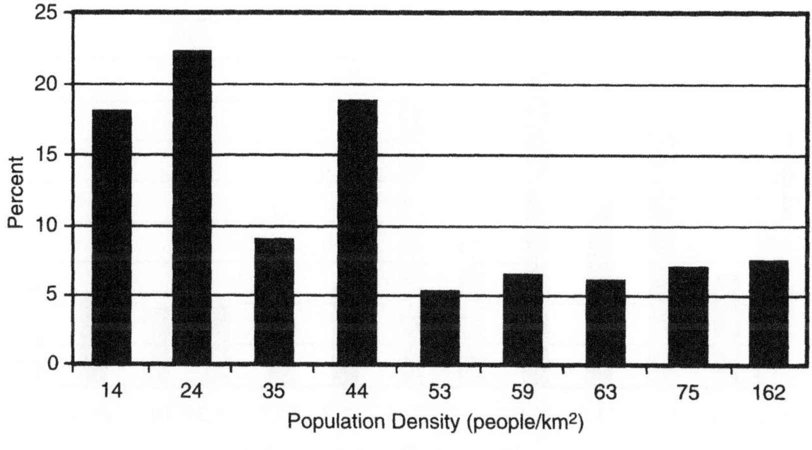

FIGURE C.26. U.S. 48 states land high population density uncertainty distribution (Initiating Event 4).

and is inversely proportional to the distance from the asteroid impact site to the shoreline. However, this calculation is based on a model that assumes deep water from the impact site to the shoreline with no continental shelf out from the shoreline or offshore reefs or barrier islands to break up the tsunami. This assumption is considered appropriate for large portions of the Pacific and Atlantic oceans, but not for the Gulf of Mexico. Also, when the wave hits the shoreline, there will be amplification effects depending on the shape of the shoreline,

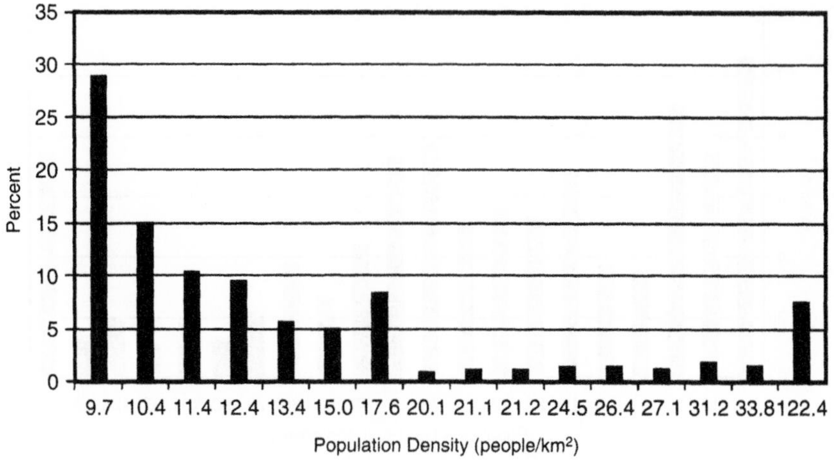

FIGURE C.27. U.S. 48 states land medium population density uncertainty distribution (Initiating Event 4).

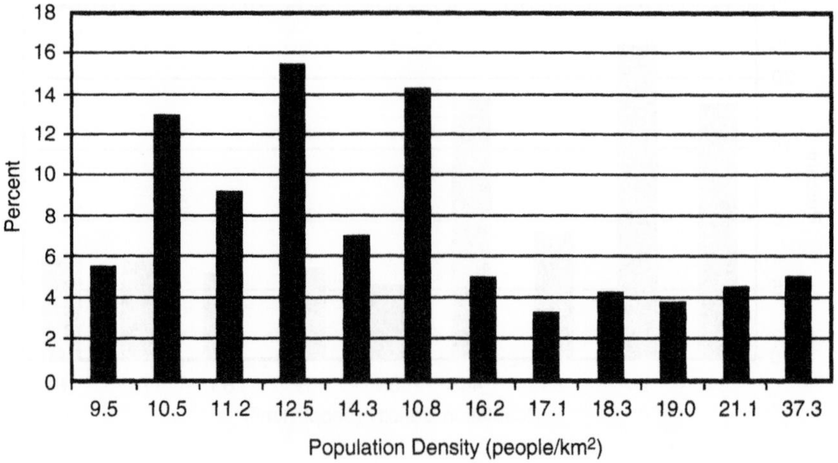

FIGURE C.28. U.S. 48 states land high population density uncertainty distribution (Initiating Event 5).

whether features such as coves and bays are present in the shoreline, and the slope of the land beneath the ocean approaching the shoreline.

For the Atlantic and Pacific oceans, the fatality fraction was assigned a value of 1.0 when the tsunami amplitude at the shoreline was calculated to be 2 m or greater. For the Gulf of Mexico, the fatality fraction was assigned a value of 1.0 when the tsunami amplitude at

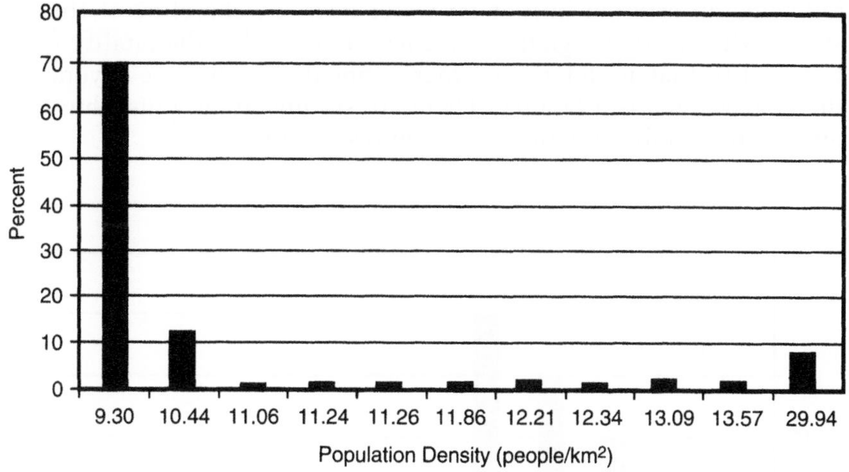

FIGURE C.29. U.S. 48 states land medium population density uncertainty distribution (Initiating Event 5).

the shoreline was calculated to be 6 m or more. The values assumed for smaller wave amplitudes at the shoreline are shown in the table below.

Calculated Wave Amplitude at the Shoreline (m)	*Fatality Fraction—Gulf of Mexico*	*Fatality Fraction—Atlantic or Pacific*
0.5	0.08	0.15
1.0	0.12	0.33
1.5	0.15	0.80
2.0	0.20	1.0
2.5	0.28	1.0
3.0	0.40	1.0
3.5	0.52	1.0
4.0	0.68	1.0
4.5	0.80	1.0
5.0	0.88	1.0
5.5	0.95	1.0
6.0	1.0	1.0
6.5	1.0	1.0
7.0	1.0	1.0

Based on the above model, an examination of Table 4.2 of Chapter 4 indicates that 20 sequences of the 51 regional sequences require uncertainty distributions with respect to the fatality fraction following an asteroid impact on water near the shoreline of the 48 states. All the other regional sequences assume that the fatality fraction is 1.0; that is, for these other regional sequences, everyone in the damage area is a fatality. There are six uncertainty distributions for tsunami fatality fractions (see Figures C.30–C.35).

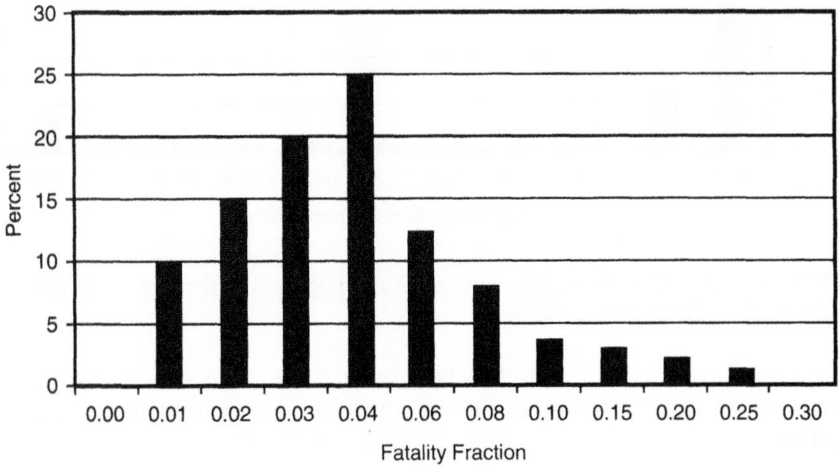

FIGURE C.30. Gulf of Mexico tsunami fatality fraction uncertainty distribution (wave amplitude 0.21 m).

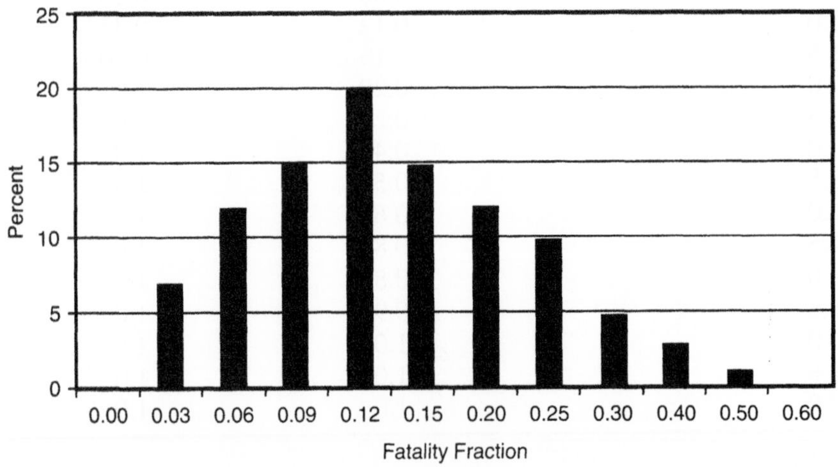

FIGURE C.31. Gulf of Mexico tsunami fatality fraction uncertainty distribution (wave amplitude 1.5 m).

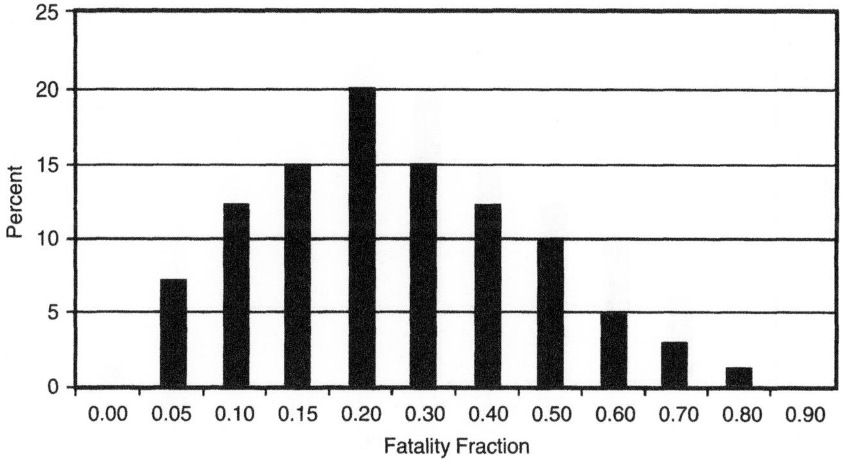

FIGURE C.32. Gulf of Mexico tsunami fatality fraction uncertainty distribution (wave amplitude 2.3 m).

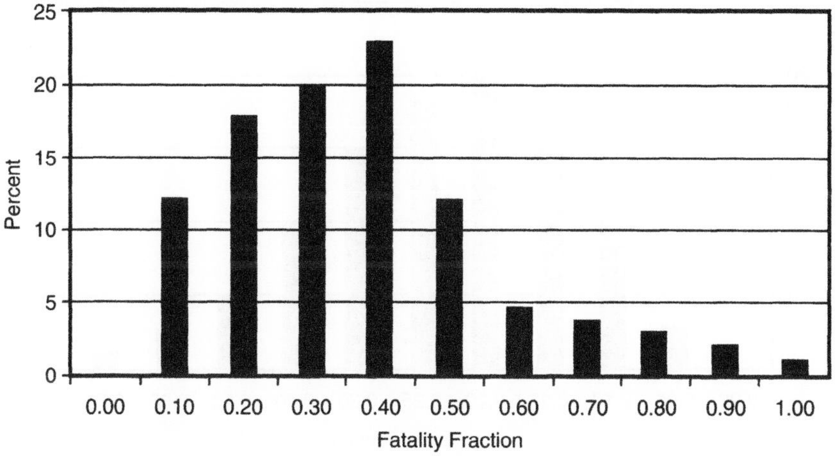

FIGURE C.33. Gulf of Mexico tsunami fatality fraction uncertainty distribution (wave amplitude 3.0 m).

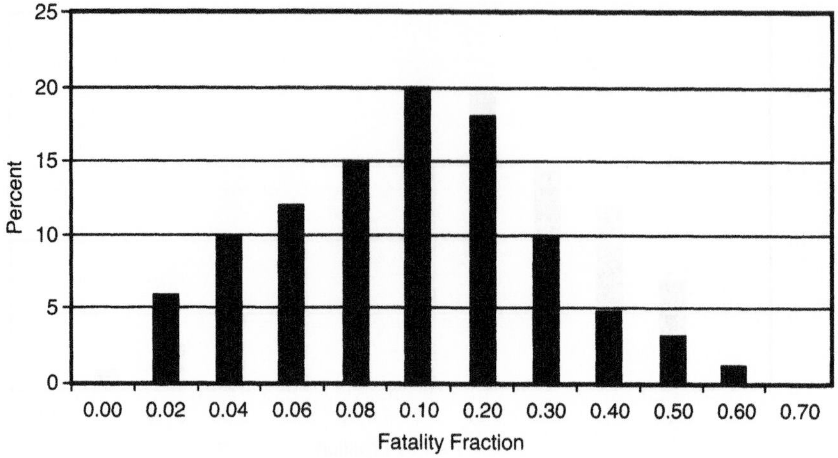

FIGURE C.34. Atlantic or Pacific tsunami fatality fraction uncertainty distribution (wave amplitude 0.52 m).

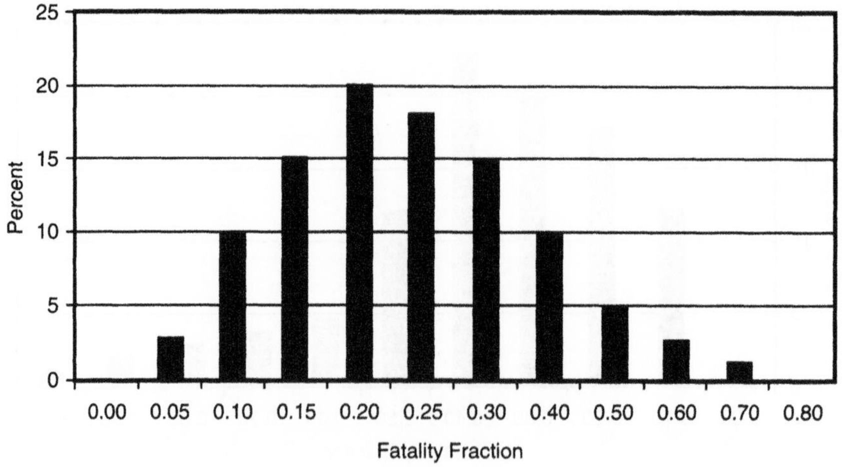

FIGURE C.35. Atlantic or Pacific tsunami fatality fraction uncertainty distribution (wave amplitude 0.69 m).

Figure	*Region*	*Wave Amplitude (m)*
C.30	Gulf of Mexico	0.21
C.31	Gulf of Mexico	1.50
C.32	Gulf of Mexico	2.30
C.33	Gulf of Mexico	3.00
C.34	Atlantic or Pacific	0.52
C.35	Atlantic or Pacific	0.69

C.1.8 Land Fatality Fraction

For asteroids with an impact energy greater than 10,000 megatons, the assumption is that everyone in the blast area is a fatality, i.e., the fatality fraction is 1.0. For Initiating Events 1 and 2, two uncertainty distributions were generated for the fatality fraction following impact on land. For Initiating Event 1, the mean value of the fatality fraction is 0.332. For Initiating Event 2, the mean value is 0.570. The uncertainty distribution for the fatality fraction for Initiating Event 1 is shown in Figure C.36. The uncertainty distribution for the fatality fraction for Initiating Event 2 is shown in Figure C.37.

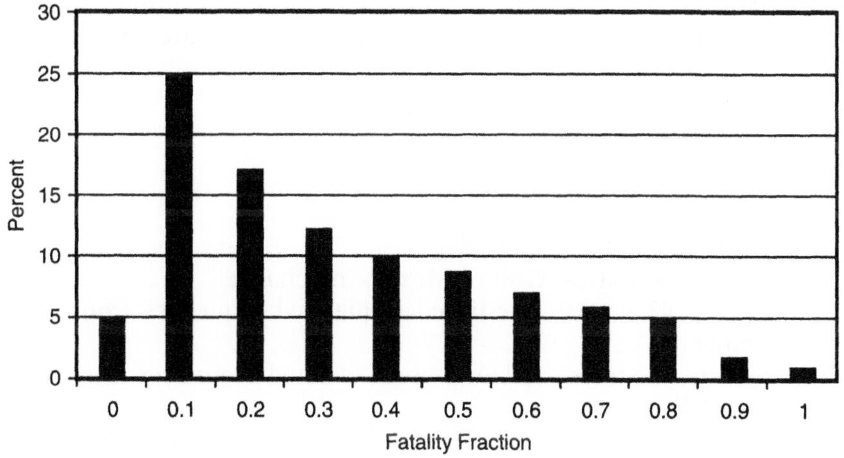

FIGURE C.36. U.S. 48 states land fatality fraction uncertainty distribution (Initiating Event 1).

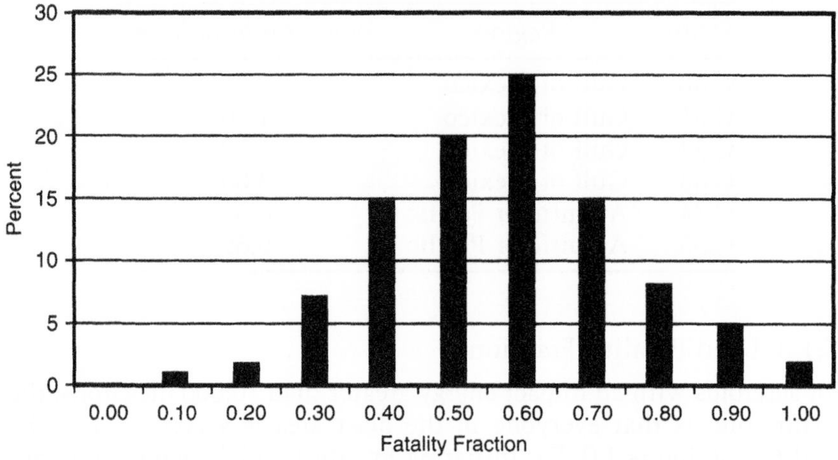

FIGURE C.37. U.S. 48 states land fatality fraction uncertainty distribution (Initiating Event 2).

C.2 Asteroid Risk Assessment for Metropolitan New Orleans, LA

The approach is to follow the same steps as above and simply indicate the differences in the two models, that is, the 48 states model and New Orleans.

C.2.1 Initiating Events

No change from the above model.

C.2.2 Impacts Water or Land

Impacts 48 states—Gulf of Mexico (no change).

Impacts 48 states—Pacific coast (deleted sequences from the model).

Impacts 48 states—Atlantic coast (deleted sequences from the model).

Impacts 48 states—land. The Earth's total land surface area is approximately 153 million km^2. The surface area of New Orleans is approximately 5980 km^2. The probability of the asteroid directly impacting New Orleans is $(0.3)(5980/153{,}000{,}000) = 1.17 \times 10^{-5}$.

C.2.3 Water Damage Area

The damage area is the entire New Orleans metropolitan area, 5980 km^2 with the resultant population at risk of 1.34 million people.

C.2.4 Land Blast Area

The damage area was confined to the New Orleans land area regardless of the impact energy of the asteroid. That is, the fatality consequences did not include populations beyond New Orleans for those asteroids resulting in larger damage areas. The table below shows the values used.

Initiating Event	*Mean Blast Area (km^2)*
1	176
2	1420
3	5980
4	5980
5	5980

C.2.5 Coastal Population Density

The coastal population was assumed to consist only of the New Orleans population, which had a density of 224 people/km^2.

C.2.6 Land Population Density

The land population was assumed to consist only of the New Orleans population, that is, a density of 224 people/km^2.

C.2.7 Water Fatality Fraction

The fatality fraction for New Orleans following an asteroid impact in the Gulf of Mexico was assumed to be twice the fatality fraction used in the 48 states model due to the unique characteristics of New Orleans. The uncertainty distributions for fatality fractions used in the 48 states model were modified as shown in Figures C.38–C.41.

Figure	*Wave Amplitude (m)*	*Mean Fatality Fraction*
C.38	0.21	0.100
C.39	1.50	0.300
C.40	2.30	0.560
C.41	3.00	0.740

C.2.8 Land Fatality Fraction

No change.

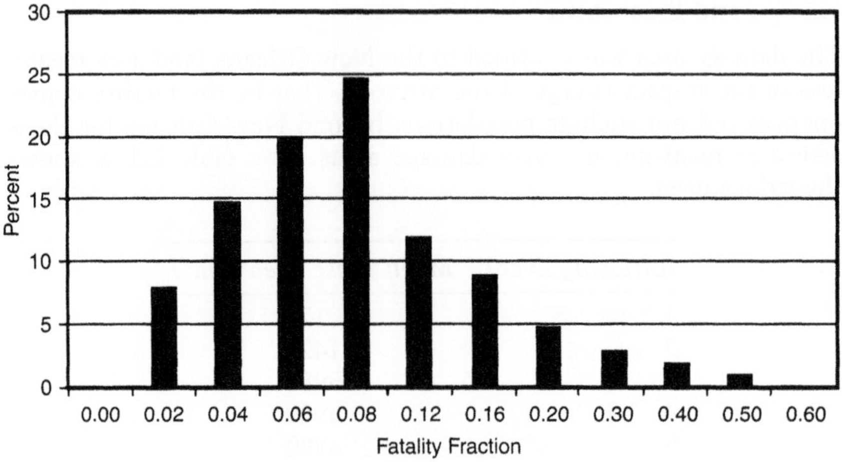

FIGURE C.38. Gulf of Mexico tsunami fatality fraction uncertainty distribution for New Orleans (wave amplitude 0.21 m).

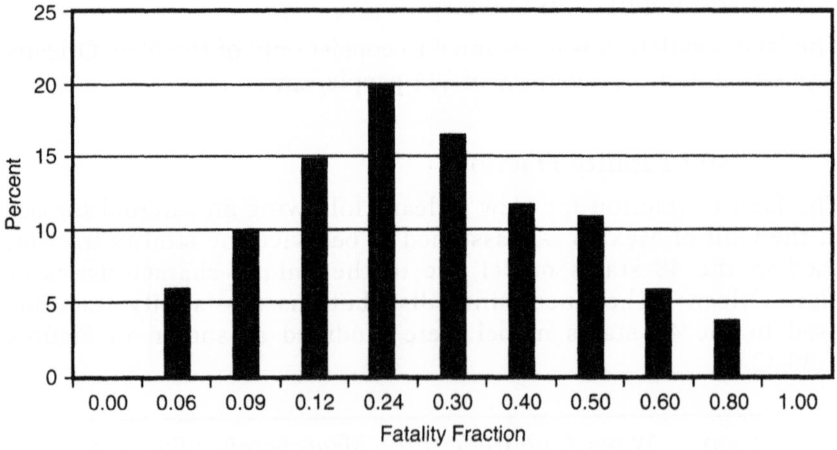

FIGURE C.39. Gulf of Mexico tsunami fatality fraction uncertainty distribution for New Orleans (wave amplitude 1.5 m).

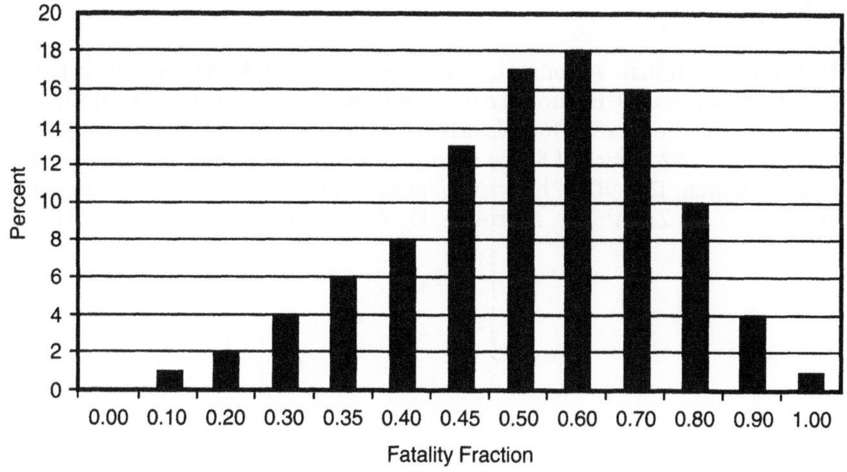

FIGURE C.40. Gulf of Mexico tsunami fatality fraction uncertainty distribution for New Orleans (wave amplitude 2.3 m).

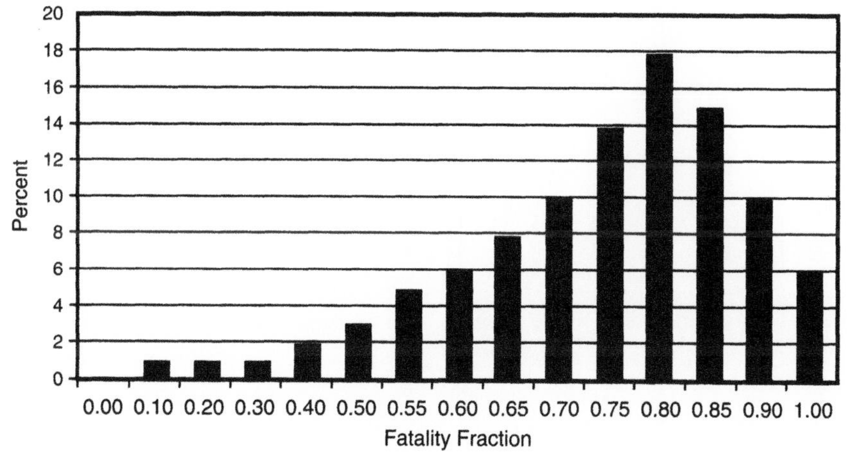

FIGURE C.41. Gulf of Mexico tsunami fatality fraction uncertainty distribution for New Orleans (wave amplitude 3.0 m).

References

[1] NASA (National Aeronautics and Space Administration). Near-Earth Object Science Definition Team, Study to Determine the Feasibility of Extending the Search for Near-Earth Objects to Smaller Limiting Diameters. NASA Office of Space Science, Solar System Exploration Division, Washington, DC, 2003. http://neo.jpl.nasa.gov/neo/neoreport030825.pdf.
[2] Toon, O. B., Zahnle, K., Morrison, D., *Rev. Geophys.,* 1997, *35*(1), 41–78.

Author Index

Subject Index

D

E

F

G

O

P

Q

R

Printed and bound by CPI Group (UK) Ltd, Croydon, CR0 4YY
03/10/2024
01040414-0006